SOLIDWORKS 2023
Basic Tools

Introductory Level Tutorials
Getting started with Parts, Assemblies and Drawings

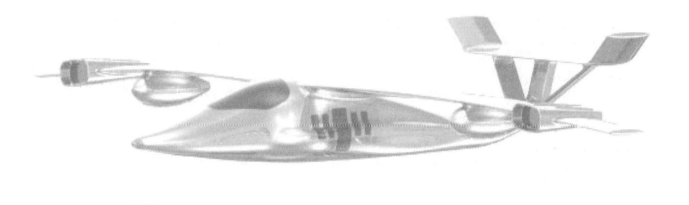

Paul Tran, CSWE, CSWI

SDC
PUBLICATIONS

SDC Publications
P.O. Box 1334
Mission, KS 66222
913-262-2664
www.SDCpublications.com
Publisher: Stephen Schroff

ISBN-13: 978-1-63057-548-9
ISBN-10: 1-63057-548-8

Printed and bound in the United States of America.

Acknowledgments

Thanks as always to my wife Vivian and my daughter Lani for always being there and providing support and honest feedback on all the chapters in the textbook.

Additionally, thanks to Kevin Douglas for writing the foreword.

I also would like to thank SDC Publications and the staff for its continuing encouragement and support for this edition of **SOLIDWORKS 2023 Basic Tools**. Thanks also to Tyler Bryant for putting together such a beautiful cover design.

Finally, I would like to thank you, our readers, for your continued support. It is with your consistent feedback that we were able to create the lessons and exercises in this book with more detailed and useful information.

Foreword

For more than two decades, I have been fortunate to have worked in the fast-paced, highly dynamic world of mechanical product development providing computer-aided design and manufacturing solutions to thousands of designers, engineers, and manufacturing experts in the western US. The organization where I began this career was US CAD in Orange County, CA, one of the most successful SOLIDWORKS resellers in the world. My first several years were spent in the sales organization prior to moving into middle management and ultimately President of the firm. In the mid-1990s is when I met Paul Tran, a young, enthusiastic Instructor who had just joined our team.

Paul began teaching SOLIDWORKS to engineers and designers of medical devices, automotive and aerospace products, high tech electronics, consumer goods, complex machinery and more. After a few months of watching him teach and interacting with students during and after class, it was becoming pretty clear – Paul not only loved to teach, but his students were the most excited with their learning experience than I could ever recall from previous years in the business. As the years began to pass and thousands of students had cycled through Paul's courses, what was eye opening was Paul's continued passion to educate as if it were his first class and students in every class, without exception, loved the course.

Great teachers not only love their subject, but they love to share that joy with students – this is what separates Paul from others in the world of SOLIDWORKS Instruction. He always has gone well beyond learning the picks & clicks of using the software, to best practice approaches to creating intelligent, innovative, and efficient designs that are easily grasped by his students. This effective approach to teaching SOLIDWORKS has translated directly into Paul's many published books on the subject. His latest effort with SOLIDWORKS 2023 is no different. Students that apply the practical lessons from basics to advanced concepts will not only learn how to apply SOLIDWORKS to real world design challenges more quickly but will gain a competitive edge over others that have followed more traditional approaches to learning this type of technology.

As the pressure continues to rise on U.S. workers and their organizations to remain competitive in the global economy, raising not only education levels but technical skills are paramount to a successful professional career and business. Investing in a learning process towards the mastery of SOLIDWORKS through the tutelage of the most accomplished and decorated educator and author in Paul Tran will provide a crucial competitive edge in this dynamic market space.

Kevin Douglas
Vice President Sales/Board of Advisors, GoEngineer

Images courtesy of C.A.K.E. Energy Corp., designed by Paul Tran

Author's Note

SOLIDWORKS 2023 Basic Tools, Intermediate Skills, and Advanced Techniques are comprised of lessons and exercises based on the author's extensive knowledge on this software. Paul has more than 30 years of experience in the fields of mechanical and manufacturing engineering; he has spent 2/3 of those years teaching and supporting the SOLIDWORKS software and its add-ins. As an active Sr. SOLIDWORKS instructor and design engineer, Paul has worked and consulted with hundreds of reputable companies including IBM, Intel, NASA, US Navy, Boeing, Disneyland, Medtronic, Guidant, Terumo, Kingston and many more. Today, he has trained more than 13,000 engineering professionals, and given guidance to nearly one half of the number of Certified SOLIDWORKS Professionals and Certified SOLIDWORKS Experts (CSWP & CSWE) in the state of California.

Every lesson and exercise in this book was created based on real world projects. Each of these projects has been broken down and developed into easy and comprehendible steps for the reader. Learn the fundamentals of SOLIDWORKS at your own pace, as you progress from simple to more complex design challenges. Furthermore, at the end of every chapter, there are self-test questionnaires to ensure that the reader has gained sufficient knowledge from each section before moving on to more advanced lessons.

Paul believes that the most effective way to learn the "world's most sophisticated software" is to learn it inside and out, create everything from the beginning, and take it step by step. This is what the **SOLIDWORKS 2023 Basic Tools, Intermediate Skills & Advanced Techniques** manuals are all about.

About the Training Files

The files for this textbook are available for download on the publisher's website at www.SDCpublications.com/downloads/978-1-63057-548-9. They are organized by the chapter numbers and the file names that are normally mentioned at the beginning of each chapter or exercise. In the **Completed Parts** folder, you will also find copies of the parts, assemblies and drawings that were created for cross referencing or reviewing purposes.

It would be best to make a copy of the content to your local hard drive and work from these documents; you can always go back to the original training files location at anytime in the future, if needed.

Who this book is for?

This book is for the beginner who is not familiar with the SOLIDWORKS program and its add-ins. It is also a great resource for the more CAD literate individuals who want to expand their knowledge of the different features that SOLIDWORKS 2023 has to offer.

The organization of the book

The chapters in this book are organized in the logical order in which you would learn the SOLIDWORKS 2023 program. Each chapter will guide you through some different tasks, from navigating through the user interface, to exploring the toolbars, from some simple 3D modeling to advancing through more complex tasks that are common to all SOLIDWORKS releases. There is also a self-test questionnaire at the end of each chapter to ensure that you have gained sufficient knowledge before moving on to the next chapter.

The conventions in this book

This book uses the following conventions to describe the actions you perform when using the keyboard and mouse to work in SOLIDWORKS 2023:

Click: means to press and release the mouse button. A click of a mouse button is used to select a command or an item on the screen.

Double Click: means to quickly press and release the left mouse button twice. A double mouse click is used to open a program or show the dimensions of a feature.

Right Click: means to press and release the right mouse button. A right mouse click is used to display a list of commands, a list of shortcuts that is related to the selected item.

Click and Drag: means to position the mouse cursor over an item on the screen and then press and hold down the left mouse button; still holding down the left button, move the mouse to the new destination and release the mouse button. Drag and drop makes it easy to move things around within a SOLIDWORKS document.

Bolded words: indicate the action items that you need to perform.

Italic words: Side notes and tips that give you additional information, or to explain special conditions that may occur during the course of the task.

Numbered Steps: indicates that you should follow these steps in order to successfully perform the task.

Icons: indicates the buttons or commands that you need to press.

SOLIDWORKS 2023

SOLIDWORKS 2023 is a program suite, or a collection of engineering programs that can help you design better products faster. SOLIDWORKS 2023 contains different combinations of programs; some of the programs used in this book may not be available in your suites.

Start and exit SOLIDWORKS

SOLIDWORKS allows you to start its program in several ways. You can either double click on its shortcut icon on the desktop or go to the Start menu and select the following: All Programs / SOLIDWORKS 2023 / SOLIDWORKS or drag a SOLIDWORKS document and drop it on the SOLIDWORKS shortcut icon.

Before exiting SOLIDWORKS, be sure to save any open documents, and then click File / Exit; you can also click the X button on the top right of your screen to exit the program.

Using the Toolbars

You can use toolbars to select commands in SOLIDWORKS rather than using the drop-down menus. Using the toolbars is normally faster. The toolbars come with commonly used commands in SOLIDWORKS, but they can be customized to help you work more efficiently.

To access the toolbars, either right click in an empty spot on the top right of your screen or select View / Toolbars.

To customize the toolbars, select Tools / Customize. When the dialog pops up, click on the Commands tab, select a Category, then drag an icon out of the dialog box and drop it on a toolbar that you want to customize. To remove an icon from a toolbar, drag an icon out of the toolbar and drop it into the dialog box.

Using the task pane

The task pane is normally kept on the right side of your screen. It displays various options like SOLIDWORKS resources, Design library, File explorer, Search, View palette, Appearances and Scenes, Custom properties, Built-in libraries, Technical alerts, and news, etc.

The task pane provides quick access to any of the mentioned items by offering the drag and drop function to all of its contents. You can see a large preview of a SOLIDWORKS document before opening it. New documents can be saved in the task pane at anytime, and existing documents can also be edited and re-saved. The task pane can be resized, closed, or moved to different locations on your screen if needed.

Table of Contents

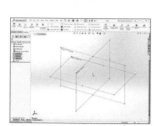
Setting the System Parameters

Basic Modeling Topics

Bottom-Up Assembly Topics

Drawing Topics

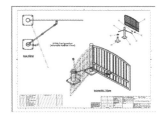

Chapter 18:   **Sheet Metal Drawings**   **18-1**

Chapter 19:   **Configurations**   **19-1**

CSWA Preparation Materials (Certified SOLIDWORKS Associate)

Glossary
Index
SOLIDWORKS 2023 Quick-Guides:
Quick Reference Guide to SOLIDWORKS 2023 Command Icons and Toolbars.

Introduction

SOLIDWORKS User Interface

The SOLIDWORKS 2023 User Interface

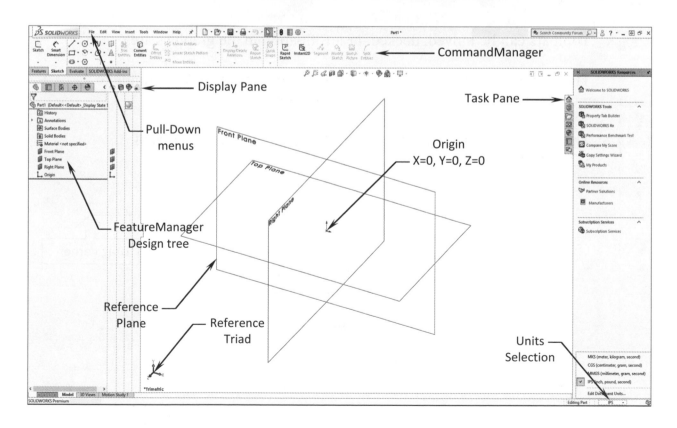

Display Pane

Pull-Down menus

FeatureManager Design tree

Reference Plane

Reference Triad

CommandManager

Task Pane

Origin
X=0, Y=0, Z=0

Units Selection

The 3 reference planes:

The Front, Top and the Right plane are 90° apart.
They share the same center point called the Origin.

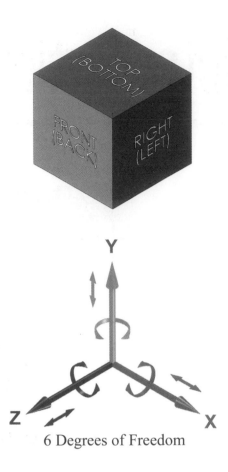

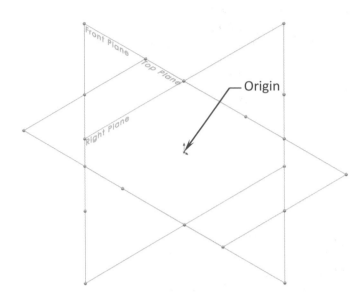

Origin

6 Degrees of Freedom

The Toolbars:

Toolbars can be moved, docked, or left floating in the graphics area.

They can also be "shaped" from horizontal to vertical, or from single to multiple rows when dragging on their corners.

The CommandManager is recommended for the newer releases of SOLIDWORKS.

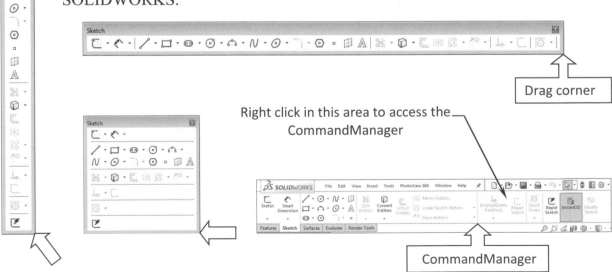

Drag corner

Right click in this area to access the CommandManager

CommandManager

If the CommandManager is not used, toolbars can be docked or left floating.

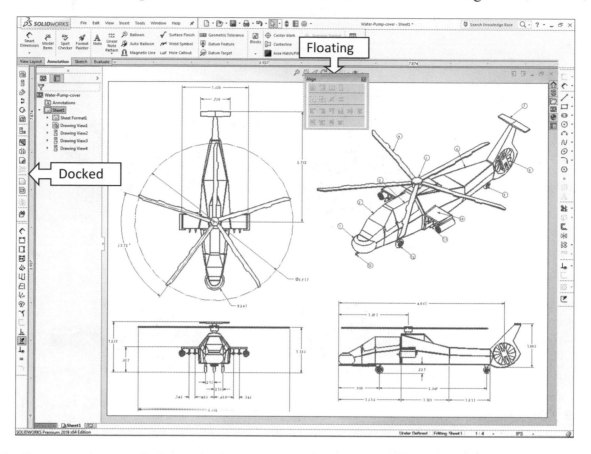

Toolbars can be toggled on or off by activating or de-activating their check boxes:

Select **Tools / Customize / Toolbars** tab.

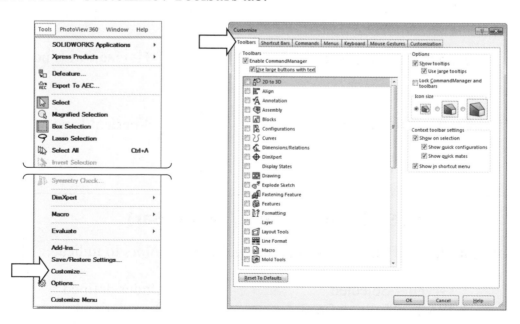

The icons in the toolbars can be enlarged when its check box is selected ☐ Large icons

The View ports: You can view or work with SOLIDWORKS model or an assembly using one, two or four view ports.

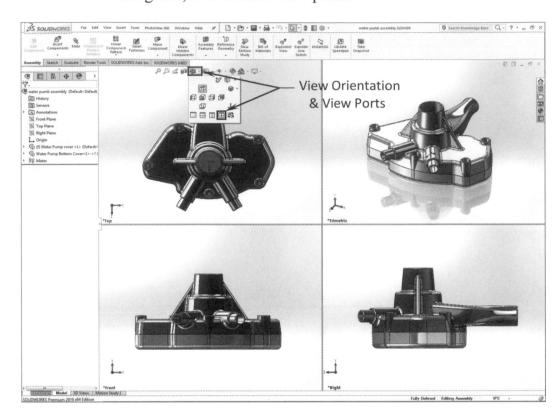

Some of the **System Feedback symbols** (Inference pointers):

Snap to **Vertex** (endpoint)

Snap to **Intersection**

Snap to **Edge** (curve)

Horizontal Line

Snap to **Mid-point**

Vertical Line

The Status Bar: (View / Status Bar)

Displays the status of the sketch entity using different colors to indicate:

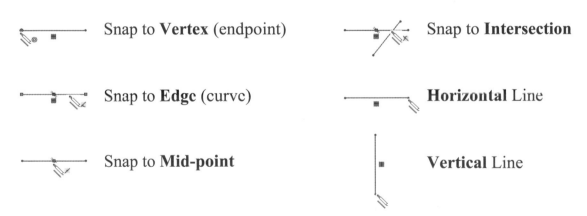

Green = Selected   **Blue** = Under defined
Black = Fully defined   **Red** = Over defined

2D Sketch examples:

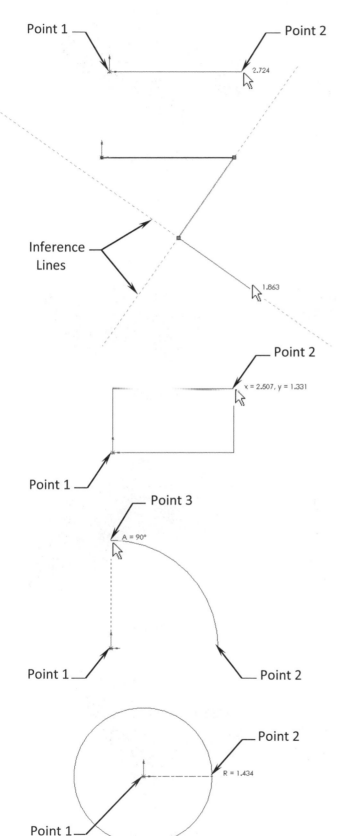

Click-Drag-Release: Single entity

(Click Point 1, hold the mouse button, drag to point 2 and release.)

Click-Release: Continuous multiple entities

(The Inference Lines appear when the sketch entities are Parallel, Perpendicular, or Tangent with each other.)

Click-Drag-Release: Single Rectangle

(Click point 1, hold the mouse button, drag to Point 2 and release.)

Click-Drag-Release: Single Centerpoint Arc

(Click point 1, hold the mouse button and drag to Point 2, release; then drag to Point 3 and release.)

Click-Drag-Release: Single Circle

(Click point 1 [center of circle], hold the mouse button, drag to Point 2 [Radius] and release.)

3D Feature examples:

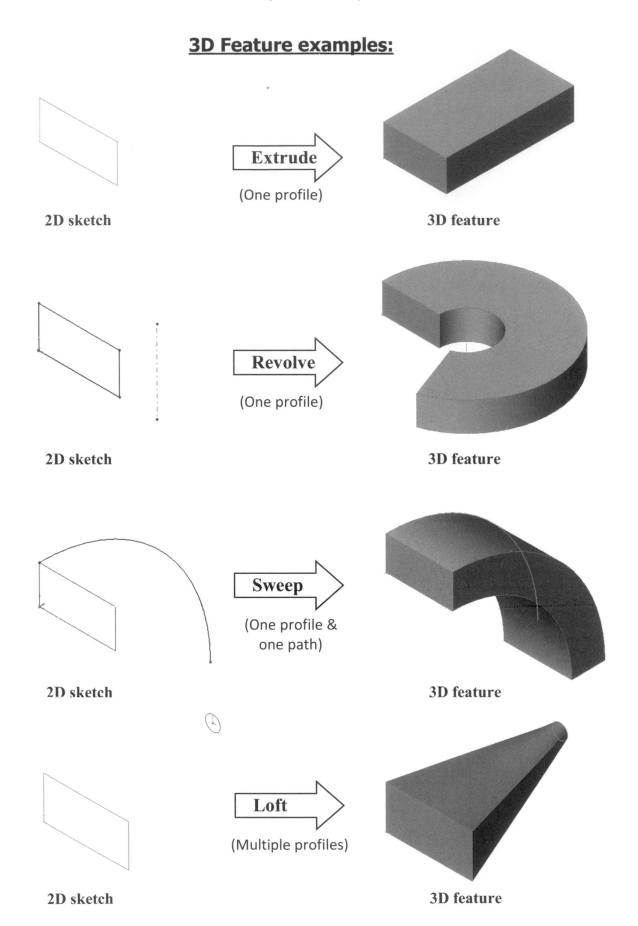

<u>Box-Select:</u> Use the Select Pointer 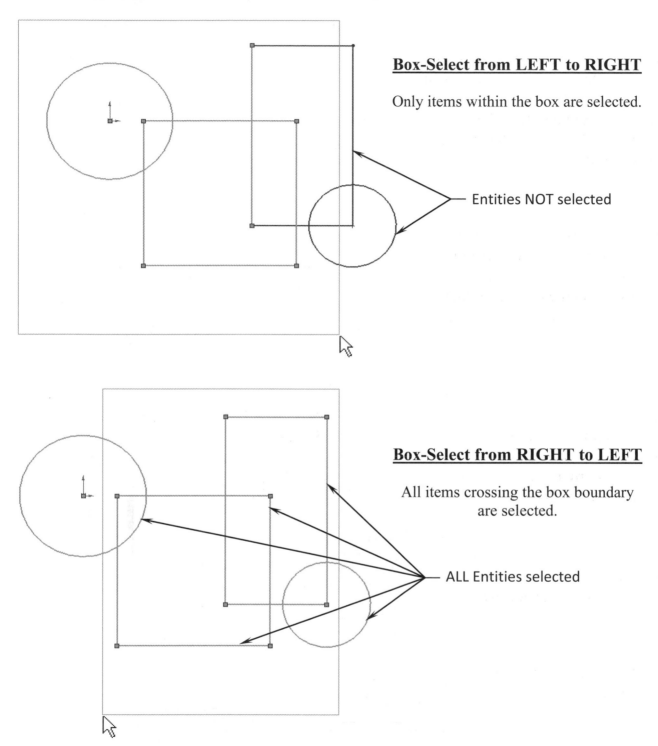 to drag a selection box around items.

Box-Select from <u>LEFT</u> to <u>RIGHT</u>

Only items within the box are selected.

— Entities NOT selected

Box-Select from <u>RIGHT</u> to <u>LEFT</u>

All items crossing the box boundary
are selected.

— ALL Entities selected

The default geometry type selected is as follows:

* Part documents – edges * Assembly documents – components * Drawing documents - sketch entities,

dims & annotations. * To select multiple entities, hold down **Ctrl** while selecting after the first selection.

The <u>Mouse Gestures</u> for Parts, Sketches, Assemblies and Drawings

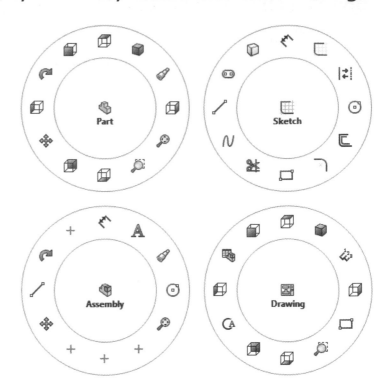

Similar to a keyboard shortcut, you can use a Mouse Gesture to execute a command. A total of 12 keyboard shortcuts can be independently mapped and stored in the Mouse Gesture Guides.

To activate the Mouse Gesture Guide, **right-click-and-drag** to see the current 12 gestures, then simply select the command that you want to use.

To customize the Mouse Gestures and include your favorite shortcuts, go to: **Tools / Customize**.

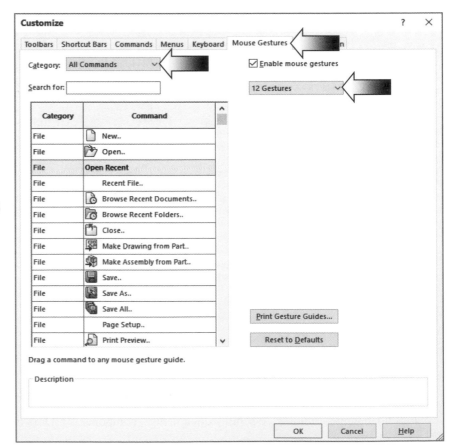

From the **Mouse Gestures** tab select **All Commands**.

Click the **Enable Mouse Gestures** checkbox.

Select the **12 Gestures** option (arrow).

Customizing Mouse Gestures

To reassign a mouse gesture:

1. With a document open, click **Tools > Customize** and select the **Mouse Gestures** tab. The tab displays a list of tools and macros. If a mouse gesture is currently assigned to a tool, the icon for the gesture appears in the appropriate column for the command.

    For example, by default, the right mouse gesture is assigned to the Right tool for parts and assemblies, so the icon for that gesture (🖱➡) appears in the Part and Assembly columns for that tool.

    To filter the list of tools and macros, use the options at the top of the tab. By default, four mouse gesture directions are visible in the Mouse Gestures tab and available in the mouse gesture guide. Select 12 gestures to view and reassign commands for 12 gesture directions.

2. Find the row for the tool or macro you want to assign to a mouse gesture and click in the cell where that row intersects the appropriate column.

    For example, to assign Make Drawing from Part 🗒 to the Part column, click in the cell where the Make-Drawing-from-Part row and the Part column intersect.

    A list of either 4 or 12 gesture directions appears as shown, depending on whether you have the 4 gestures, or 12 gestures option selected.

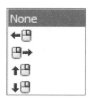 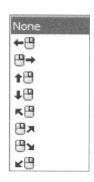  

4 gestures   **12 gestures**

Some tools are not applicable to all columns, so the cell is unavailable, and you cannot assign a mouse gesture. For example, you cannot assign a mouse gesture for Make Drawing from Part in the Assembly or Drawing columns.

3. Select the mouse gesture direction you want to assign from the list. The mouse gesture direction is reassigned to that tool and its icon appears in the cell.

4. Click OK.

Designed with SOLIDWORKS 2023

CHAPTER 1

System Options

System Options
Setting up the System Options

One of the first things to do after installing SOLIDWORKS is to set up the system options to use as the default settings for all documents.

System Options include the following:

Input dimension value, Face highlighting…

Drawing views controls, edge, and hatch display.

System colors, errors, sketch, text, grid, etc.

Sketch display, Automatic relation.

Edges display and selection controls.

Performance and Large assembly mode.

Area hatch/fill and hatch patterns.

Feature Manager and Spin Box increment controls.

View rotation and animation.

Backup files and locations, etc.

The settings are all set and saved in the **system registry**. While not part of the document itself, these settings affect all documents, including current and future documents.

This chapter will guide you through some of the options settings for use with this textbook; you may need to modify them to ensure full compatibility with your applications or company's standards.

System Options

The **General** Options

To start setting up the system options, go to: **Tools / Options.**

The first section is the **General** options.

Enable only the checkboxes as shown in the General Options dialog box.

Next, go down the list and select the **MBD** option. Enable the checkbox shown below.

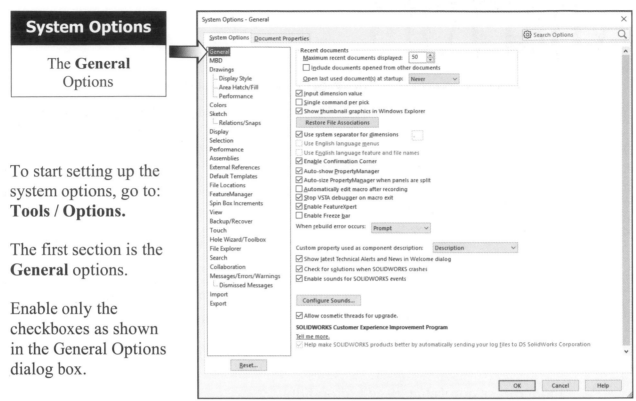

System Options

The **MBD** Options

This checkbox sets the options for Model Based Definition such as Editing the 3D-PDF Templates, adding the Company's Logo, Inserting Custom Text Fields, etc.

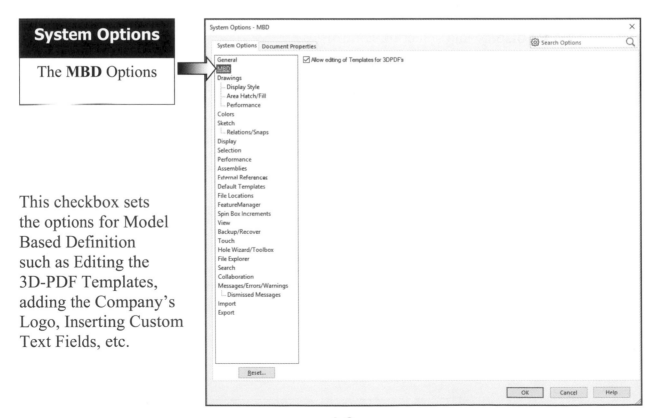

System Options

The **Drawings** Options

Enable only the check boxes as indicated in this dialog box.

(These settings are intended for use with this training manual only, you may need to modify them to meet your company's requirements).

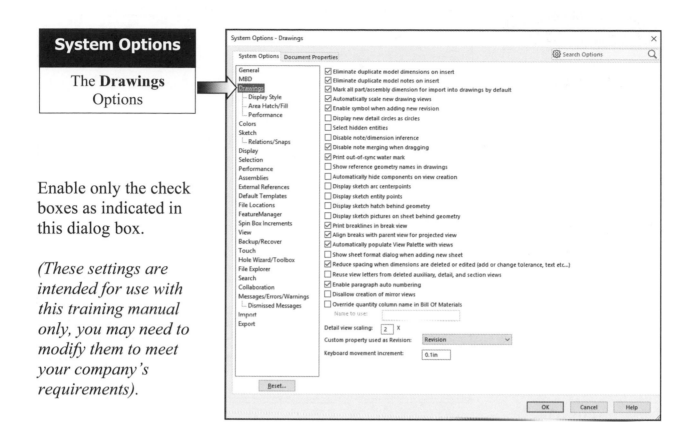

System Options

The **Display Style** Options

Go down the list and follow the sample settings shown in the dialog boxes to set up your System Options.

For more information on these settings click the Help button at the lower right corner of the dialog box (arrow).

System Options

The **Area Hatch/Fill** Options

The Area Hatch/Fill option sets the hatch pattern and Spacing (Scale).

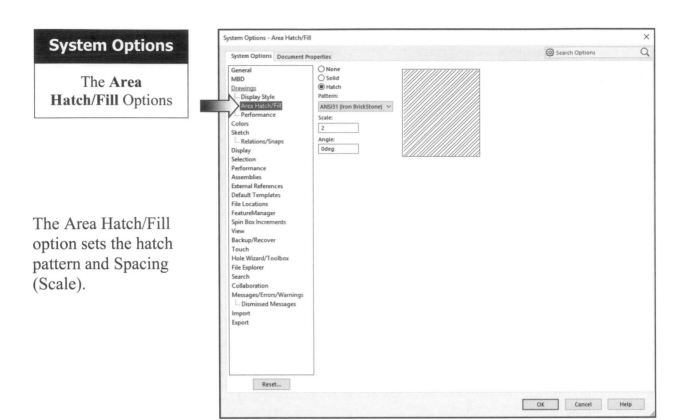

System Options

The **Performance** Options (Drawing)

Preview and High Quality display options take extra memory, which affects the performance of the computer. (Use the Automatically Load Components Light-weight option in the large assembly mode.)

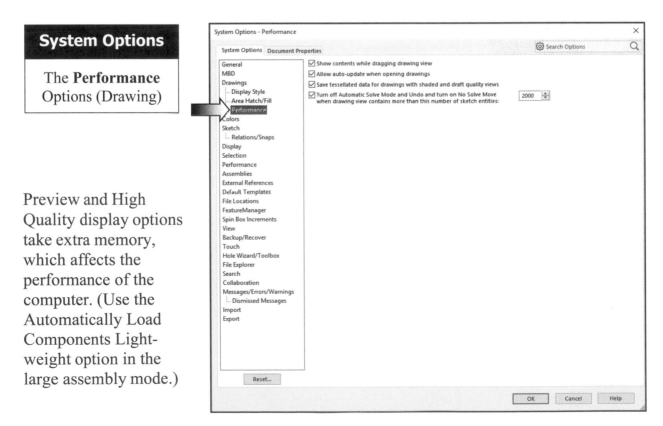

System Options

The **Colors** Options

The Colors options set the colors of the background and the Feature-Manager Design tree.

For Background-Appearance select the PLAIN option, and for Viewport Background color click Edit and select the White color.

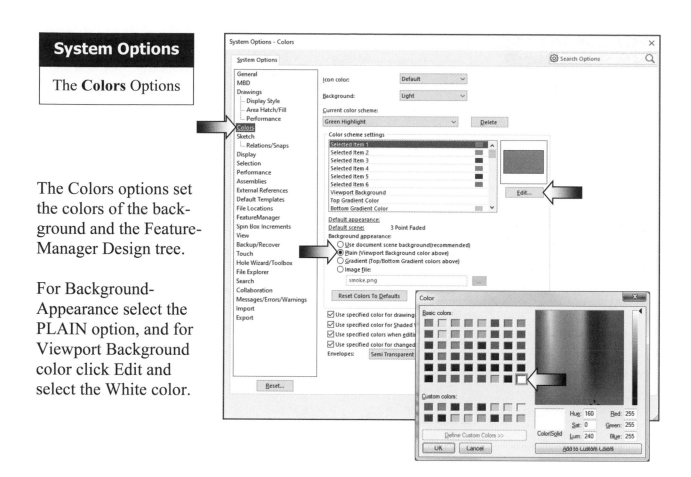

System Options

The **Sketch** Options

The Sketch options control the displays of the sketch entities and their orientation.

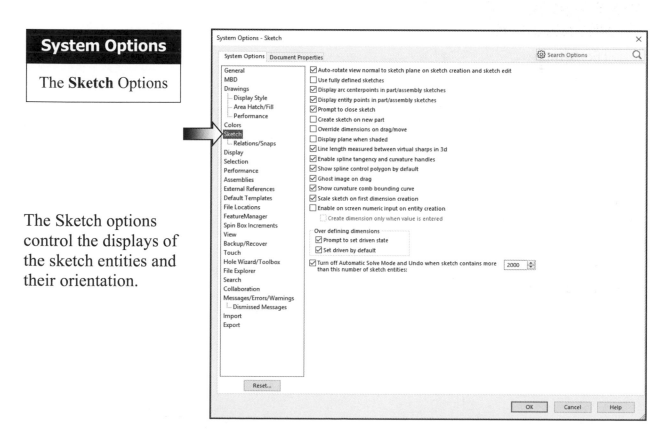

System Options

The **Relations / Snaps** Options

Enable all Automatic-Relation options except for the Grid option.

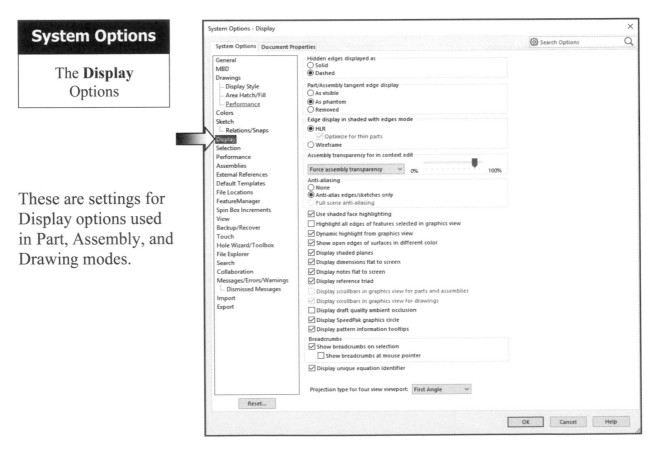

System Options

The **Display** Options

These are settings for Display options used in Part, Assembly, and Drawing modes.

System Options

The **Selection**
Options

Parameters for selection
options of edges in
Wireframe, Hidden
Lines Visible, Hidden
Lines Removed, and
Shaded modes.

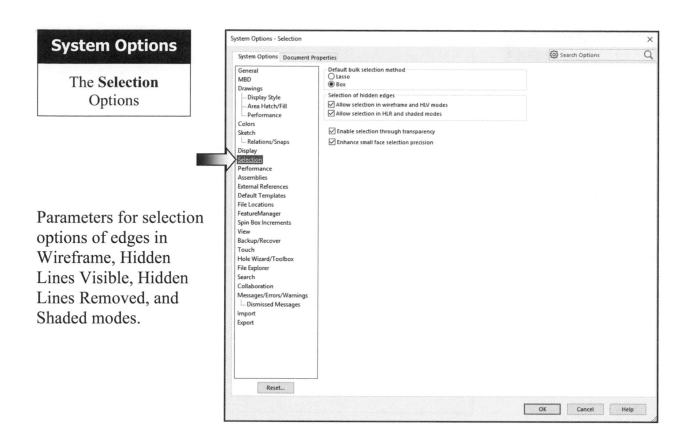

System Options

The **Performance**
Options (Assembly)

The Preview and High
Quality display options
require extra memory,
which affects the
performance of the
computer. (Enable the
Automatically Optimize
Resolved Mode, Hide
Lightweight Mode.)

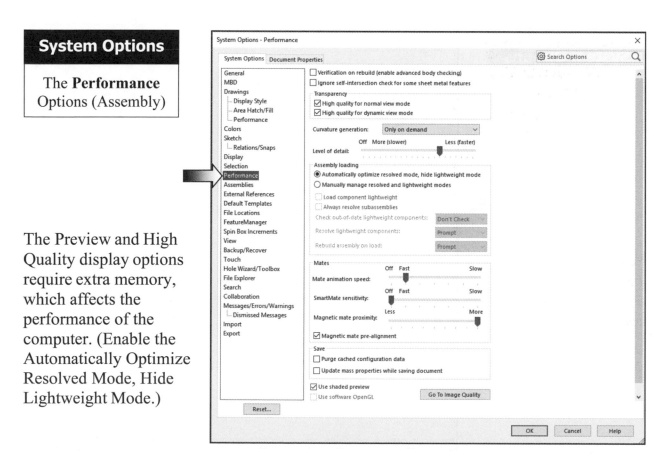

System Options

The **Assemblies** Options

Use the Lightweight Mode and Large Assembly Settings to help speed up the process of opening and saving large assemblies.

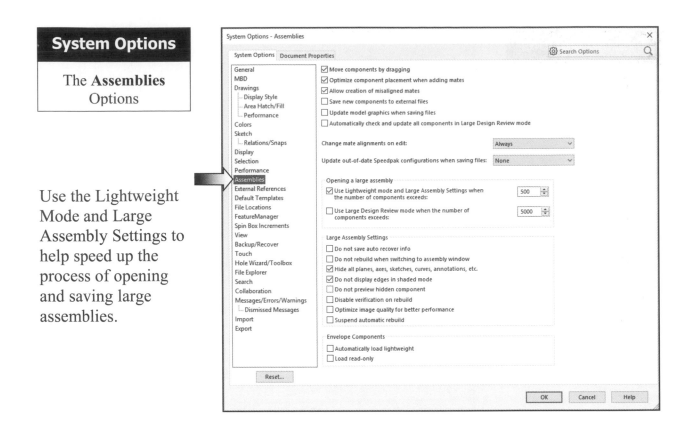

System Options

The **External References** Options

Select the options that help locate the file references and update component names.

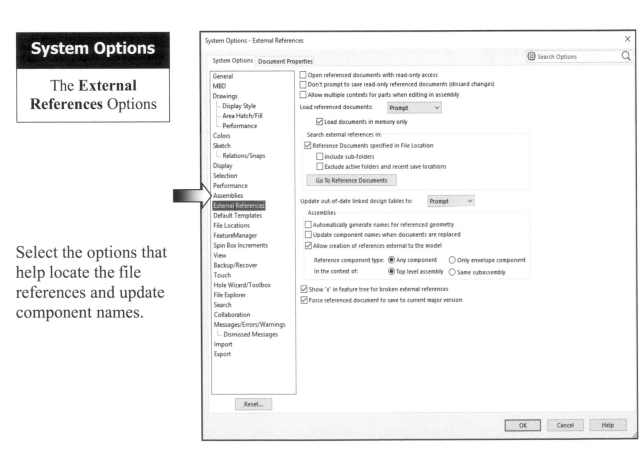

System Options

The **Default Templates** Options

Set the default Template for Parts, Assemblies, and Drawings.

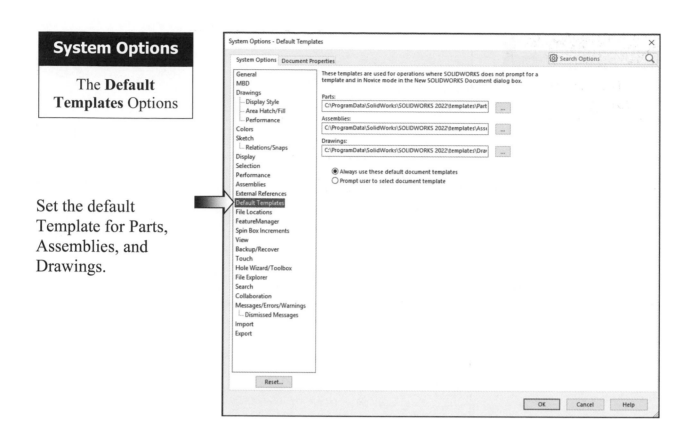

System Options

The **File Locations** Options

Locate and add the files and folders for: Document Templates, Design Library, Holes Table Templates, Weldments, etc.

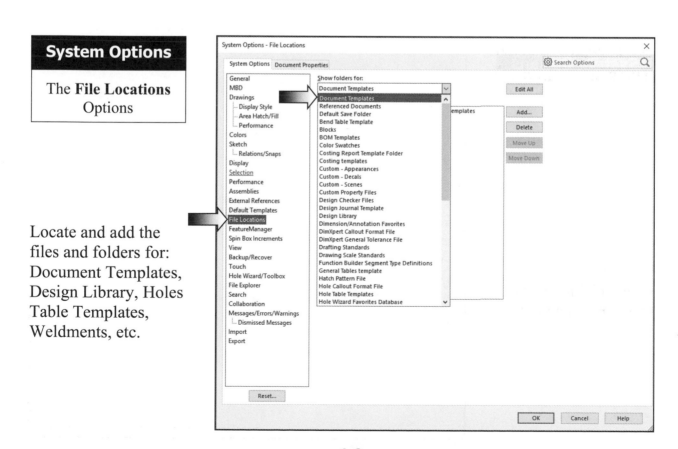

System Options

The **Feature-Manager** Options

Select the Folders
that you wish to display
on the FeatureManager
Design tree.

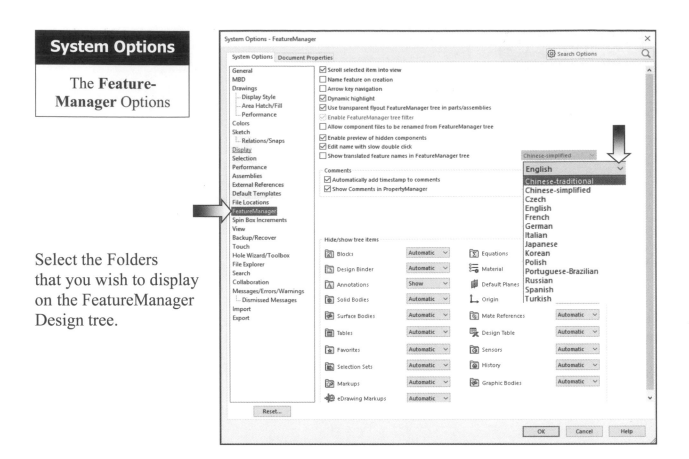

System Options

The **Spin Box Increments** Options

Set the default
increments for
the Modify Spin
Box (when creating
or Editing dimensions).

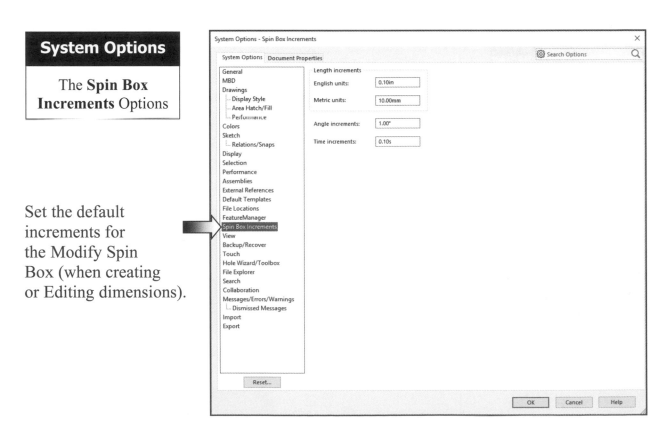

System Options

The **View** Options

Set the Rotation,
Mouse Speed,
and Transitions
of the model views.

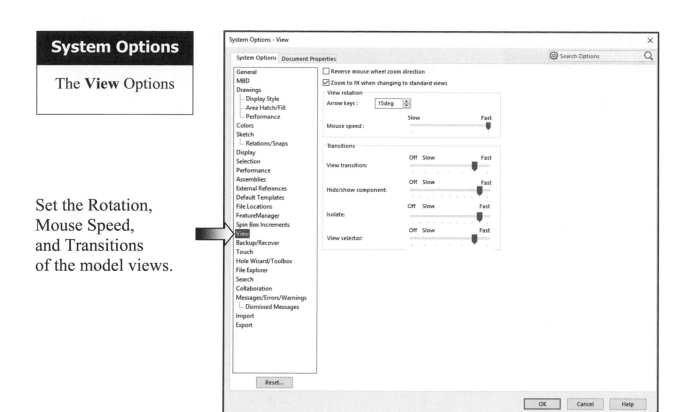

System Options

The **Backups /
Recover** Options

Set the Auto-
Recover Location,
Time and the
Number of
Backup Copies.

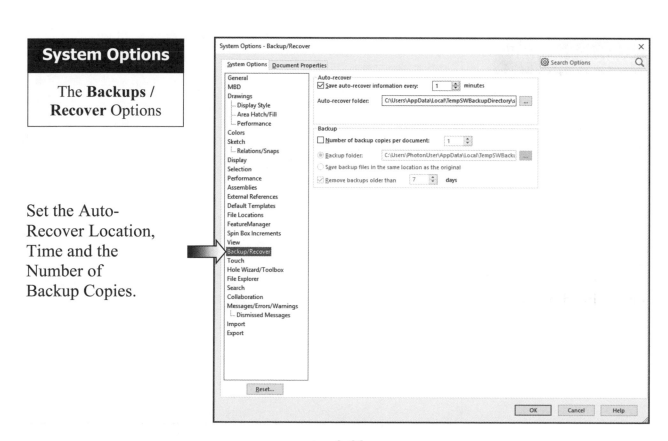

System Options

The **Touch** Options

With a Touch-enabled computer, you can use flick touch and multi-touch gestures in SOLIDWORKS 2023.

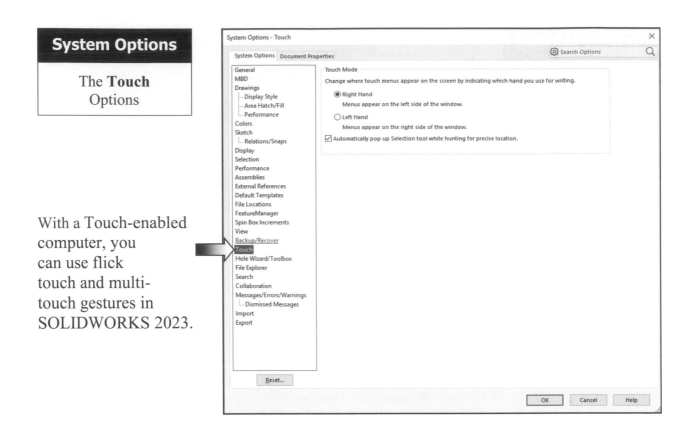

System Options

The **Hole-Wizard/Toolbox** Options

Locate the Hole-Wizard and Toolbox folder.

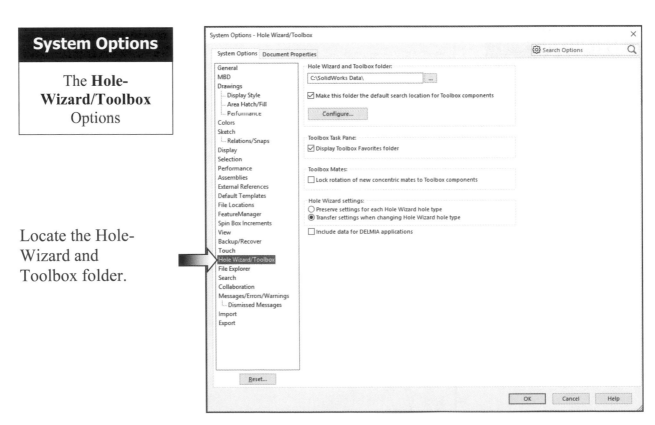

System Options

The **File Explorer** Options

Enable the File Locations for accessing the SOLIDWORKS documents.

System Options

The **Search** Options

Set the Search options.

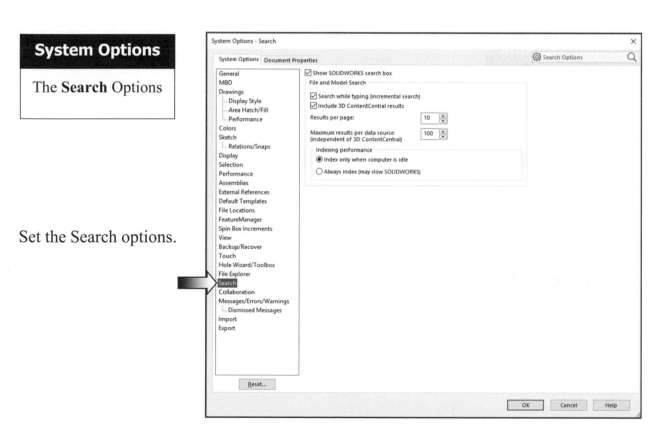

System Options

The **Collaboration**
Options

Enable / Disable
the Multi-User
Environment.

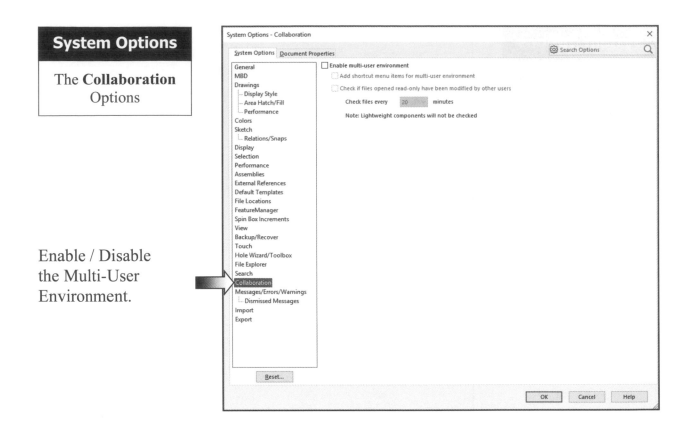

System Options

The **Messages
/Errors/Warnings**
Options

Set the Error and
Warning messages.

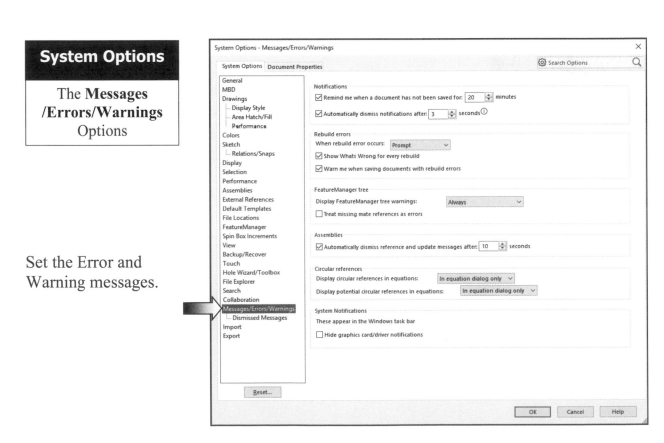

System Options

The **Import**
Options

Sets the options for
importing documents.

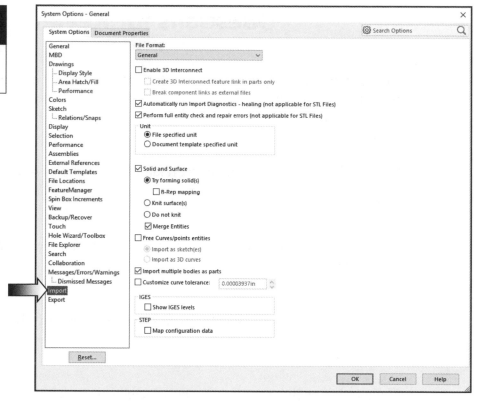

System Options

The **Export**
Options

Sets the options for
exporting documents.

Continue with
setting up the
Document Properties
in Chapter 2...

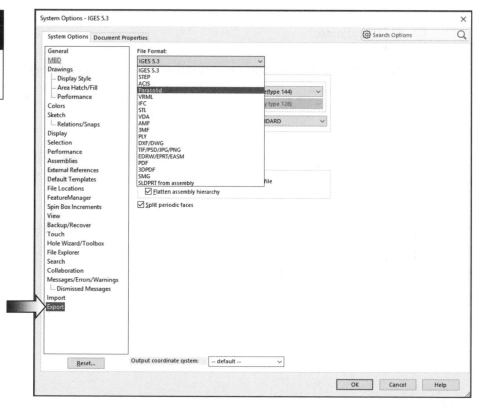

Questions for Review

1. The settings in the System Options affect all documents including the current and the future documents.
   a. True
   b. False

2. The sketch displays and automatic relations are samples of System Options.
   a. True
   b. False

3. The background colors cannot be changed or saved in the System Options.
   a. True
   b. False

4. The mouse wheel zoom direction can be reversed and saved as a default option.
   a. True
   b. False

5. Default document templates can be selected and saved in the System Options.
   a. True
   b. False

6. The spin box increment value is fixed in the System Options; it cannot be changed.
   a. True
   b. False

7. The number of backup copies for a document can be specified and saved in the System Options.
   a. True
   b. False

8. The System Options can be copied using the SOLIDWORKS Utility / Copy Settings Wizard.
   a. True
   b. False

1. TRUE   2. TRUE
3. FALSE 4. TRUE
5. FALSE 6. FALSE
7. TRUE   8. TRUE

CHAPTER 2

Document Templates

Document Templates
Setting up the Document Properties

After setting up the System Options, the next step is to set up the document template where drafting standards and other parameters can be set, saved, and used over and over again as a template. The Document Properties include the following:

Drafting Standard (ANSI, ISO, DIN, JIS, etc.).

Dimension, Note, Balloon, Fonts Sizes, and Annotation display.

Arrowhead sizes.

Grid spacing and grid display.

Units (Inches, Millimeters, etc.) and Decimal places.

Feature Colors, Wireframe, and Shading colors.

Material Properties.

Image quality controls.

Plane display controls.

These parameters are all set and saved in the templates; all settings affect only the **current document** (C:\Program Data\SOLIDWORKS\SOLIDWORKS 2023\ Templates) OR (C:\ProgramFiles\SOLIDWORKSCorp\SOLIDWORKS\ Lang\ English\Tutorial).

The following are examples of various document settings which are intended for use with this textbook only; you may need to modify them to ensure full compatibility with your applications.

Document Properties

The **Drafting Standard** options

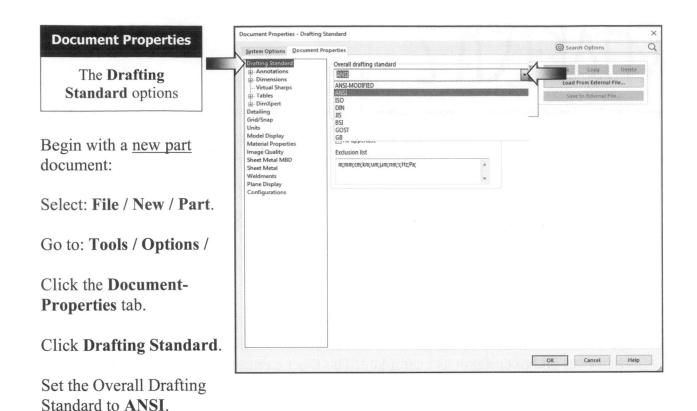

Begin with a <u>new part</u> document:

Select: **File / New / Part**.

Go to: **Tools / Options /**

Click the **Document-Properties** tab.

Click **Drafting Standard**.

Set the Overall Drafting Standard to **ANSI**.

<u>Note</u>: *If your Units are in Millimeters, skip to the Units options on page 2-13 and set your units to IPS, then return to where you left off and continue with your template settings.*

Document Properties

The **Annotations** options

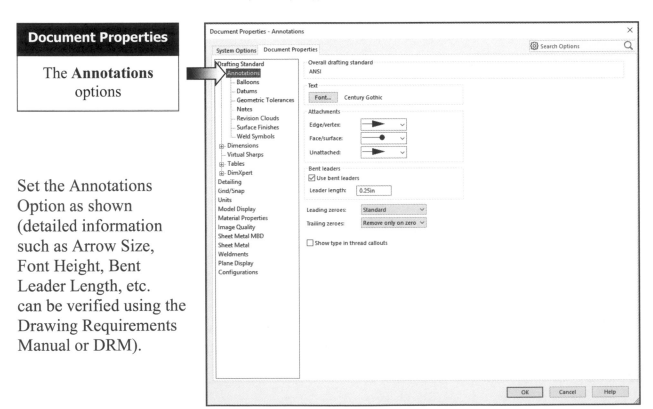

Set the Annotations Option as shown (detailed information such as Arrow Size, Font Height, Bent Leader Length, etc. can be verified using the Drawing Requirements Manual or DRM).

Document Properties

The **Annotations / Balloons** options

Set the Balloons standard to ANSI and the other options as shown.

Click the Help button at any time to access the information on these topics.

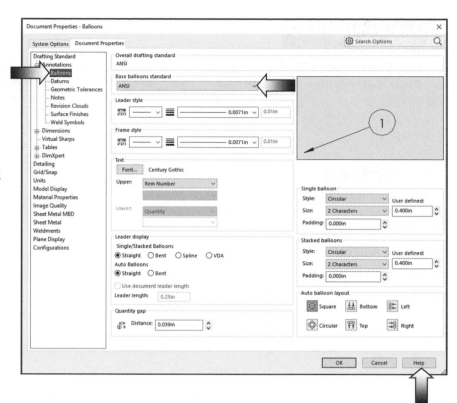

Document Properties

The **Annotations / Datums** options

Set the Datums standard to ANSI and other options as shown.

Note:

1982 Datum symbol

1994 Datum symbol

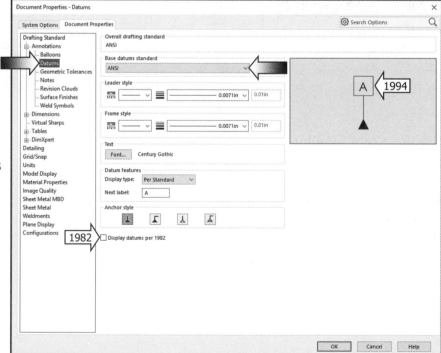

Document Properties

The **Annotations /
Geo. Tol.** options

Set the Geometric-
Tolerance standard
to ANSI.

The Font selection
should match the
other options for
all annotations.

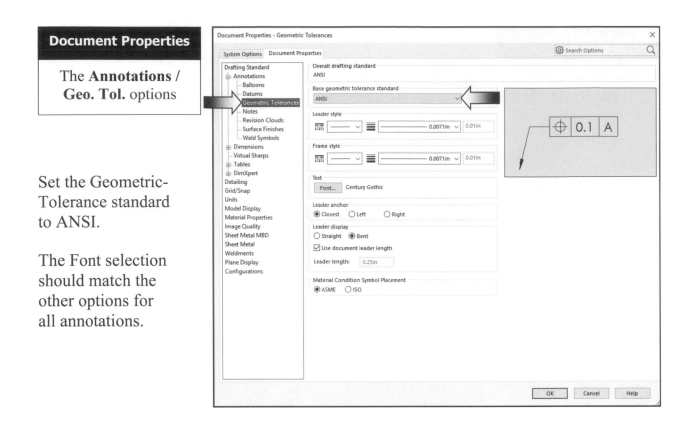

Document Properties

The **Annotations /
Notes** options

Set the Notes standard
to ANSI.

Use the same settings
for Font and Leader
display.

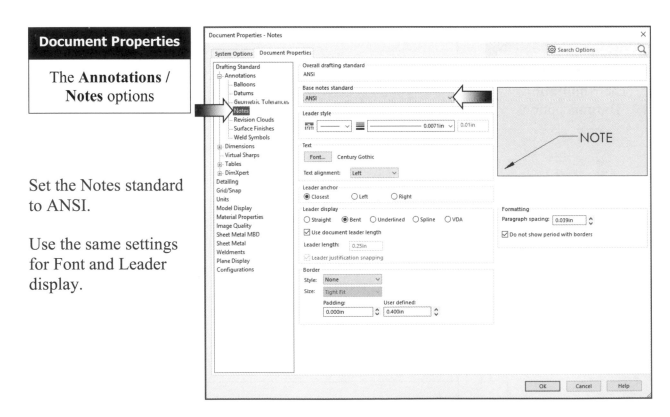

Document Properties

The **Revision Clouds**
options

Use Revision Clouds to
call attention to geometry
changes in a drawing.

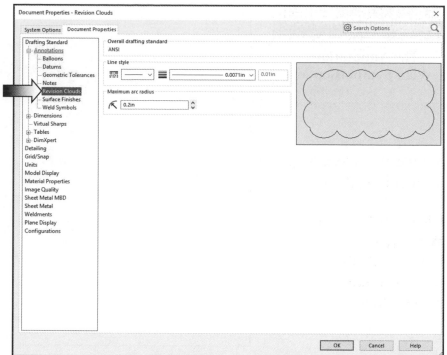

Document Properties

The **Surface Finishes**
options

Set the Surface Finish
standard to ANSI and
set the Leader Display
to match the ones in
the previous options.

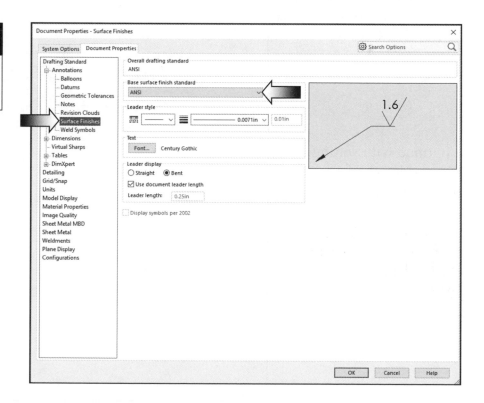

Document Properties

The **Annotations /
Weld Symbols** options

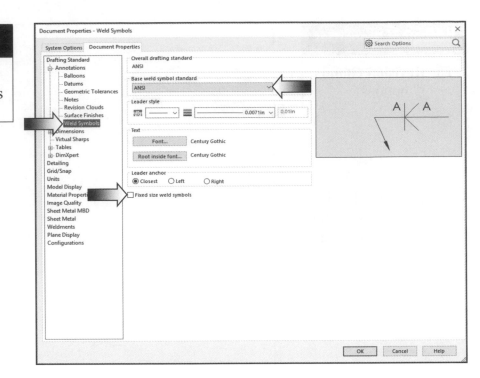

Set the Weld Symbols
standard to ANSI.

Clear the Fixed Size
Weld Symbols check-
box to scale the size
of the symbol to the
symbol font size.

Document Properties

The **Dimensions**
options

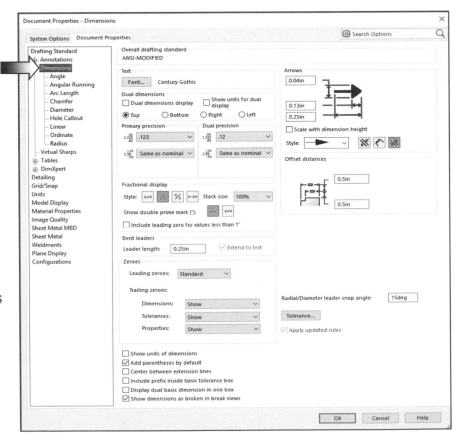

Continue with setting
the options shown in
the next dialog boxes.

The document-level
drafting standard settings
for all dimensions are
set here.

The **Dimensions / Angle** options

Set the Angle Dimension standard to ANSI and set the other options shown.

Set the number of decimal places for angular dimensions based on your company's standard practices.

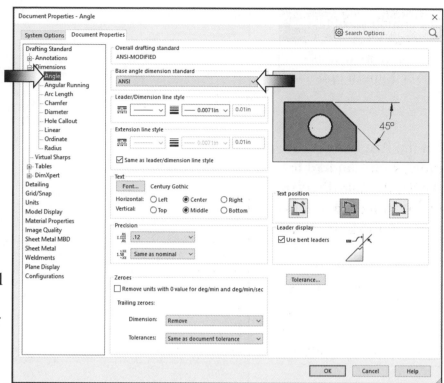

The **Angular Running** options

A set of dimensions measured from a zero-degree dimension in a sketch or drawing.

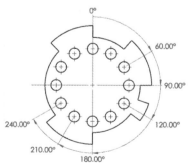

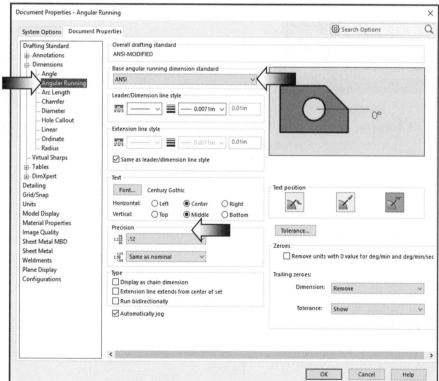

Set the Angular Precision to either 0 or 2 decimals (depends on your company's standard).

Document Properties

The **Dimensions /
Arc Length** options

Set the Arc Length
dimension standard to
ANSI, Primary Precision
to 3 decimals, and Dual
Precision to 2 decimals.

The Arc Length
dimension is created by
holding the Control key
and clicking the arc and
both of its endpoints.

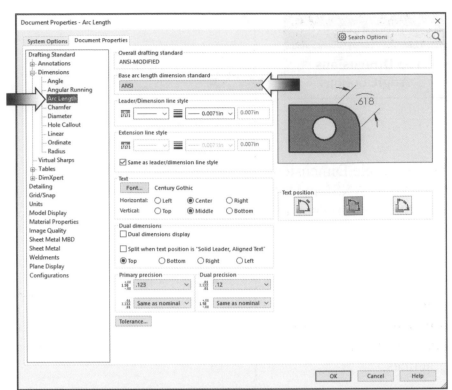

Document Properties

The **Dimensions /
Chamfer** options

Set the Chamfer-
Dimension standard to
ANSI and the Primary-
Precision to 3 decimals.

Set other options shown.

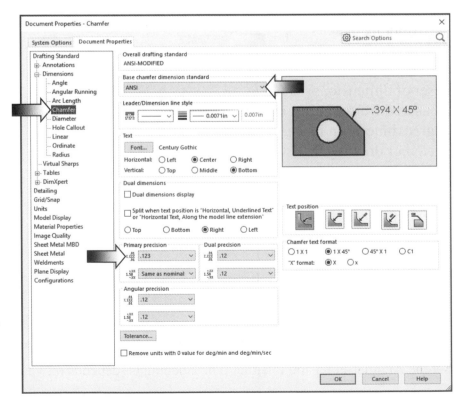

Document Properties

The **Dimensions /
Diameter** options

Set the Diameter-
Dimension standard
to ANSI.

Set the Leader Style
and Thickness here.

Set other options shown.

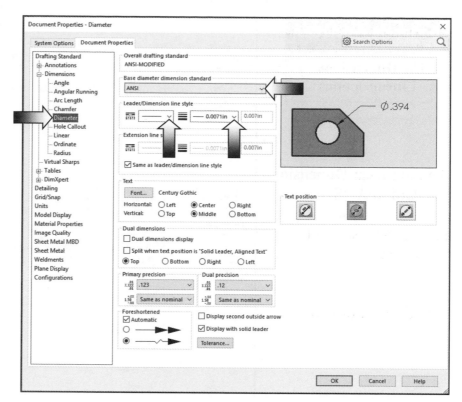

Document Properties

The **Dimensions /
Hole Callout** options

Set the Hole Callout
Dimension standard
to ANSI.

Set the Text justification
positions to Center and
Middle.

Set other options shown.

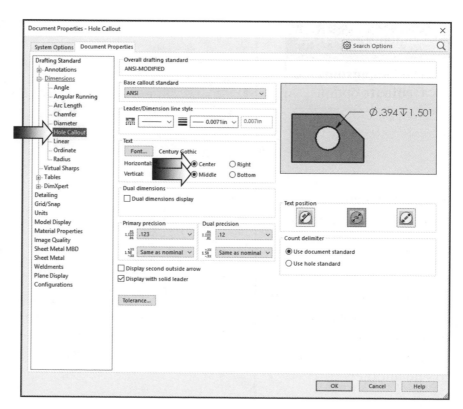

Document Properties

The **Dimensions /
Linear** options

Set the Linear Dimension
standard to ANSI.

Enable the Use Bent-
Leader checkbox.

Set other options shown.

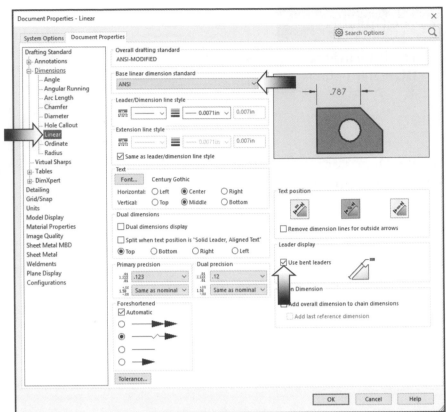

Document Properties

The **Dimensions /
Ordinate** options

Set the Ordinate-
Dimension standard to
ANSI.

Enable the Automatically
Jog Ordinates when
dimensions are
overlapping or too close
to one another.

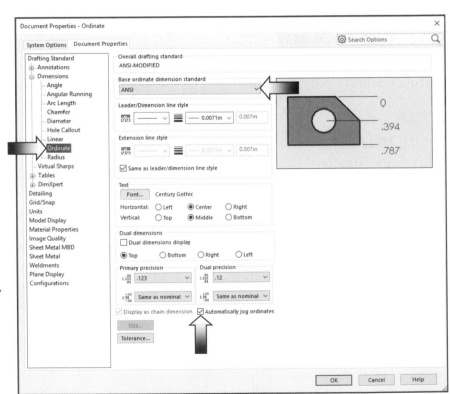

Document Properties

The **Dimensions /
Radius** options

Set the Radius-
Dimension standard
to ANSI.

Enable the Display with
Solid Leader Checkbox.

Set other options shown.

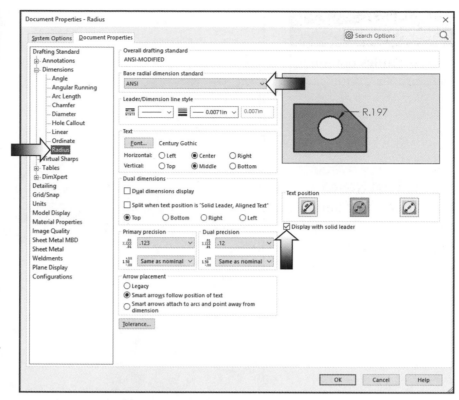

Document Properties

The **Virtual Sharps**
options

Select the Plus symbol
for Virtual Sharp
(the intersection
between the two
entities).

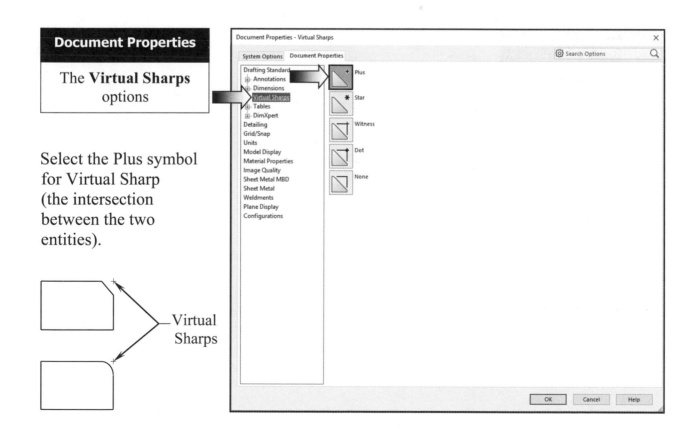

Virtual
Sharps

Document Properties

The **Tables** options

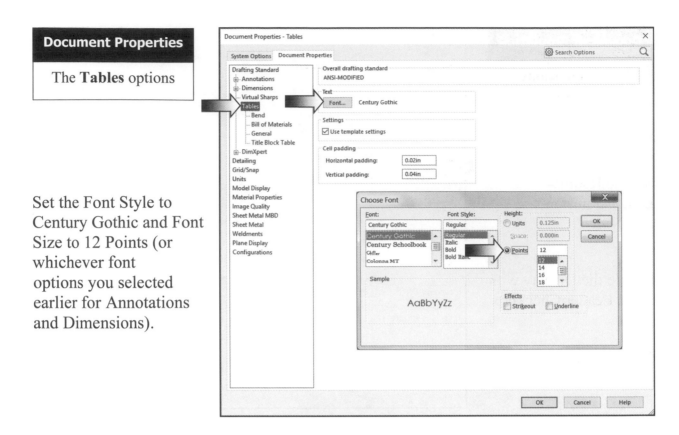

Set the Font Style to Century Gothic and Font Size to 12 Points (or whichever font options you selected earlier for Annotations and Dimensions).

Document Properties

The **Bend** options

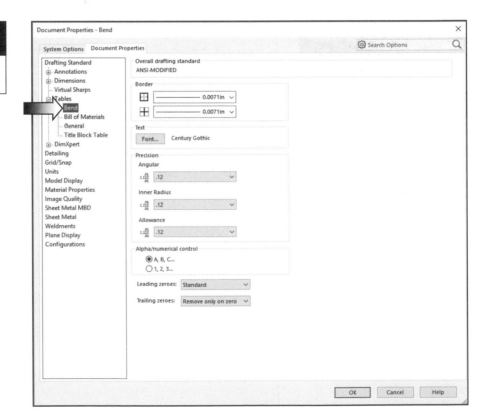

You can specify document-level drafting settings for bend tables. Available for drawings only.

Document Properties

The **Bill of Materials** options

Set the border for the Bill of Materials (BOM).

Enable the Automatic Update of BOM checkbox.

Set other options shown.

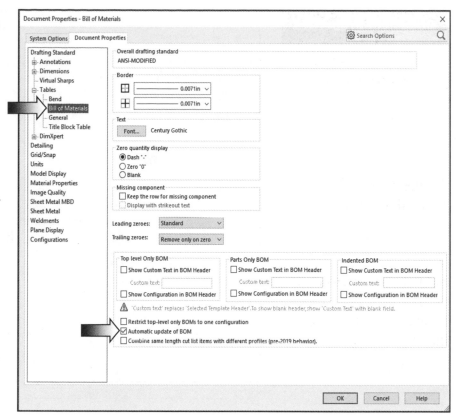

Document Properties

The **General** options

Create a General Table to use in a drawing. The user inputs data in the cell manually.

Set the line thickness for the border of the table here.

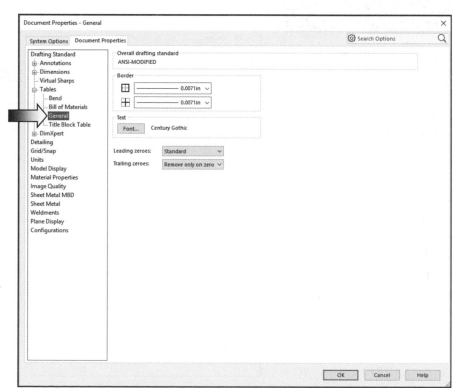

Document Properties

The **Title Block Table** options

Set the border for the Title Block Table.

Set the Thickness and the Font for the Title Block Table to match the BOM border.

Document Properties

The **DimXpert** options

Set the DimXpert to your company's specifications.

Block Tolerance is a common form of tolerancing used with inch units.

General Tolerance is a common form of tolerancing used with metric units in conjunction with the ISO standard.

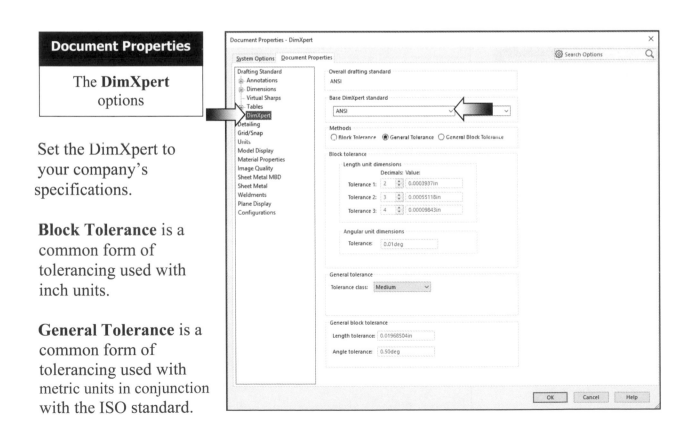

Document Properties

The **Size Dimension** options

Set per company Std.

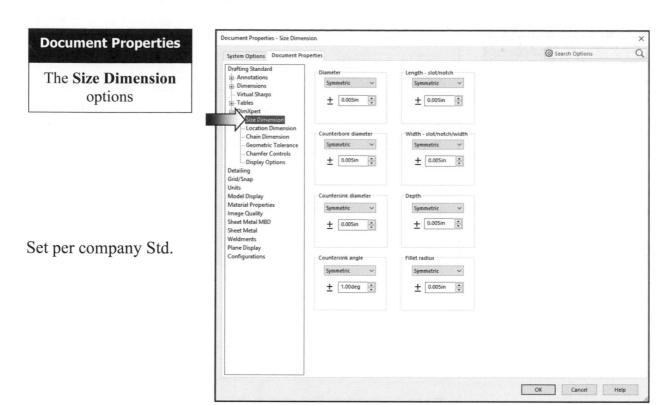

Document Properties

The **Location-Dimension** options

Set per company Std.

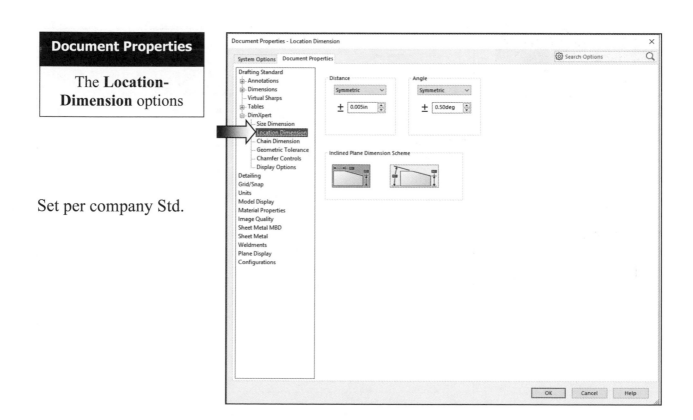

Document Properties

The **Chain Dimension** options

Set per company Std.

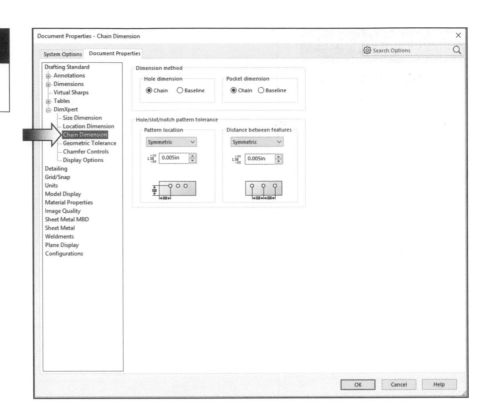

Document Properties

The **Geometric-Tolerance** options

Set per company Std.

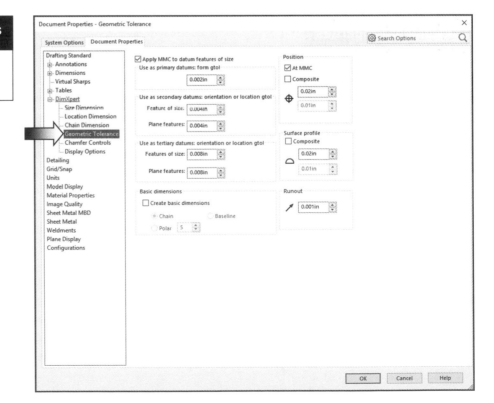

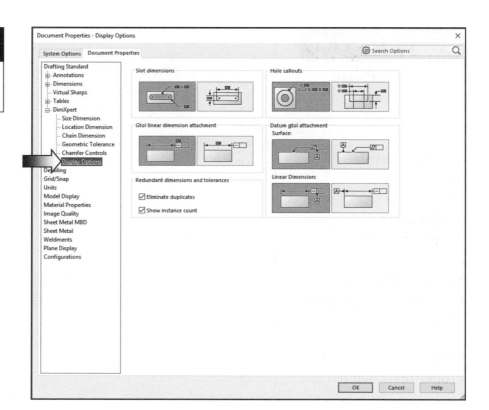

Document Properties

The **Chamfer-Controls** options

Set per company Std.

Document Properties

The **Display** options

Set per company Std.

Document Properties

The **Detailing** options

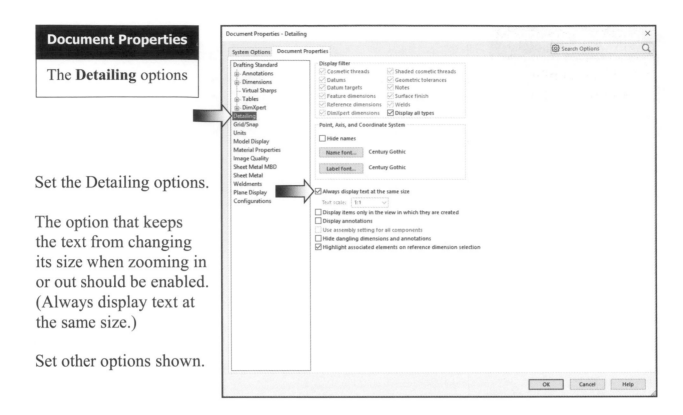

Set the Detailing options.

The option that keeps the text from changing its size when zooming in or out should be enabled. (Always display text at the same size.)

Set other options shown.

Document Properties

The **Grid/Snap** options

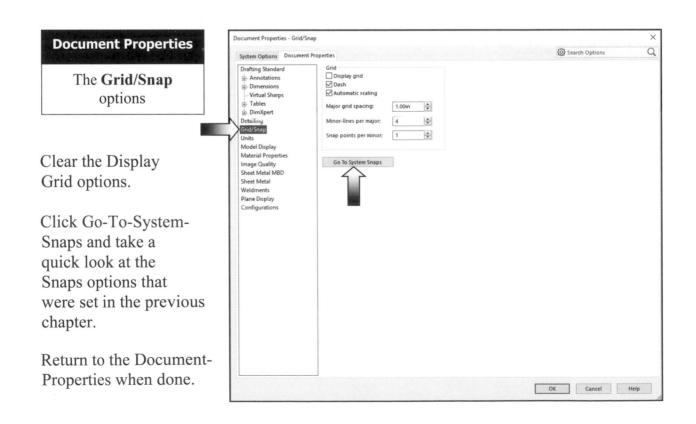

Clear the Display Grid options.

Click Go-To-System-Snaps and take a quick look at the Snaps options that were set in the previous chapter.

Return to the Document-Properties when done.

Document Properties

The **Units** options

Set the Unit options to IPS and the number of decimals as indicated.

(The Dual Dimension options are set in the Dimension section.)

Set other options shown.

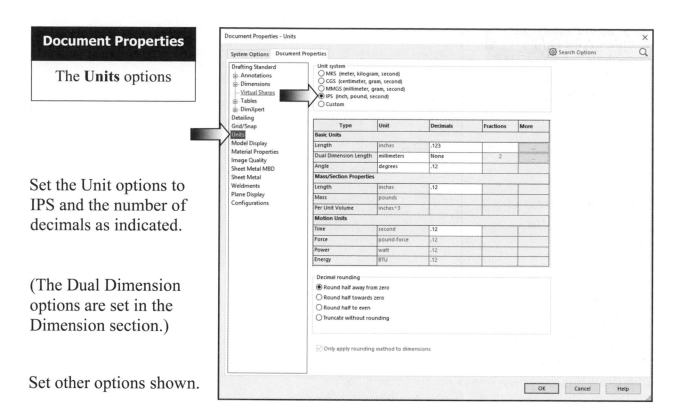

Document Properties

The **Model Display** options

Keep all default Colors as they are; do not change.

Uncheck the Store Appearance, Decal, and Scene… to reduce the file size.

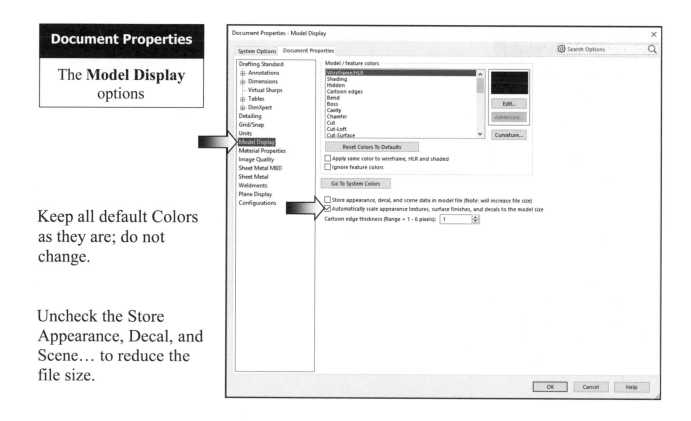

Document Properties

The **Material Properties** options

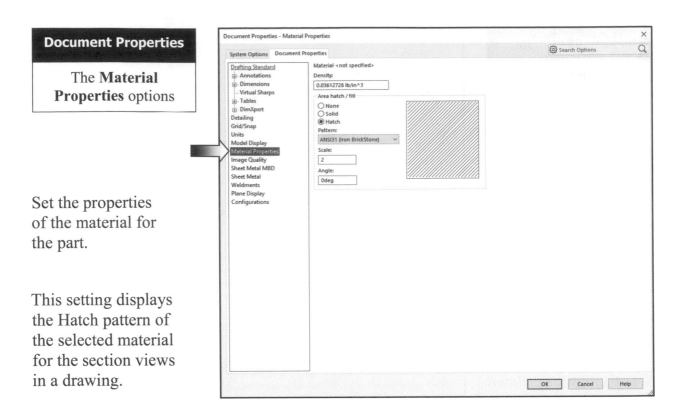

Set the properties of the material for the part.

This setting displays the Hatch pattern of the selected material for the section views in a drawing.

Document Properties

The **Image Quality** options

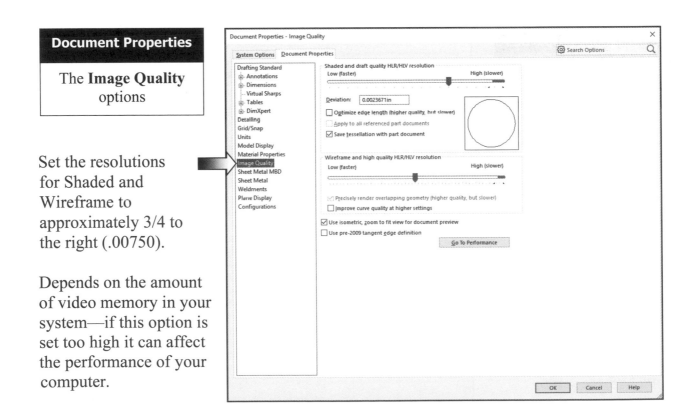

Set the resolutions for Shaded and Wireframe to approximately 3/4 to the right (.00750).

Depends on the amount of video memory in your system—if this option is set too high it can affect the performance of your computer.

Document Properties

The **Sheet Metal MBD** options

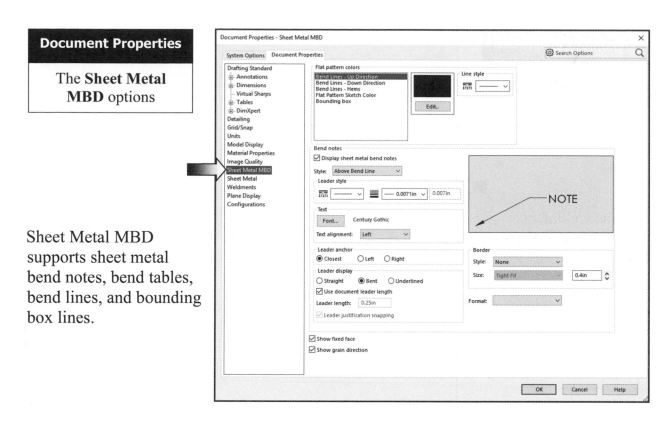

Sheet Metal MBD supports sheet metal bend notes, bend tables, bend lines, and bounding box lines.

Document Properties

The **Sheet Metal** options

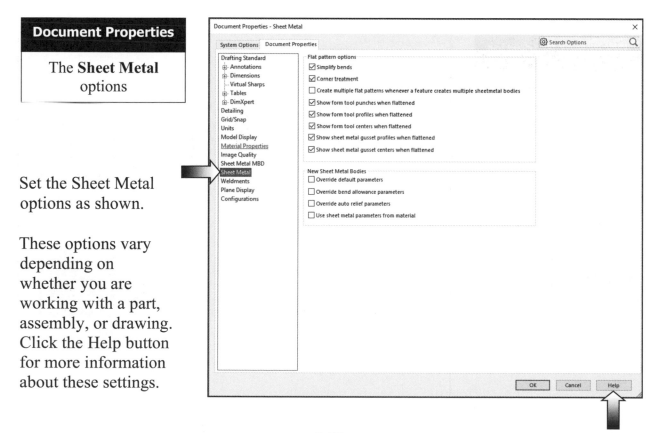

Set the Sheet Metal options as shown.

These options vary depending on whether you are working with a part, assembly, or drawing. Click the Help button for more information about these settings.

Document Properties

The **Weldments**
options

Sets the weldments
behaviors per document
basis.

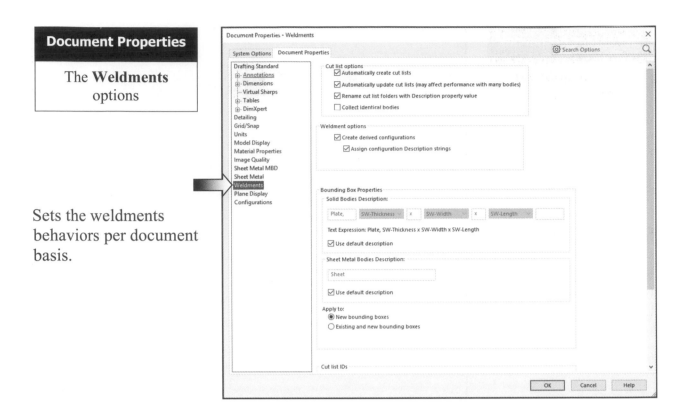

Document Properties

The **Plane Display**
options

Change the Plane
Colors and its
Transparency only
as needed.

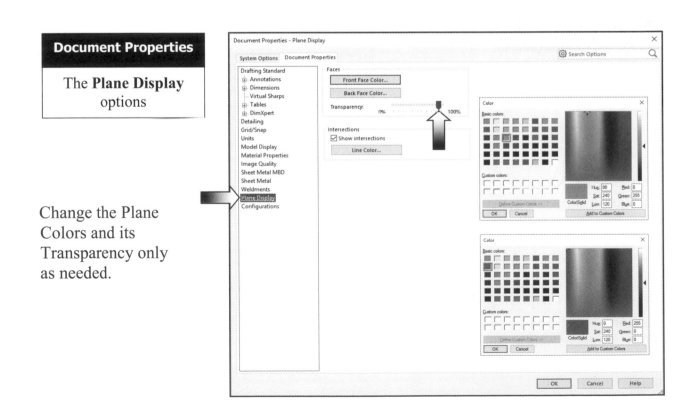

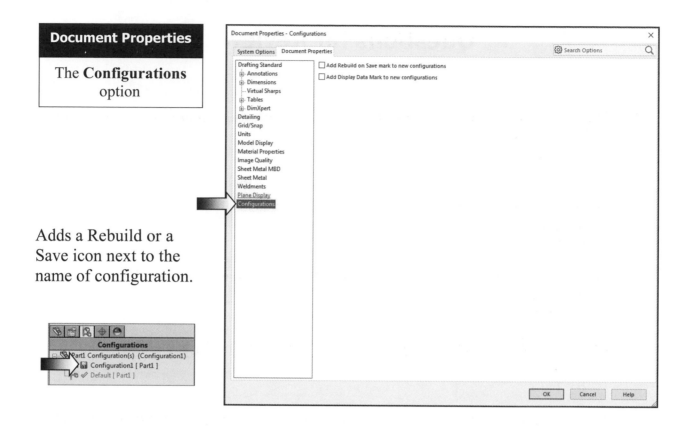

Document Properties

The **Configurations** option

Adds a Rebuild or a Save icon next to the name of configuration.

Saving the settings as a Part Template:

These settings should be saved within the Document Template for use in future documents.

Click **File / Save As.**

Change the Save-As-Type to **Part Templates** (*.prtdot or part document template).

Enter **Part-Inch.prtdot** for the file name.

Save either in the default Templates folder *(C:\ProgramData\SOLIDWORKS Corp\ SOLIDWORKS\Data\ Templates).*

Click **Save**.

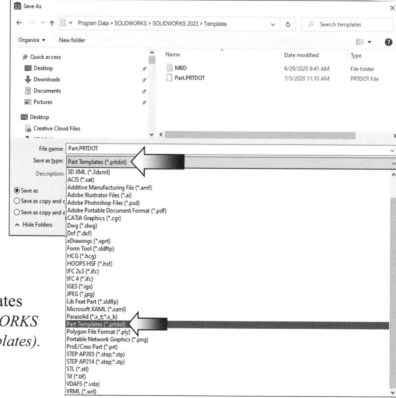

Questions for Review

1. The ANSI drafting standard (American National Standards Institute) can be set and saved in the System Options.
   a. True
   b. False

2. The size of dimension arrows can be controlled globally from the Document Templates.
   a. True
   b. False

3. The balloon's size and shape can be set and saved in the Document Templates.
   a. True
   b. False

4. Dimension and note fonts can be changed and edited both locally and globally.
   a. True
   b. False

5. The Grid option is only available in the drawing environment, not in the part or assembly.
   a. True
   b. False

6. The maximum number of decimal places can be set to 10 digits.
   a. True
   b. False

7. The feature colors can be pre-set and saved in the Document Templates.
   a. True
   b. False

8. The display quality of the model can be adjusted using the settings in the Image Quality option.
   a. True
   b. False

9. The plane colors and transparency can be set and saved in the Document Templates.
   a. True
   b. False

9. TRUE
7. TRUE   8. TRUE
5. FALSE   6. FALSE
3. TRUE   4. TRUE
1. FALSE   2. TRUE

CHAPTER 3

Basic Solid Modeling — Extrude Options

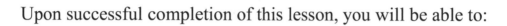

Basic Solid Modeling
Extrude Options

Upon successful completion of this lesson, you will be able to:

* Sketch on planes and/or planar surfaces

* Use the sketch tools to construct geometry

* Add the geometric relations or constraints

* Add/modify dimensions

* Explore the different extrude options

The following 5 basic steps will be demonstrated throughout this exercise:

* Select the sketch plane (or sketch face)

* Activate Sketch toolbar

* Sketch the profile using the sketch tools

* Define the profile with dimensions or relations

* Extrude the profile

Be sure to review the self-test questionnaires at the end of the lesson prior to moving to the next chapter.

Basic Solid Modeling
Extrude Options

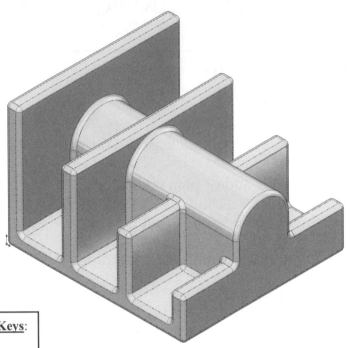

View Orientation Hot Keys:

Ctrl + 1 = Front View
Ctrl + 2 = Back View
Ctrl + 3 = Left View
Ctrl + 4 = Right View
Ctrl + 5 = Top View
Ctrl + 6 = Bottom View
Ctrl + 7 = Isometric View
Ctrl + 8 = Normal To
        Selection

Dimensioning Standards: **ANSI**

Units: **INCHES** – 3 Decimals

Tools Needed:

 Insert Sketch

 Line

 Circle

 Add Geometric Relations

 Dimension

 Sketch Fillet

 Trim Entities

 Boss / Base Extrude

Parent & Child explained:

The sample part below has <u>1 Parent</u> feature (the Base) and <u>3 Child</u> features (the Boss, the Holes, and the Fillets).

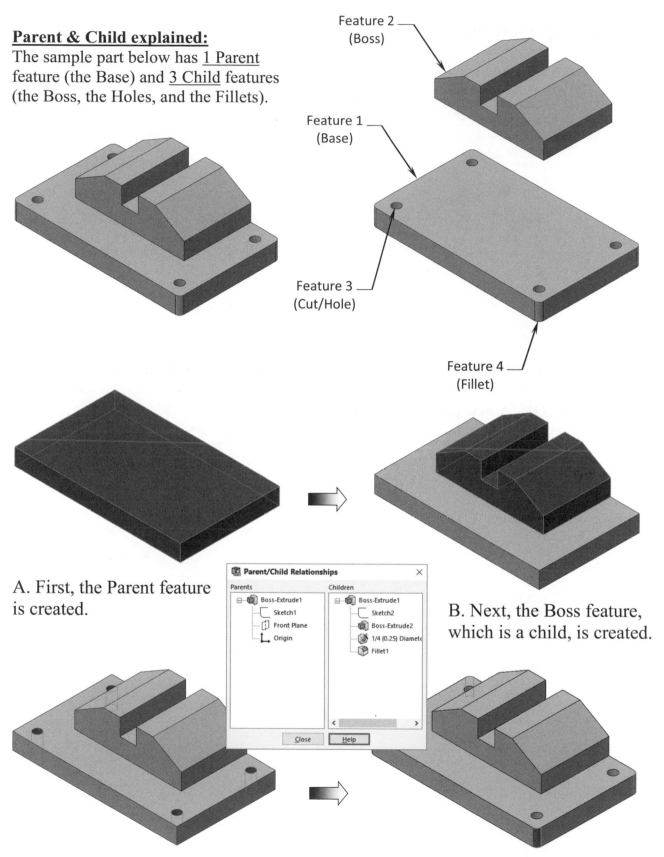

Feature 2
(Boss)

Feature 1
(Base)

Feature 3
(Cut/Hole)

Feature 4
(Fillet)

A. First, the Parent feature is created.

B. Next, the Boss feature, which is a child, is created.

C. The features that remove material such as Extruded Cuts or Holes are created next.

D. Finally, the Fillets and Chamfers features are added last.

1. Starting a new Part:

From the **File** menu, select **New / Part**, or click the **New** icon.

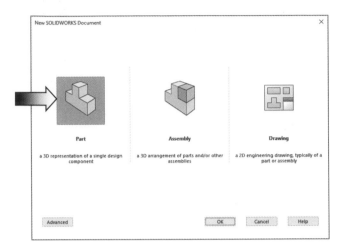

Select the **Part** template from either the Templates or Tutorial folders.

Click **OK**. A new part template is open.

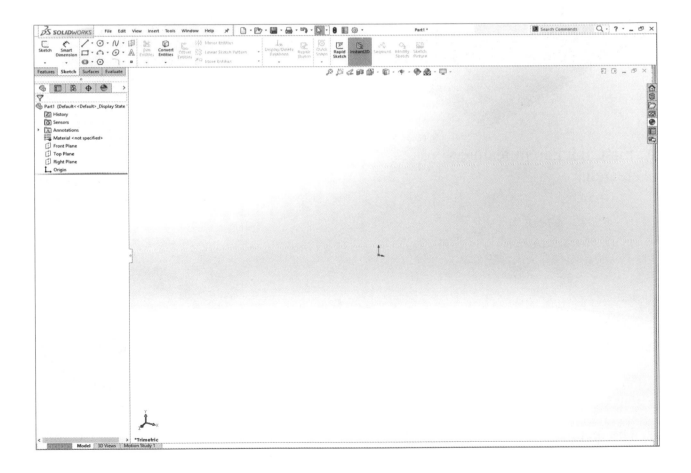

2. Changing the Scene:

From the View (Heads-up) toolbar, click the **Apply Scene** button (arrow) and select the **Plain White** option (arrow).

By changing the scene color to **Plain White,** we can see the colors of the sketch entities and their dimensions a little better.

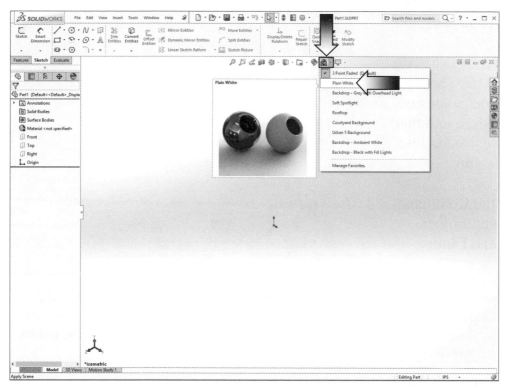

To show the Origin, click the **View, Hide /Show** drop-down menu and select **Origins**.

The Blue color Origin is the Zero position of the part and the Red Origin is the Zero position of the sketch.

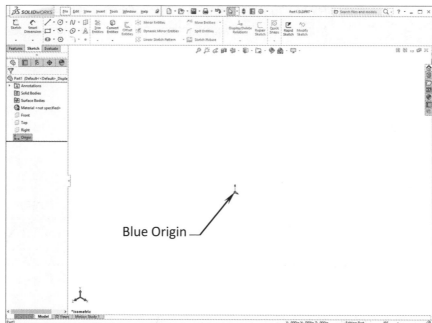

3. Starting a new Sketch:

Select the <u>Front</u> plane from the Feature-Manager tree and click the **Sketch** drop-down arrow and select the **Sketch** command to start a new sketch.

A sketch is usually created first, relations and dimensions are added after, and then it gets extruded into a 3D feature.

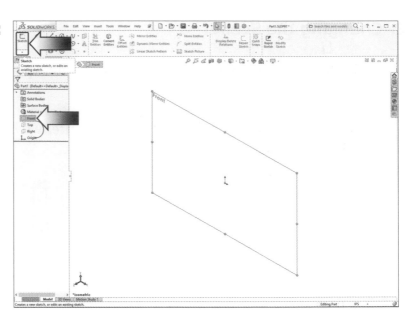

From the Command-Manager toolbar, select the **Line** command.

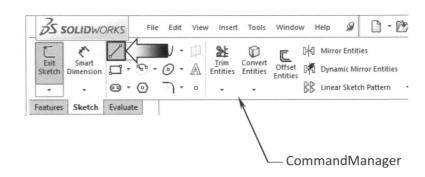

CommandManager

Mouse-Gesture

OPTIONAL:
Right-Drag to display the Mouse-Gesture wheel and select the Line command from it. (See the Introduction section, page XVIII for details on customizing the Mouse Gesture.)

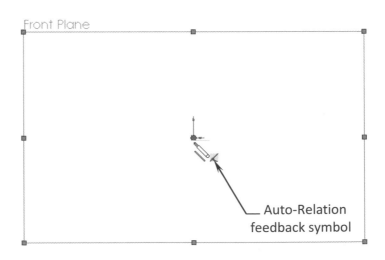

Hover the mouse cursor over the Origin point; a yellow feedback symbol appears to indicate a relation (Coincident) is going to be added automatically to the 1st endpoint of the line. This endpoint will be locked at the zero position.

Auto-Relation feedback symbol

<u>**Note:**</u> *There are 3 different ways to use the line tool:* **Single Line**, **Multiple Lines** *(connecting),*
and **Line-to-Arc** *(transitioning from a line to a tangent arc).*

The following steps are examples to demonstrate how to use all embedded
functions of the Line command. We will delete the example and go back to the
lesson on page 3-8.

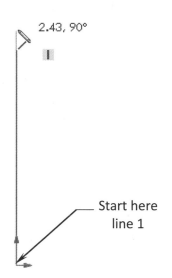

A. To sketch a <u>single line</u>, use the **click + hold + drag**
method:

With the Line tool already selected, click on the
Origin point to start the line (press and hold the
mouse button), drag the cursor upward, look for
the **vertical relation** symbol, and release the
mouse button.

Only 1 line is drawn; the mouse cursor <u>is dis-
connected</u> from the line.

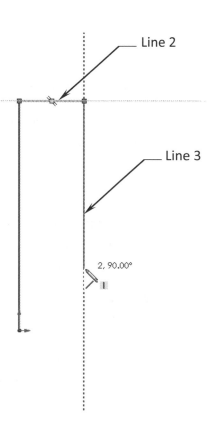

B. To sketch <u>multiple lines</u>, use the
click + release + click method.

Position the mouse cursor at the **top end point**
of the first line and click to start the line (do not
hold the mouse button this time).
Move the cursor to the right and look for the
Horizontal relation symbol and click to complete
the line.

The next line <u>is connecting</u> to the previous line.
Move the mouse cursor downward, look for the
vertical relation symbol and click to complete
the line.

The next line is connecting to the previous line.

C. To change from a **Line** to a **Tangent Arc,**
use the **click + release + press A** method.

Start a new line at the bottom of the 3rd line
and move the mouse cursor outward approx.
as shown.

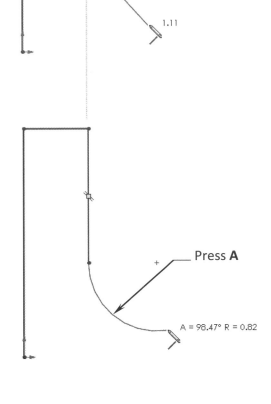

Move
downward

1.11

D. Press <u>once</u> on the **A** key and then move
the cursor slightly to the left or to the right.
The line is changed into a **Tangent Arc**.

Move the cursor back and forth, up and
down to see how the Tangent Arc maintains
its tangent relation with the line that it is
connecting to.

Press **A**

A = 98.47° R = 0.82

Based on the starting point, the arc
changes its direction from negative
to positive when the cursor moves
pass its zero point.

Move the mouse cursor towards
the lower right side and click to
complete the arc.

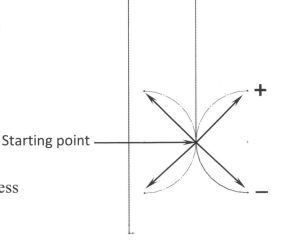

+

Starting point

−

<u>Delete</u> all sketch entities before
moving on to step number 4 (press
Control+A then push Delete).

4. Using the Click + Hold + Drag technique:

Select the **Line** command. Click the Origin point and *hold* the mouse button to start the line at point 1, *drag upwards* to point 2, then release the mouse button.

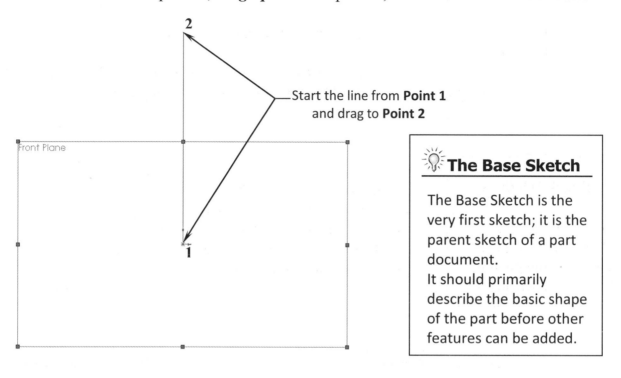

Start the line from **Point 1** and drag to **Point 2**

💡 The Base Sketch

The Base Sketch is the very first sketch; it is the parent sketch of a part document.
It should primarily describe the basic shape of the part before other features can be added.

Continue adding other lines using the *Click-Hold-Drag* technique.

The relations like Horizontal and Vertical are added automatically to each sketch line. Other relations like Collinear and Equal are added manually.

The size and shape of the profile will be corrected in the next few steps.

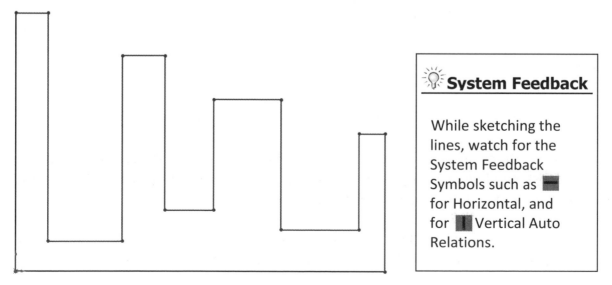

💡 System Feedback

While sketching the lines, watch for the System Feedback Symbols such as ▬ for Horizontal, and for ▌ Vertical Auto Relations.

5. Adding Geometric Relations*:

Click **Add Relation**  under Display/Delete Relations - OR - select **Tools / Relations / Add**.

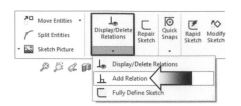

Select the **4 lines** shown below.

Click **Equal** from the Add Geometric Relation dialog box. This relation makes the length of the four selected lines equal.

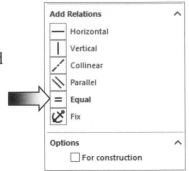

Equal Relations

Adding the EQUAL relations to these lines eliminates the need to dimension each line.

Geometric relations can be created manually or automatically. The next few steps in this chapter will demonstrate how geometric relations are added manually.

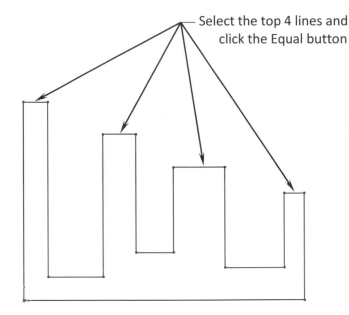

Select the top 4 lines and click the Equal button

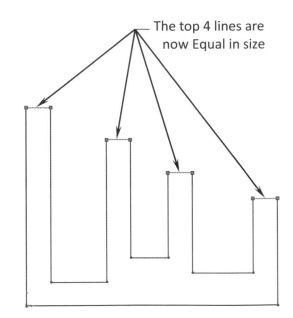

The top 4 lines are now Equal in size

* Geometric relations are one of the most powerful features in SOLIDWORKS. They are used in the sketch level to control the behaviors of the sketch entities when they are moved or rotated and to keep the associations between one another.

(When applying geometric relations between entities, one of them should be a 2D entity and the other can either be a 2D sketch entity, a model edge, a plane, an axis, or a curve).

6. Adding a Collinear relation**:

Select the **Add Relation** 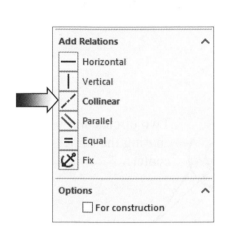 command again.

Select the **3 lines** shown below.

Click **Collinear** from the Add Geometric Relations dialog box.

Click **OK** to accept the relation.

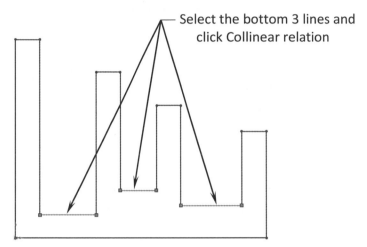

Select the bottom 3 lines and click Collinear relation

Collinear Relations

Adding a Collinear relation to these lines will force them to lie at the same straight line.
Only one dimension is needed to drive the height of all 3 lines.

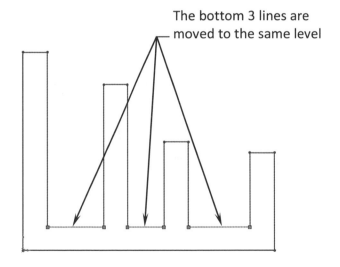

The bottom 3 lines are moved to the same level

** Collinear relations can be used to constrain the geometry as follows:

Collinear between a line and another line(s) (2D and 2D).

Collinear between a line(s) to a linear edge of a model (2D and 3D).

Geometric Relation Examples

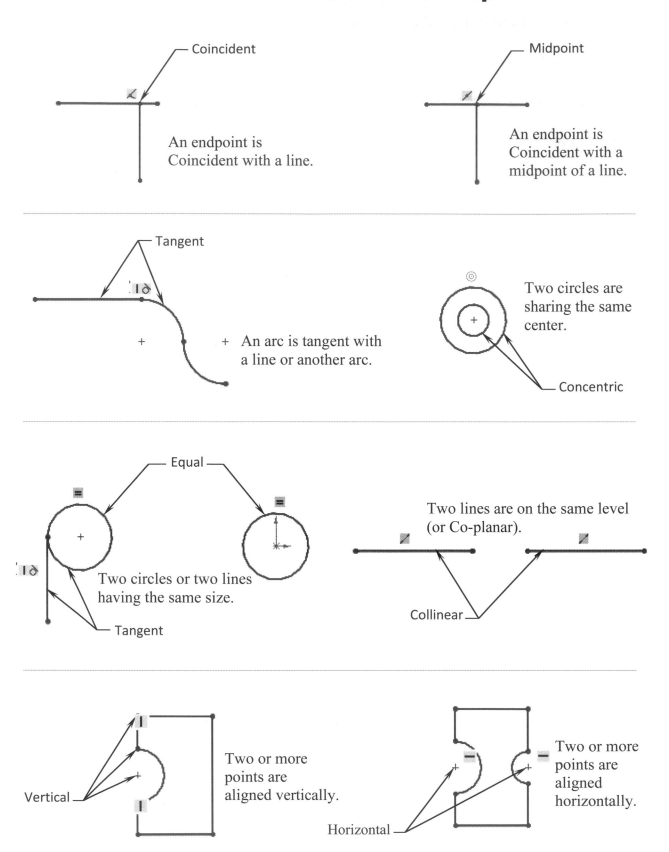

Coincident

An endpoint is
Coincident with a line.

Midpoint

An endpoint is
Coincident with a
midpoint of a line.

Tangent

An arc is tangent with
a line or another arc.

Two circles are
sharing the same
center.

Concentric

Equal

Two circles or two lines
having the same size.

Tangent

Two lines are on the same level
(or Co-planar).

Collinear

Vertical

Two or more
points are
aligned vertically.

Two or more
points are
aligned
horizontally.

Horizontal

7. Adding the horizontal dimensions:

Select 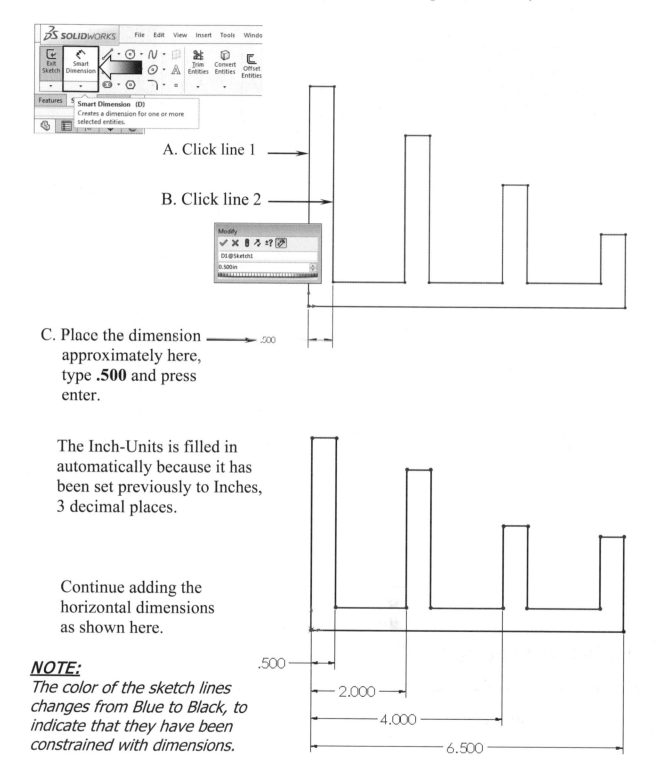 from the Sketch toolbar - OR – press the letter D on the keyboard, and Add the dimensions shown below (follow the 3 steps A, B and C).

A. Click line 1 ⟶

B. Click line 2 ⟶

C. Place the dimension ⟶ .500
approximately here,
type **.500** and press
enter.

The Inch-Units is filled in
automatically because it has
been set previously to Inches,
3 decimal places.

Continue adding the
horizontal dimensions
as shown here.

NOTE:
The color of the sketch lines
changes from Blue to Black, to
indicate that they have been
constrained with dimensions.

8. Adding the Vertical dimensions:

With the Smart-Dimension tool still selected, select line 1 and line 2, then place the dimension approximately as shown; change the value to **.500 in**.

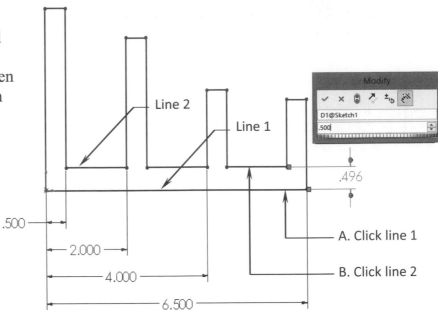

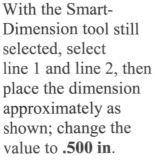

A. Click line 1

B. Click line 2

Continue adding other dimensions until the entire sketch turns into the Black color.

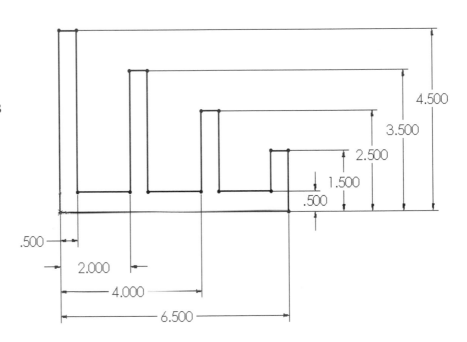

The Status of a Sketch:

The current status of a sketch is displayed in the lower right corner of the screen.

Fully Defined	=	**Black**	Fully Defined
Under Defined	=	**Blue**	Under Defined
Over Defined	=	**Red**	Over Defined

Sketch Relation Symbols

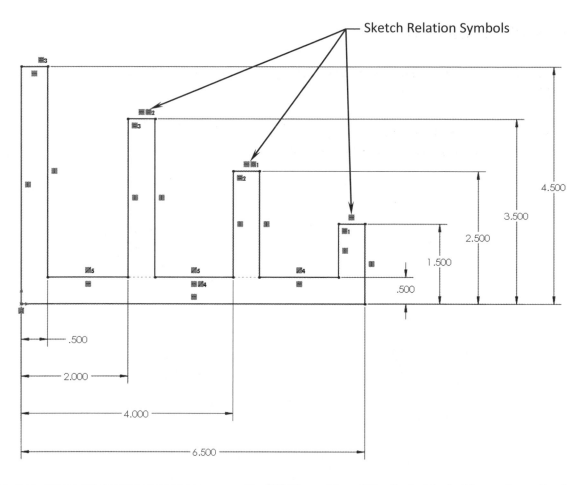

.500

2.000

4.000

6.500

1.500

2.500

3.500

4.500

.500

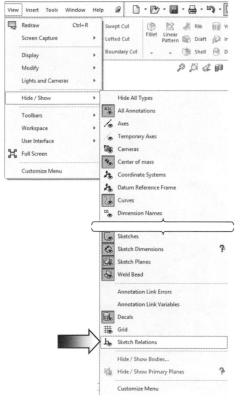

9. Hiding the Sketch Relation Symbols:

The Sketch Relation Symbols indicate which geometric relation a sketch entity has, but they get quite busy as shown.

To hide or show the Sketch Relation Symbols, go to the **View, Hide/Show** menus and click off the **Sketch Relations** option.

Sketch Relation Symbols at a Glance

▬	Horizontal relation	▮	Vertical relation
═	Equal relation	◩	Coincident relation
◩	Tangent relation	◪	Collinear relation

10. Extruding the Base:

The **Extrude Boss/Base** command is used to define the characteristic of a 3D linear feature.

Switch to the **Features** tab and click 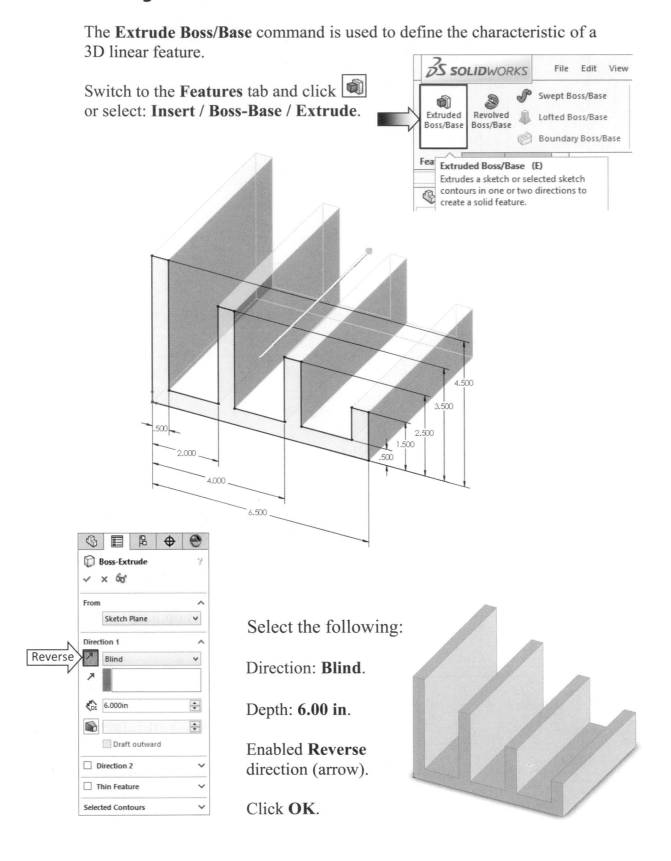 or select: **Insert / Boss-Base / Extrude**.

Select the following:

Direction: **Blind**.

Depth: **6.00 in**.

Enabled **Reverse** direction (arrow).

Click **OK**.

11. Sketching on a Planar Face:

Select the <u>face</u> as indicated.

Click 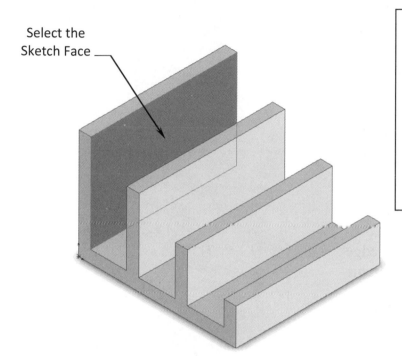 or select **Insert/Sketch** and press the shortcut key **Ctrl+7** to change to the Isometric view.

Select the **Circle** command from the Sketch Tools toolbar.

Select the
Sketch Face

> **Planar Surfaces**
>
> A planar surface of the model can also be used as a Sketch Plane.
>
> The Sketch will then be extruded normal to the selected surface.

Position the mouse cursor near the center of the selected face, click and drag outward to make a circle.

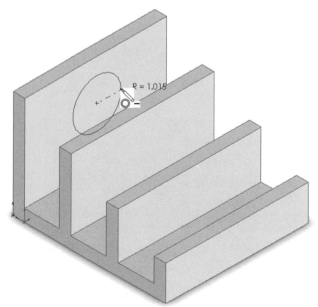

R = 1.015

While sketching the circle, the system displays the radius value next to the mouse cursor.

Dimensions are added after the profile is created.

Select the **Smart Dimension**

command 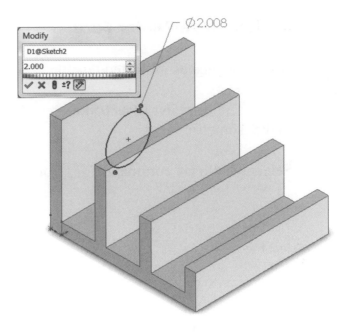 and add a
diameter dimension to the
circle. (Click on the edge of
the circle and move the cursor
outward at approximately 45
degrees, and click to place
the dimension.)

To add the location dimensions,
click the edge of the circle and
the edge of the model, place
the dimension, then correct
the value.

Continue adding the location
dimensions as shown to fully
define the sketch.

Select the Line command
and sketch the **3 lines** as shown
below. Snap to the hidden edge
of the model when it highlights.

The color of the sketch should
change to black at this point
(Fully Defined).

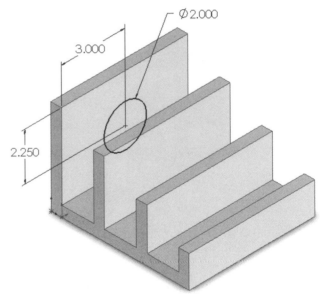

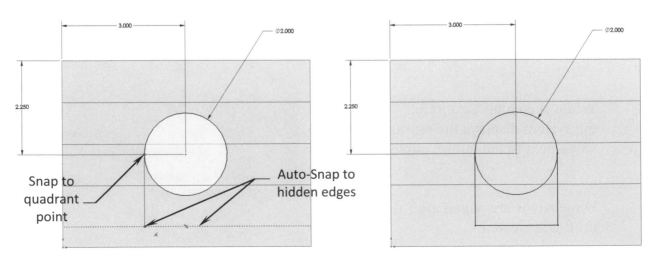

12. Using the Trim Entities command:

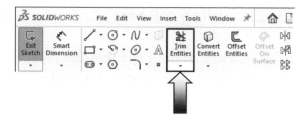

Select the **Trim Entities** command from the Sketch toolbar (arrow).

Click the **Trim to Closest** option (arrow). When the pointer is hovered over a sketch entity, this trim command will highlight the entity prior to trimming them to the next intersection.

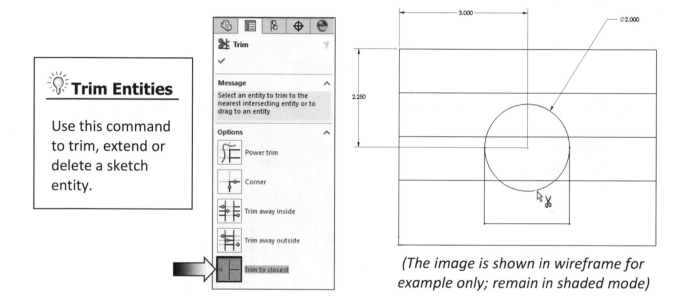

> 💡 **Trim Entities**
>
> Use this command to trim, extend or delete a sketch entity.

(The image is shown in wireframe for example only; remain in shaded mode)

Hover the pointer over the lower portion of the circle; the portion that is going to be trimmed will highlight. Click the mouse button to trim.

The bottom portion of the circle is trimmed, leaving the sketch as one-continuous-closed-profile, suitable to extrude and make a 3D feature.

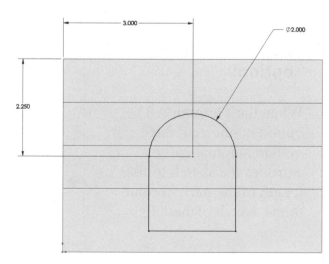

Next, we are going to look at some of the extrude options available in SOLIDWORKS.

13. Extruding a Boss:

Switch to the **Features** tab and click 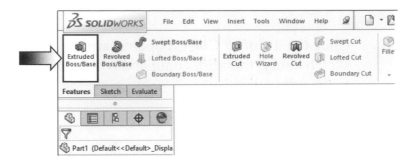 or select:
Insert / Boss-Base / Extrude.

Extrude Options...

Explore each extrude option to see the different results. Press Undo to go back to the original state after each one.

A **Using the Blind option:**

When extruding with the Blind option, the following conditions are required:

* Direction

* Depth dimension

Drag the direction arrow on the preview graphics to define the direction, and then enter a dimension for the extrude depth.

Direction & Depth

Blind
Condition

B **Using the Through All option:**

When the Through All option is selected, the system automatically extrudes the sketch to the length of the part, normal to the sketch plane.

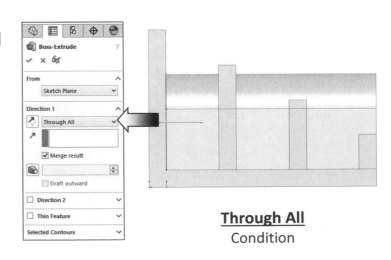

Through All
Condition

C Using the Up To Next option:

With the Up To Next option selected, the system extrudes the sketch to the very next group of surface(s), and blends it to match the geometry of the surface(s).

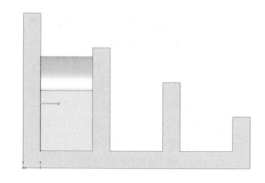

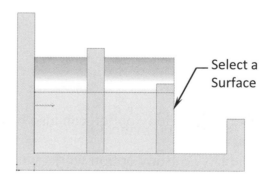

Up To Next
Condition

D Using the Up To Vertex option:

This option extrudes the sketch from its plane to a vertex, specified by the user, to define its depth.

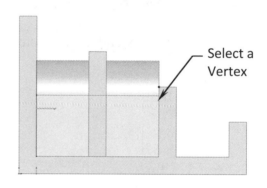

Select a Vertex

Up To Vertex
Condition

E Using the Up To Surface option:

This option extrudes the sketch from its plane to a single surface to define its depth.

Select a Surface

Up To Surface
Condition

F Using the Offset From Surface option:

This option extrudes the sketch from its plane to a selected face, and then offsets at a distance specified by the user.

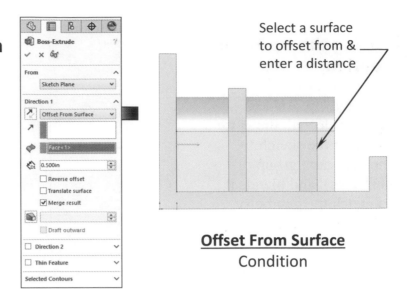

Select a surface to offset from & enter a distance

Offset From Surface
Condition

G Using the Up To Body option (optional):

This option extrudes the sketch from its sketch plane and blends to a selected body.

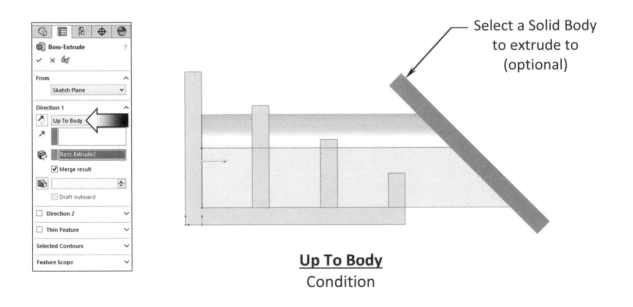

Select a Solid Body to extrude to (optional)

Up To Body
Condition

The Up To Body option can also be used in assemblies or multi-body parts.

The Up To Body option works with either a solid body or a surface body.

It is also useful when making extrusions in an assembly to extend a sketch to an uneven surface.

H Using the Mid Plane option:

This option extrudes the sketch from its plane equally in both directions.

Enter the Total Depth dimension when using the Mid-Plane option.

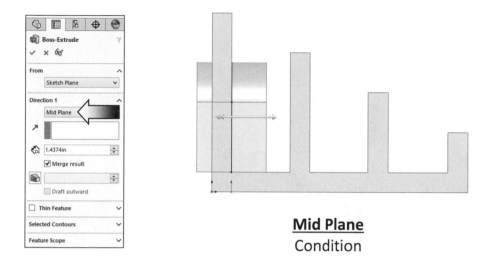

Mid Plane
Condition

After you are done with exploring all the extrude options, change the Direction 1 to **Through All**.

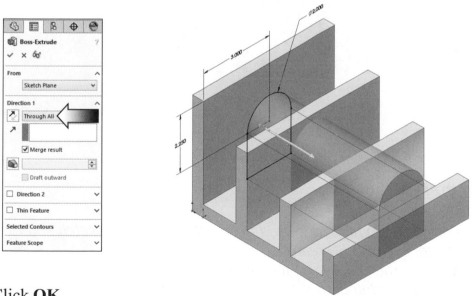

Click **OK**.

The system extrudes the sketch to the outermost surface as the result of the Through All end condition.

The overlapped material between the first and the second extruded features is removed automatically.

If the Merge Result checkbox is cleared, all interferences will be kept and the model became a multibody part.

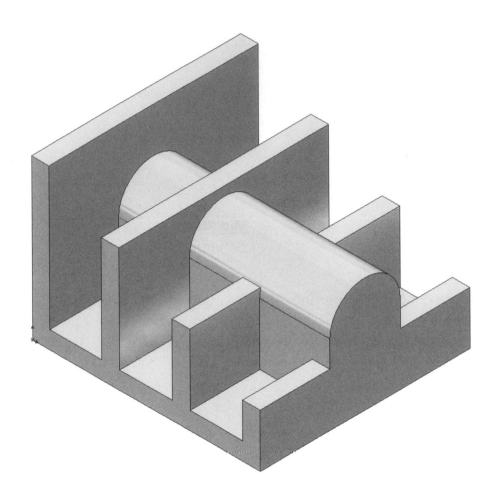

Extrude summary:

* The Extrude Boss/Base command is used to add thickness to a sketch and to define the characteristic of a 3D feature.

* A sketch can be extruded in both directions at the same time, from its sketch plane.

* A sketch can also be extruded as a solid or a thin feature.

14. Adding the model fillets:

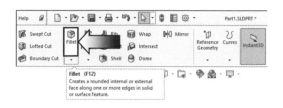

Fillet/Round creates a rounded internal or external face on the part. You can fillet all edges of a face, select groups of faces, edges, or edge loops.

The **radius** value stays in effect until you change it.
Therefore, you can select any number of edges or faces in the same operation.

Click 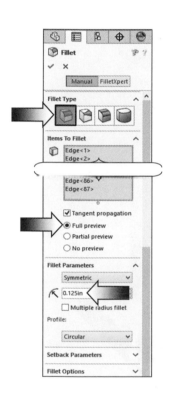 or select **Insert / Features / Fillet/Round**.

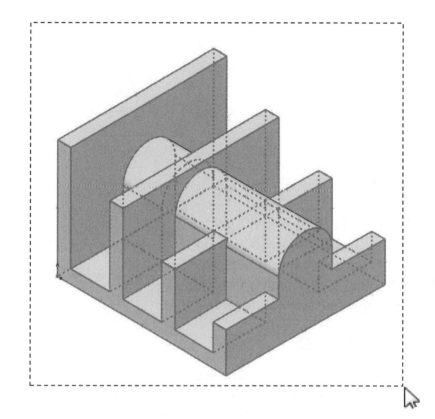

Select the **Constant Size Fillet** button (arrow).

Either "drag-select" to highlight all edges of the model, or press the shortcut key **Control+A** (select all).

Enter **.125 in**. for radius size.

Enable the **Full Preview** checkbox.

Click **OK**.

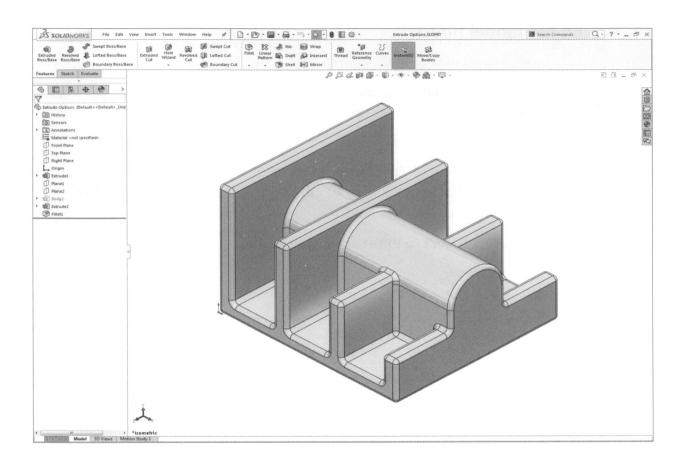

In the Training Files folder, in the
<u>Built Parts folder</u>, you will also find
copies of the parts, assemblies, and
drawings that were created for cross
referencing or reviewing purposes.

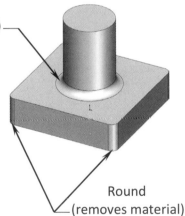

Fillet
(adds material)

Round
(removes material)

* Fillets and Rounds:

*Using the same Fillet command, SOLIDWORKS "knows"
whether to add material (Fillet) or remove material
(Round) to the faces adjacent to the selected edge.*

15. Saving your work:

Select **File / Save As**.

Change the file type to **Part** file (.sldprt).

Enter **Extrude Options** for the name of the file.

Click **Save**.

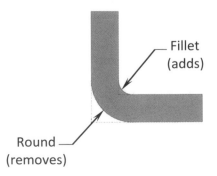

Fillet
(adds)

Round
(removes)

Questions for Review

1. To open a new sketch, first you must select a plane from the FeatureManager tree.
   a. True
   b. False

2. Geometric relations can be used only in the assembly level, not in the part level.
   a. True
   b. False

3. The current status of a sketch is displayed in the lower right area of the screen as Under defined, Fully defined, or Over defined.
   a. True
   b. False

4. Once a feature is extruded, its extrude direction cannot be changed.
   a. True
   b. False

5. A planar face can also be used as a sketch plane.
   a. True
   b. False

6. The Equal relation only works for Lines, not Circles or Arcs.
   a. True
   b. False

7. After a dimension is created, its value cannot be changed.
   a. True
   b. False

8. When the UP TO SURFACE option is selected, you must choose a surface as the end-condition to extrude up to.
   a. True
   b. False

9. UP TO VERTEX is not a valid Extrude option.
   a. True
   b. False

9. FALSE
7. FALSE   8. TRUE
5. TRUE   6. FALSE
3. TRUE   4. FALSE
1. TRUE   2. FALSE

Using the Search Commands:

The Search Commands lets you find and run commands
from SOLIDWORKS Search or locate commands in the user interface.

These features make it easy to find and run any SOLIDWORKS
command:

> The results are filtered as you type and typically find the
> command you need within a few keystrokes.

> When you run a command from the results list for a query, Search Commands
> remembers that command and places it at the top of the results list when you
> type the same query again.

> Search shortcuts let you assign simple and familiar keystroke sequences to
> commands you use more regularly.

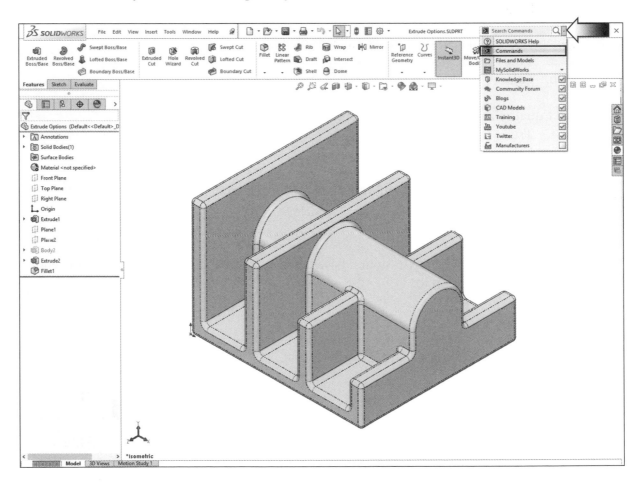

Click the drop-down arrow to see the search options (arrow).

1. Search Commands in Features Mode:

The example below shows how you might use Search Commands to find and run the **Lasso Selection** command in the <u>Feature Mode</u>.

With the part still open, start typing the command **Lasso Selection** in Search Commands. As soon as you type the first few letters of the word Lasso, the results list displays only those commands that include the character sequence **"lasso,"** and **Lasso Selection** appears near the top of the results list.

Click **Show Command Location** ; a **red arrow** indicates the command in the user interface (do not move the mouse while the system is searching).

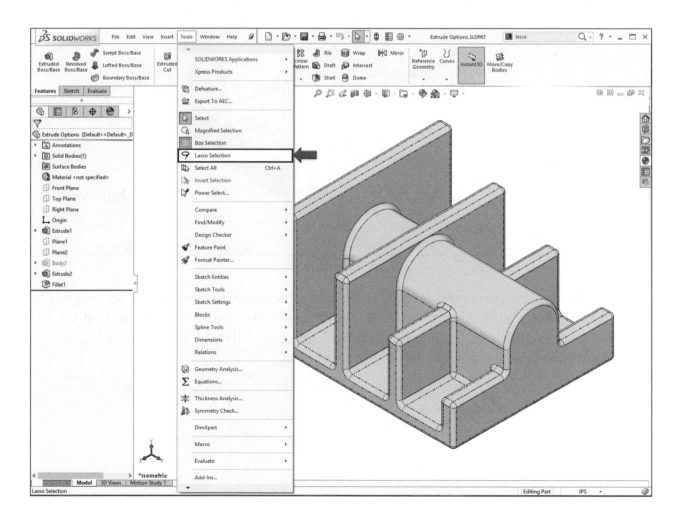

2. Search Commands in Sketch Mode:

The example below shows how you might use Search Commands to find and run the **Dynamic Mirror** command in the Sketch Mode.

Using the same part, open a **new sketch** on the side face of the model as noted.

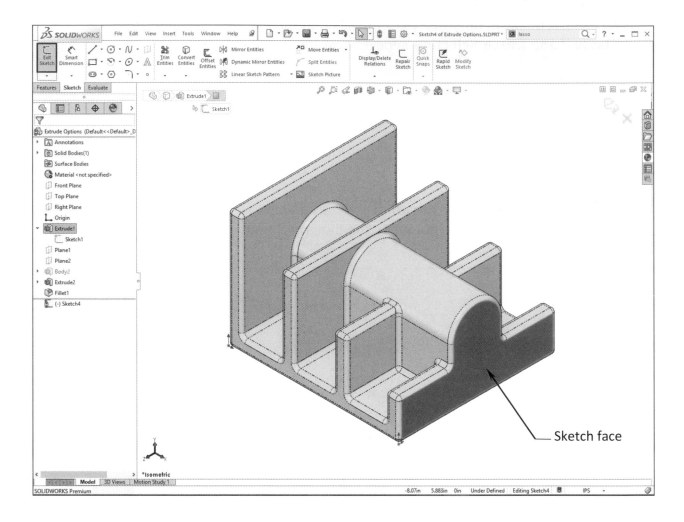

Sketch face

Start typing the command **Dynamic Mirror** in Search Commands. As soon as you type the first few letters of the word Dynamic, the results list displays only those commands that include the character sequence **"dyna,"** and **Dynamic command** appears near the top of the results list.

Click **Show Command Location** 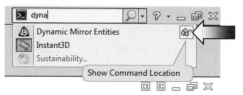; a red arrow indicates the command in the user interface (You will cancel the search operation if the mouse is moved during this time).

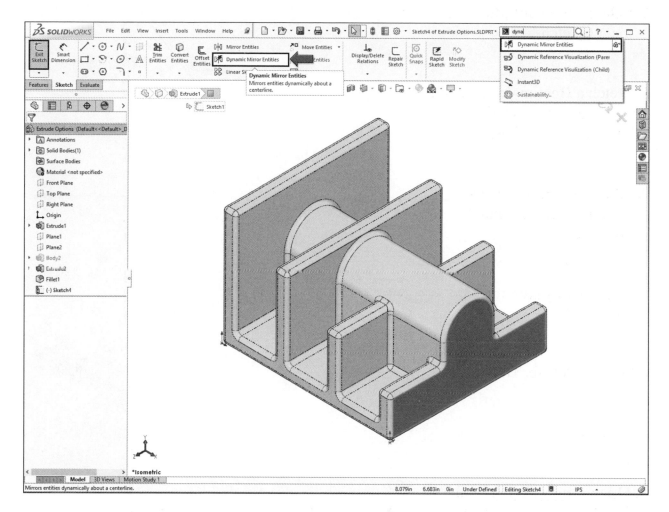

Additionally, a Search Shortcut can be assigned to any command to help find it more quickly (see Customize Keyboard in the SOLIDWORKS Help for more info):

1. Click **Tools / Customize**, and select the **Keyboard** tab.
2. Navigate to the command to which you want to assign a search shortcut.
3. In the Search Shortcut column for the command, type the shortcut letter you want to use, and then click OK.

Save and close all documents.

Exercise 1: Extrude Boss & Hole Wizard

NOTE: The exercise gives you the opportunity to apply what you have learned from the lesson. There will be enough instruction provided for you to create the model but some of the steps may require you to plan ahead of time on how you should constrain the geometry such as: use only geometric relations, or use only dimensions, or use both.

1. Select the Front plane and open a <u>new sketch</u>.

Sketch a **Rectangle** and add the dimensions and relations needed to fully define the sketch.

2. Change to the Features tab and click **Extruded Boss Base**.

Select **Mid-plane** for Direction 1.

Enter **3.00in** for Depth.

Click **OK**.

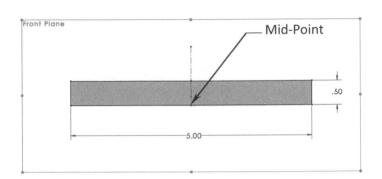

3. Open a <u>new sketch</u> on the <u>face in the back</u> of the part.

Sketch the profile shown.

Add the dimensions and relations as indicated to fully define the sketch.

4. Change to the Features tab and press **Extruded Boss Base**.

For Direction 1, select the **Blind** option.

For Depth, enter **2.00in**.

Click **Reverse** (arrow) to extrude towards the front.

Click **OK**.

5. Adding the Holes:

Click the **Hole Wizard** command on the Features tab.

Set/select the following:

* Hole Type: **Hole**
* Standard: **ANSI Inch**
* Type: **All Drill Sizes**
* Size: **1/4**
* End Condition: **Through All**

Click the **Position tab** (arrow).

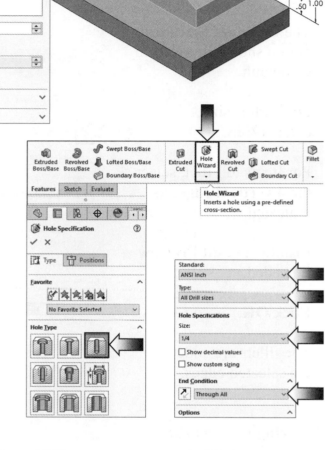

Select the <u>face</u> as indicated to activate the preview graphics.

Place 4 holes and add the dimensions/relations as noted to fully define the sketch.

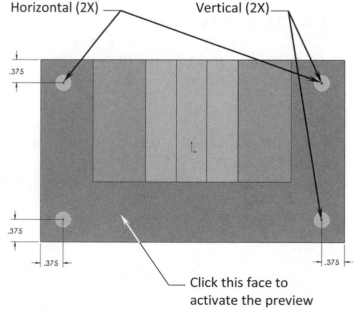

Horizontal (2X) ——— Vertical (2X) ———

.375

.375

.375   .375

Click this face to activate the preview

Click **OK** to accept and exit the Hole Wizard command.

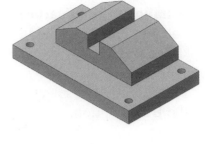

6. Adding Fillets:

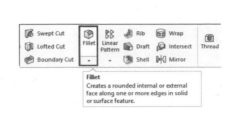

Click the **Fillet** command on the Features tab.

The **Constant Size** (arrow) option should be selected by default.

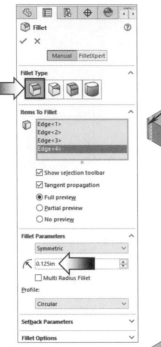

For Radius size, enter **.125in**.

For Items to Fillet, select the **4 vertical edges** as indicated.

Enable the **Full Preview** if needed.

Click **OK**.

The fillets are added to the selected edges.

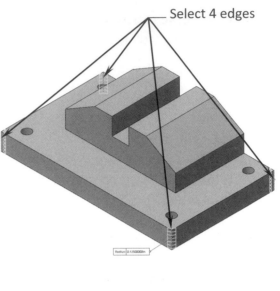

Select 4 edges

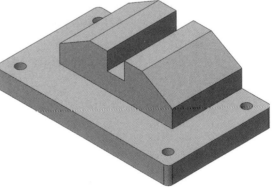

7. Saving your work:

Select **File, Save As**.

Enter **Extrudes_Exe2** for the file name.

Click **Save**.

(Try out the option **Round Corners**).

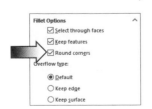

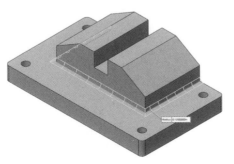

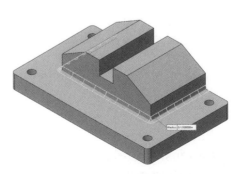

Without Round Corners   With Round Corners

<u>Exercise 2:</u> **Extrude Boss & Extrude Cut**

<u>NOTE:</u> *In an exercise, there will be less step-by-step instruction than those in the lessons which will give you a chance to apply what you have learned in the previous lesson and create the model on your own.*

1. Dimensions are in inches, 3 decimal places.
2. Use Mid-Plane end condition for the Base feature.
3. The part is symmetrical about the Front plane.
4. Use the instructions on the following pages if needed.

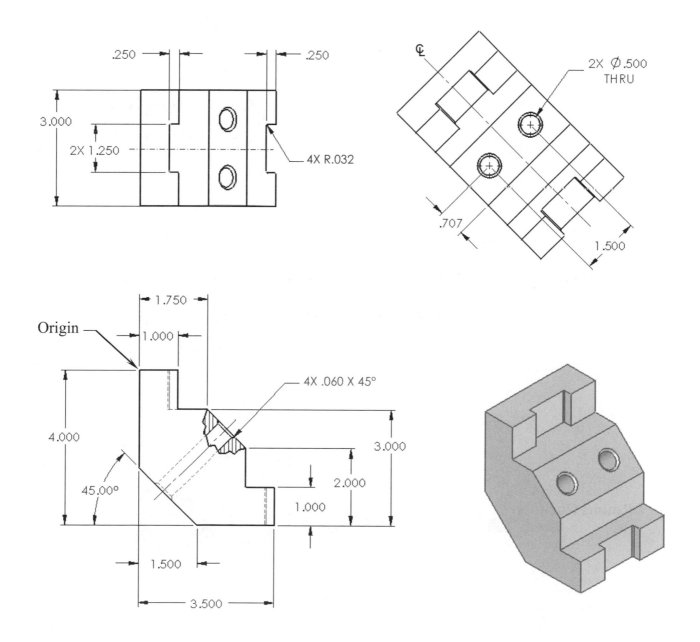

1. Starting with the base sketch:

Select the <u>Front</u> plane and open a **new sketch**.

Starting at the top left corner, using the **line** command, sketch the profile below.

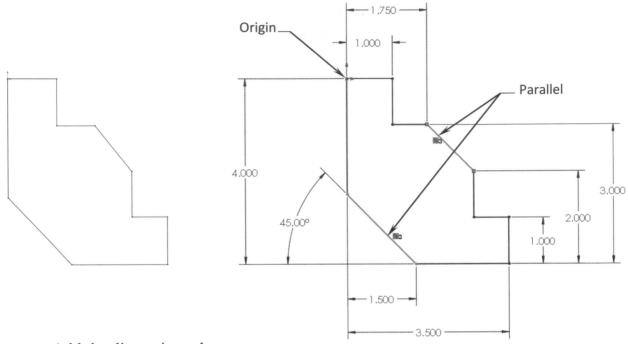

Add the dimensions shown.

Add the Parallel relation to fully define the sketch.

Extrude Boss/Base with **Mid Plane** and **3.000"** in depth.

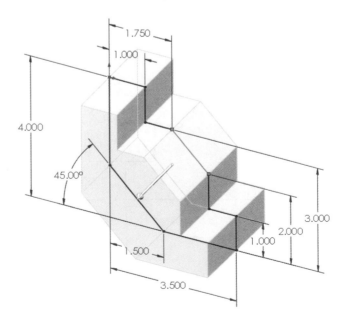

2. Adding the through holes:

Select the <u>face</u> as indicated and click the Normal-To button.

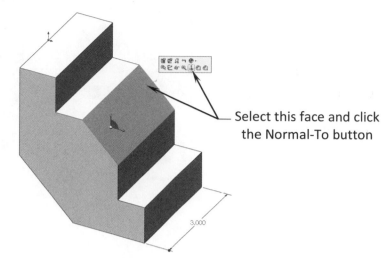

Select this face and click the Normal-To button

This command rotates the part normal to the screen.

The hotkey for this command is **Ctrl+8**.

Open a **new sketch** and draw a **centerline** that starts from the origin point.

Sketch **2 circles** on either side of the centerline.

Add the diameter and location dimensions shown. Push Escape when done.

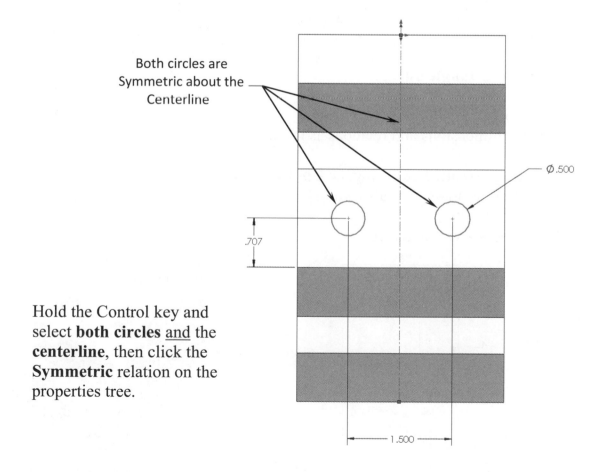

Both circles are Symmetric about the Centerline

Ø.500

.707

1.500

Hold the Control key and select **both circles** <u>and</u> the **centerline**, then click the **Symmetric** relation on the properties tree.

Create an extruded cut using the **Through-All** end condition.

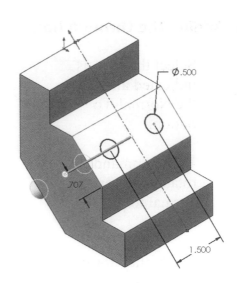

3. Adding the upper cut:

Select the upper face and click the Sketch button to open a **new sketch**.

Sketch a **centerline** that starts at the Origin.

Both lines are Symmetric about the Centerline

Sketch a **rectangle** as shown.

Add the dimensions and relations as indicated.

Create an **extruded cut** using the **Up-To-Vertex** condition (up-to-surface also works).

Select the **Vertex** indicated.

Select Vertex

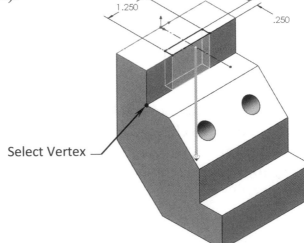

Click **OK**.

4. Adding the lower cut:

Select the <u>lower face</u> of the part and open a **new sketch**.

Sketch a **rectangle** on this face.

Add a **Collinear** <u>and</u> an **Equal** relation to the **lines** and the **edges** as noted.

The line is Collinear <u>and</u> Equal with the edge on both sides

Extrude a cut using the **Through All** condition.

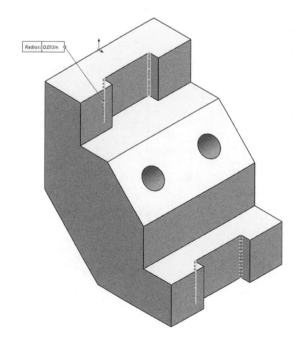

5. Adding fillets:

Select the **Fillet** command from the Features toolbar.

Enter **.032in**. for radius size.

Select the **4 vertical edges** on the inside of the 2 cuts.

Keep all other options at their default settings.

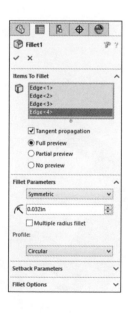

Click **OK**.

6. Adding chamfers:

Click **Chamfer** under the Fillet drop-down.

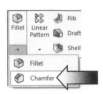

Enter **.060** for depth.

Select the **4 circular edges** of the 2 holes (both front and back).

Click **OK**.

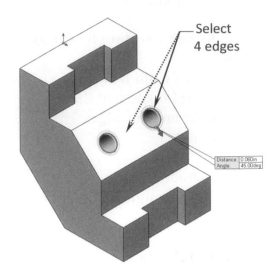

Select 4 edges

7. Saving your work:

Click **File / Save As**.

Enter **Extrudes_Exe2** for the file name.

Select a location to save the file.

Click **Save**.

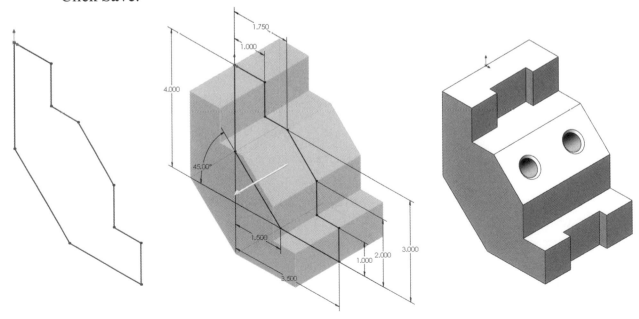

CHAPTER 4

Basic Solid Modeling — Extrude & Revolve

Basic Solid Modeling
Extrude & Revolve

Upon successful completion of this lesson, you will be able to:

* Perform basic modeling techniques

* Sketch on planar surfaces

* Add dimensions

* Add geometric relations or constraints

* Use extrude Boss/Base to add material

* Use Extruded Cut to remove material

* Create revolved features

* Create Fillets and Chamfers

The components created in this lesson will be used again later in an assembly chapter to demonstrate how they can be copied several times and constrained to form a new assembly, checked for interferences, and viewed to see the dynamic motion of an assembly.

Be sure to review self-test questionnaires at the end of the lesson, prior to going to the next chapter. They will help to see if you have gained enough knowledge that is required to move forward to the following chapters.

Basic Solid Modeling
Link Components

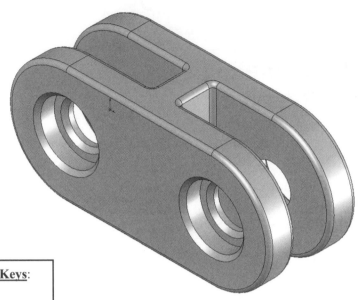

Dimensioning Standards: **ANSI**

Units: **INCHES** – 3 Decimals

Tools Needed:

Insert Sketch	Line	Straight Slot
Circle	Mirror	Dimension
Add Geometric Relations	Fillet	Chamfer
Extruded Boss/Base	Extruded Cut	Boss/Base Revolve

1. Sketching the first profile:

Select the <u>Front</u> plane from the FeatureManager tree.

Click 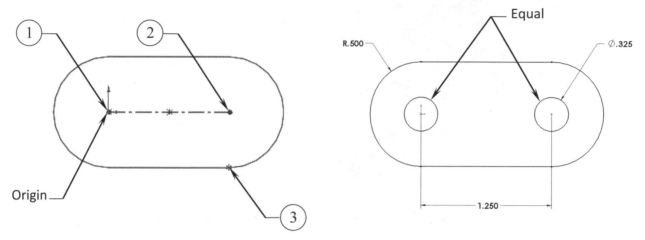 (Insert Sketch) and select the **Straight Slot** command .

Sketch a **slot** following the 3 clicks as indicated below. Clicks 1 and 2 define the length of the slot, and click 3 defines the slot's radius.

Add **2 circles** on the same centers as the arcs.

Add the dimensions and relations shown.

2. Extruding the first solid:

Switch to the **Features** tab and click **Extruded Boss-Base** or select **Insert, Boss-Base, Extrude**.

End condition: **Mid Plane**

Extrude depth: **.750 in**.

Click **OK**.

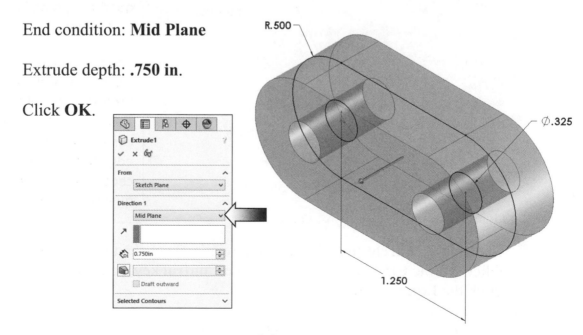

3. Creating the Bore holes:

Select the front <u>face</u> as indicated and click **Sketch** (or select: Insert / Sketch).

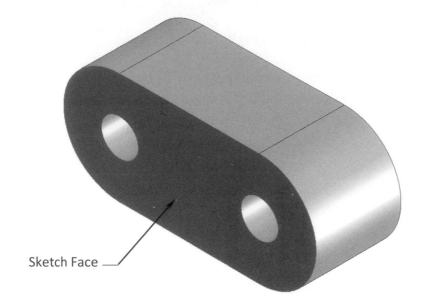

> ### Planar Surfaces
>
> The planar surfaces of the model can also be used as sketch planes; sketch geometry can be drawn directly on these surfaces.

Sketch Face

(The Blue origin is the Part-origin and the Red origin is the Sketch-origin).

Sketch a **circle** starting at the center of the existing hole.

Add **Ø.500** dimension as shown.

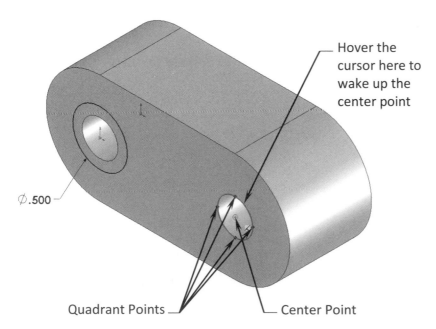

Hover the cursor here to wake up the center point

Ø.500

Quadrant Points

Center Point

> ### Wake-Up Entities*
>
> To find the center (or the quadrant points) of a hole or a circle:
>
> * In the Sketch mode, select one of the sketch tools (Circle in this case), then hover the mouse cursor over the circular edge to "wake-up" the center point & its quadrant points.

Sketch a 2nd **Circle** and add an **Equal** relation between the 2 circles.
Click-off the Circle command. Hold the **Control** key, select the **2 Circles** and click the **Equal** relation (arrow).

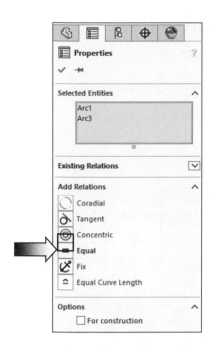

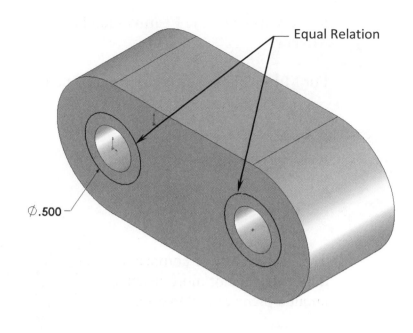

Equal Relation

⌀.500

4. Cutting the Bore holes:

Switch to the **Features** tab and click **Extruded Cut** or select **Insert, Cut, Extrude**.

End Condition: **Blind**   Extrude Depth: **.150 in**.

Click **OK**.

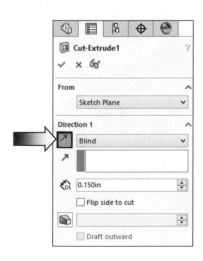

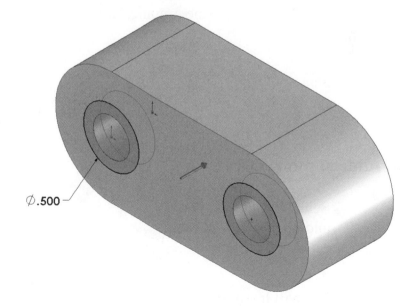

⌀.500

5. Mirroring the Bore holes:

Click **Mirror** on the **Features** toolbar or select: **Insert, Pattern/Mirror, Mirror**.

For Mirror Plane, select the <u>Front</u> plane from the FeatureManager tree.

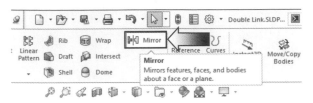

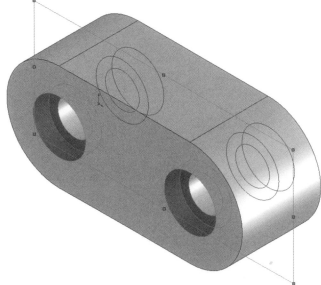

The Feature Mirror command duplicates one or more features about a plane or a planar face.

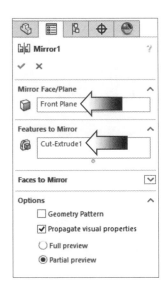

For Features to Mirror, select the **Cut-Extrude1** from the FeatureManager tree, or click one of the Bore holes from the graphics area.

Click **OK**.

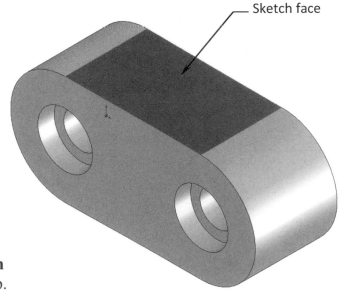

Sketch face

6. Adding more Cuts:

Select the <u>face</u> as noted and open a **new sketch**.

We will learn to use the **sketch mirror** entities in the next step.

Sketch a **vertical centerline** and a **horizontal centerline** as shown.

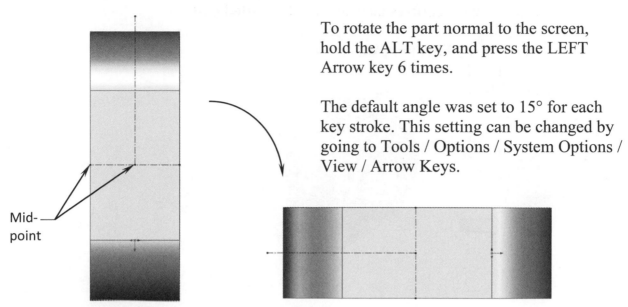

To rotate the part normal to the screen, hold the ALT key, and press the LEFT Arrow key 6 times.

The default angle was set to 15° for each key stroke. This setting can be changed by going to Tools / Options / System Options / View / Arrow Keys.

Mid-point

Select the vertical centerline and click the **Dynamic Mirror** command (or click **Tools, Sketch Tools, Dynamic Mirror**).

Sketch a **Corner Rectangle** on one side of the centerline, it will get mirrored to the opposite side automatically.

Click-off the Dynamic mirror button. Add a **Symmetric** relation to the 3 lines as noted.

Add the dimensions shown to fully define the sketch.

Click **OK**.

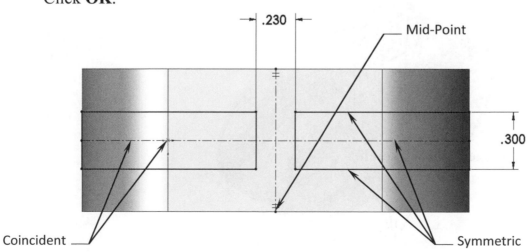

.230

Mid-Point

.300

Coincident

Symmetric

7. Extruding a Through All cut:

Switch to the **Features** tab and click **Extruded Cut**.

Direction 1:   **Through All**.

Click **OK**.

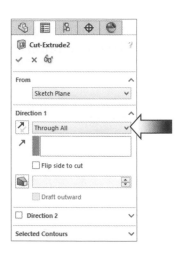

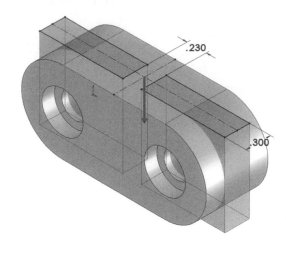

8. Adding the .032" fillets:

Click **Fillet** or select **Insert, Features, Fillet/Round**.

Enter **.032in.** for radius.

Select the <u>edges</u> as indicated below.

Click **OK**.

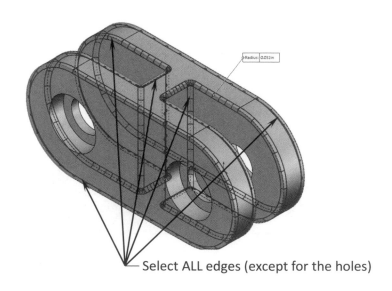

Select ALL edges (except for the holes)

9. Adding the .032" chamfers:

Click **Chamfer**  or select **Insert / Features / Chamfer.**

Enter **.032 in.** for the depth of the chamfer and select the <u>edges</u> of the 4 holes.

Click **OK**.

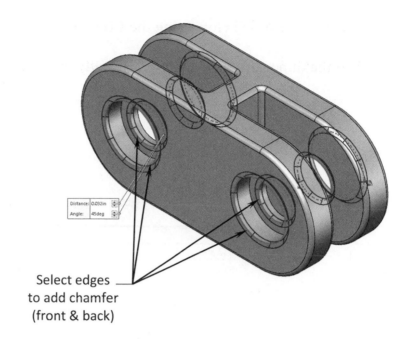

Select edges
to add chamfer
(front & back)

10. Saving your work:

Select **File, Save As, Double Link, Save**.

11. Creating the Sub-Components:

The 1st sub-component is the **Alignment Pin**.

Select the <u>Right</u> plane from the FeatureManager Tree.

Click or select **Insert / Sketch**.

Sketch the profile below using the **Line** tool.

Add the dimensions shown below to fully define the sketch.

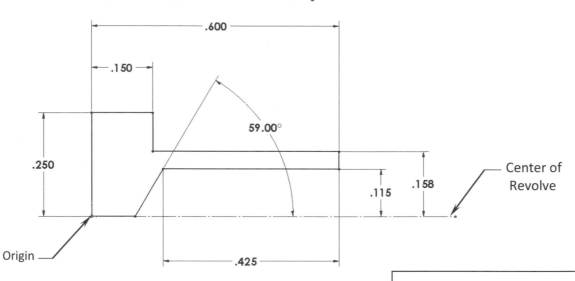

12. Revolving the base feature:

Click  or select **Insert, Boss-Base, Revolve**.

Revolve Type: **Blind**.

Revolve Angle: **360 deg**. (default).

Click **OK**.

> ☼ **Center of Revolve**
>
> A centerline is used when revolving a sketch profile.
>
> A model edge, an axis, or a sketch line can also be used as the center of the revolve feature.

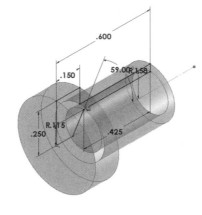

13. Adding chamfers:

Click **Chamfer** 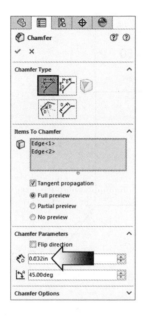 or select **Insert, Features, Chamfer**.

Enter **.032** for Distance.

Enter **45 deg**. for Angle.

Select the 2 Edges as indicated (one outside and one inside).

Click **OK**.

Select 2 edges

14. Saving your work:

Select **File, Save As, Alignment Pin, Save**.

15. The 2ⁿᵈ Sub-Components:

The 2ⁿᵈ sub-component is the **Pin Head**:

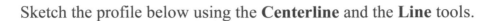

Select the <u>Front</u> plane from the FeatureManager Tree.

Click 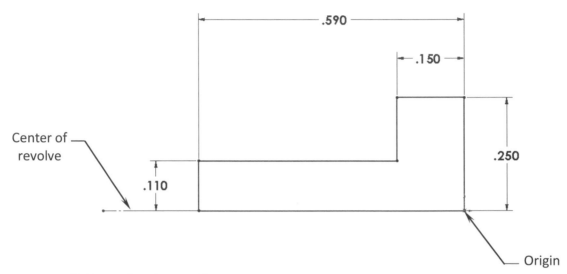 or select **Insert, Sketch**.

Sketch the profile below using the **Centerline** and the **Line** tools.

Add the dimensions shown below to fully define the sketch.

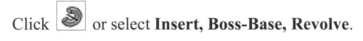

16. Revolving the base feature:

Click 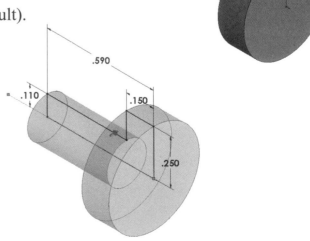 or select **Insert, Boss-Base, Revolve**.

Revolve Type: **Blind**.

Revolve Angle: **360 deg**. (default).

Click **OK**.

17. Adding chamfer:

Click **Chamfer**  or select **Insert, Features, Chamfer**.

Enter **.032** for Distance.

Enter **45 deg**. for Angle.

Select the <u>Edges</u> as indicated (one on the left, one on the right).

Click **OK**.

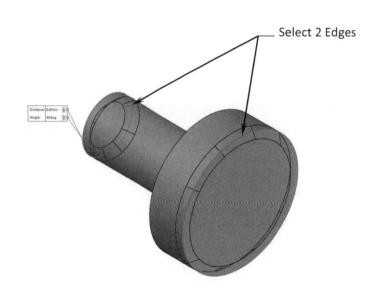

Select 2 Edges

18. Saving your work:

Select **File, Save As, Pin Head, Save**.

19. Creating the 3ʳᵈ Sub-Components:

The 3ʳᵈ sub-component is the **Single Link**:

Select <u>Front</u> plane from FeatureManager Tree.

Click 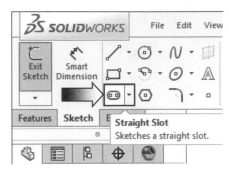 Insert Sketch and select the **Straight Slot** command.

Enable the **Add Dimensions** checkbox (arrow below).

This command requires 3 clicks. Start at the origin for point 1, move the cursor horizontally and click point 2, then move downward to make the point 3.

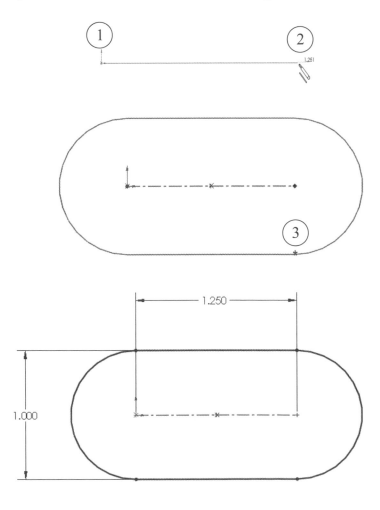

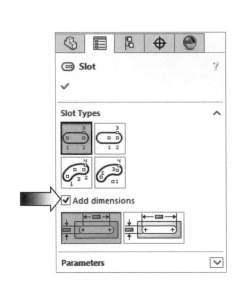

Click **OK** to exit the
Straight Slot command
and modify the values to
match the ones shown
in the image.

20. Extruding the base:

Switch to the **Features** tab and click **Extruded Boss-Base** or select: **Insert, Boss-Base, Extrude**.

End condition: **Mid Plane**

Extrude depth: **.750 in**.

Click **OK**.

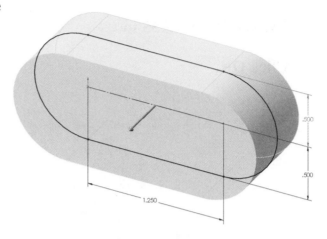

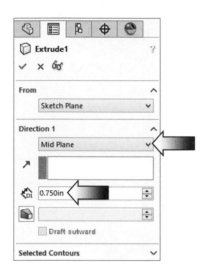

21. Sketching the Recess Profiles:

Select the <u>face</u> indicated and open a **new sketch** .

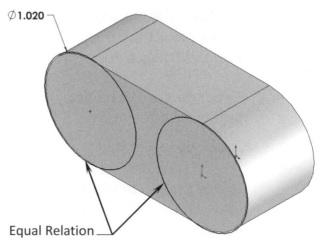

Sketch Face

Sketch **2 Circles** from the centers of the circular edges (with the Circle command already selected, hover over the circular edge to see its center).

Ø1.020

Add a **Ø1.020 in.** dimension to one of the circles. (The circles are slightly larger than the part.)

Add an **Equal** relation between the 2 circles.

Equal Relation

22. Extruding a blind cut:

Click **Extruded Cut** or select **Insert, Cut, Extrude**.

End condition: **Blind**

Extrude Depth: **.235 in.**

Click **OK**.

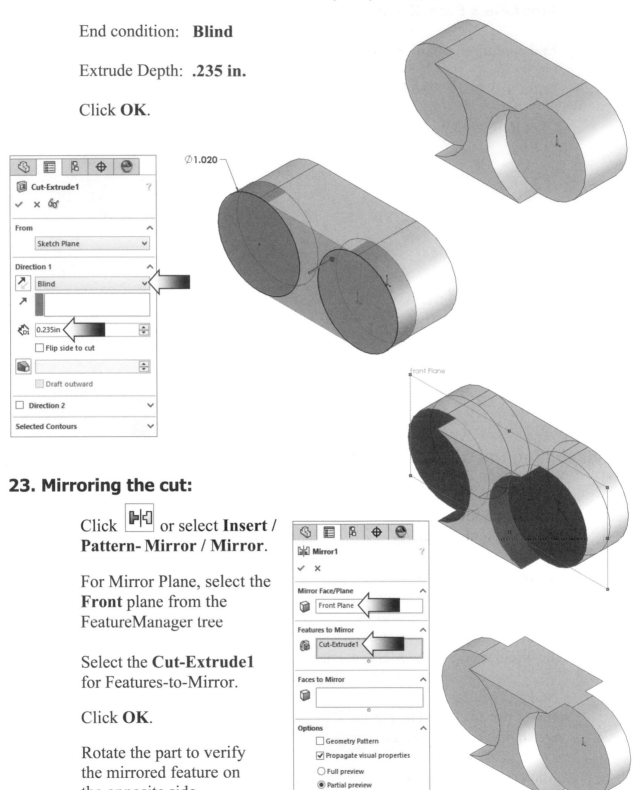

23. Mirroring the cut:

Click [icon] or select **Insert /
Pattern- Mirror / Mirror**.

For Mirror Plane, select the
Front plane from the
FeatureManager tree

Select the **Cut-Extrude1**
for Features-to-Mirror.

Click **OK**.

Rotate the part to verify
the mirrored feature on
the opposite side.

24. Adding the Holes:

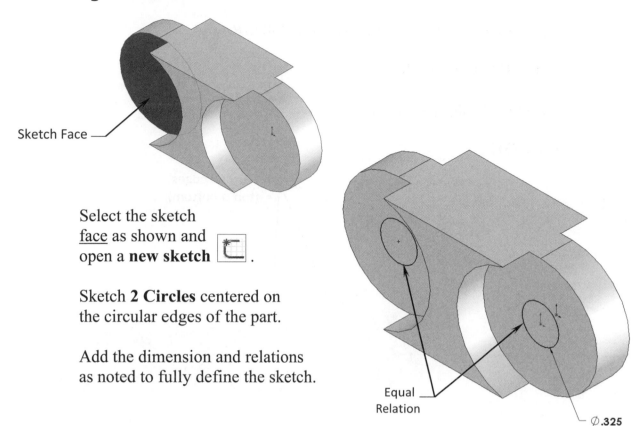

Sketch Face

Select the sketch
<u>face</u> as shown and
open a **new sketch** .

Sketch **2 Circles** centered on
the circular edges of the part.

Add the dimension and relations
as noted to fully define the sketch.

Equal
Relation

Ø.325

25. Cutting the holes:

Click **Extruded Cut** or select **Insert, Cut, Extrude**.

End condition: **Through All**

Click **OK**.

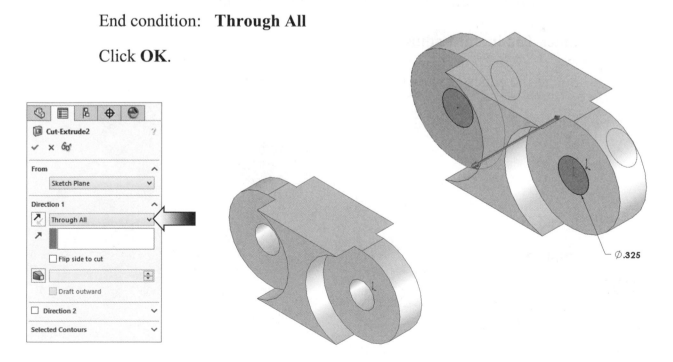

Cut-Extrude2

From
Sketch Plane

Direction 1
Through All

Flip side to cut

Draft outward

Direction 2

Selected Contours

Ø.325

26. Adding the .100" fillets:

Click **Fillet** or select **Insert, Features, Fillet/Round**.

Enter **.100 in.** for Radius.

Select the 8 edges as indicated below.

Click **OK**.

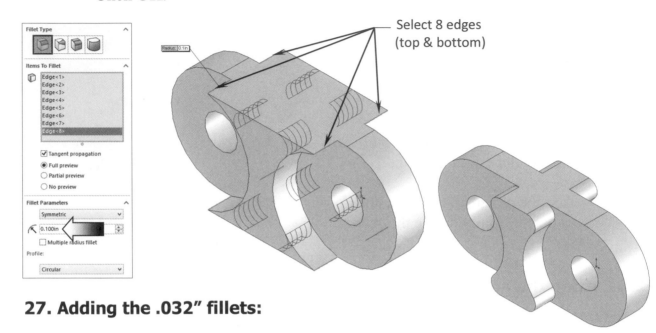

Select 8 edges
(top & bottom)

27. Adding the .032" fillets:

Click **Fillet** or select **Insert, Features, Fillet/Round**.

Enter **.032 in.** for Radius.

Select the edges as shown.

Click **OK**.

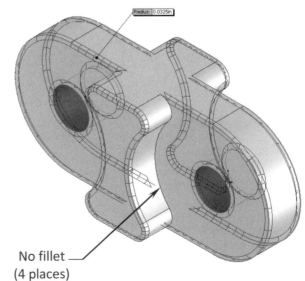

<u>NOTE:</u>
*There are no fillets on
the 4 edges as indicated.*

No fillet
(4 places)

28. Saving your work:

Click **File, Save As, Single Link, Save.**

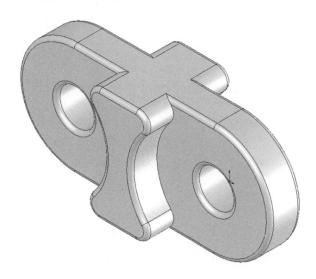

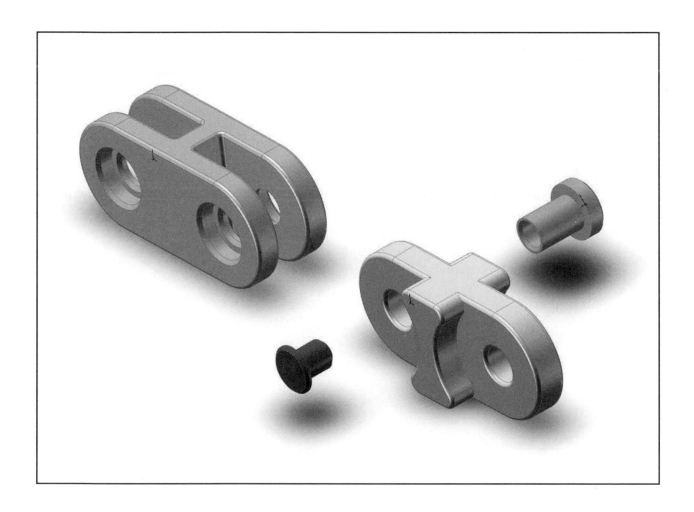

Questions for Review

1. Tangent relations only work with the same type of entities such as an arc to an arc, not between a line and an arc.
   a. True
   b. False

2. The first feature in a part is the parent feature, not a child.
   a. True
   b. False

3. The dimension arrows can be toggled to flip inward or outward by clicking on its handle points.
   a. True
   b. False

4. The Shaded-with-Edges option cannot be used in the part mode, only in the drawing mode.
   a. True
   b. False

5. The Concentric relation makes the diameters of the 2 circles equal.
   a. True
   b. False

6. More than one model edge can be selected and filleted at the same time.
   a. True
   b. False

7. To revolve a sketch profile, a centerline should be selected as the axis of the revolution.
   a. True
   b. False

8. After a sketch is revolved, its revolved angle cannot be changed.
   a. True
   b. False

7. TRUE   8. FALSE
5. FALSE 6. TRUE
3. TRUE   4. FALSE
1. FALSE 2. TRUE

Exercise: Extrude Boss & Extrude Cut

1. Create the solid model using the drawing provided below.
2. Dimensions are in Inches, 3 decimal places.
3. Tangent relations between the transitions of the Arcs should be used.
4. The Ø.472 holes are Concentric with R.710 Arcs.
5. Use the instructions on the following pages,
   if needed.

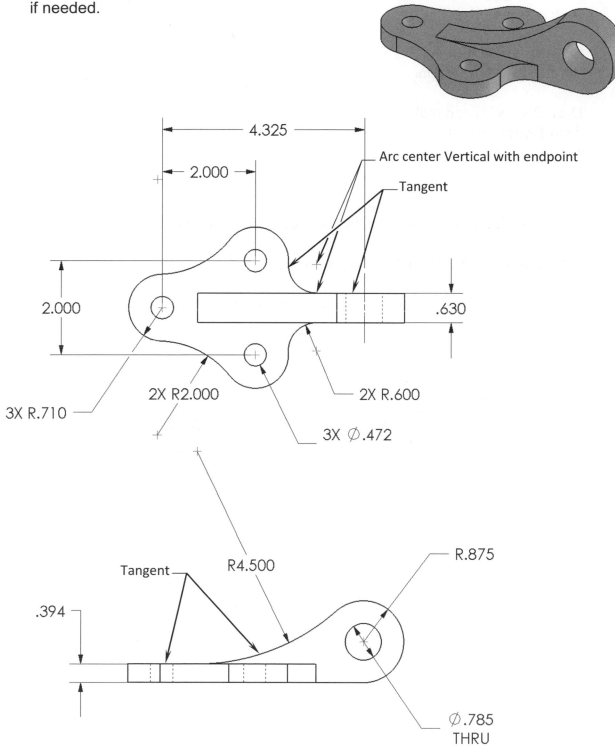

4.325

2.000

Arc center Vertical with endpoint

Tangent

2.000

.630

3X R.710

2X R2.000

2X R.600

3X Ø.472

Tangent

R4.500

R.875

.394

Ø.785
THRU

1. Creating the Base sketch:

There are many ways to create this part, but let us try this basic method first.

Select the <u>Top</u> plane and open a **new sketch**.

Sketch **2 centerlines** starting from the Origin.

Draw **3 Circles** and make them **Equal**. Add the dimensions shown.

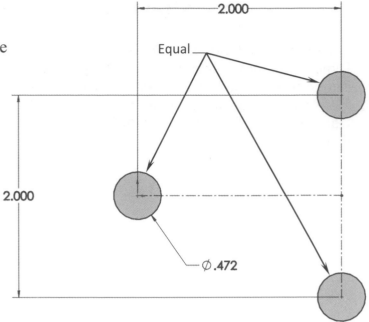

Next, sketch **3 more circles** on the same centers as the first three.

Add an **Equal** relation to the 3 larger circles.

Add the **Ø1.420** dimension to fully define the sketch.

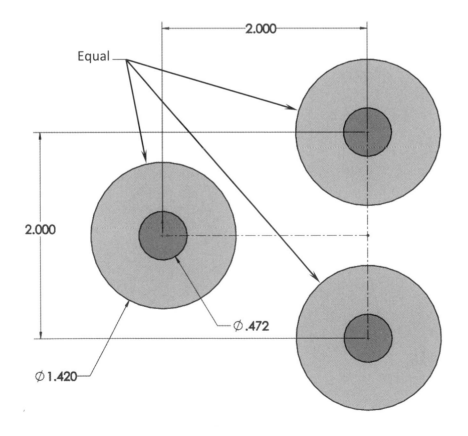

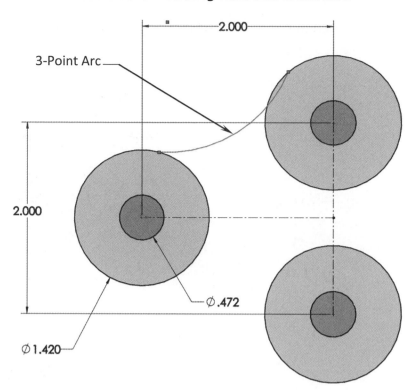

Sketch a **3-Point Arc** approximately as shown.

Add a **Tangent** relation between the 3-Point Arc and the Circle (both sides).

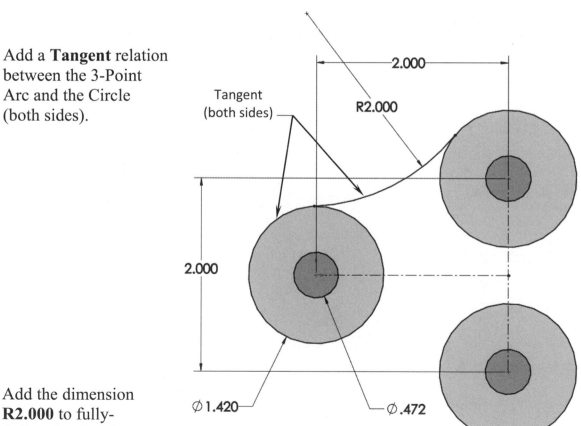

Add the dimension **R2.000** to fully-define the sketch.

Sketch another **3-Point Arc** approximately as shown.

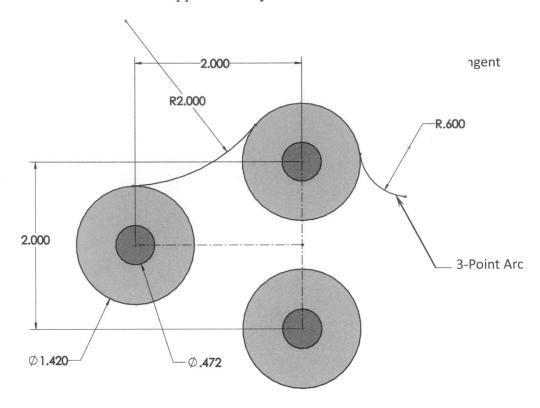

Mirror the 2 arcs about the horizontal centerline as indicated.

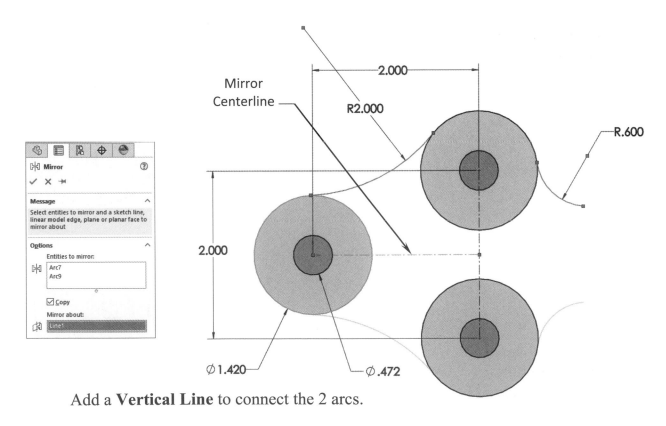

Add a **Vertical Line** to connect the 2 arcs.

Add a **Vertical** relation between the center of the arc and the endpoint of the vertical line.

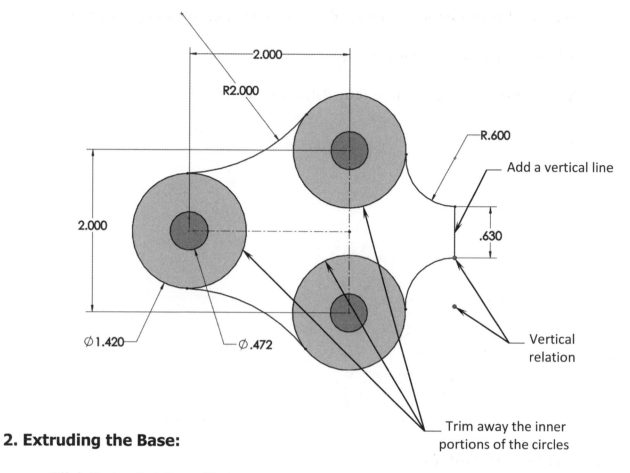

2. Extruding the Base:

Click **Extruded Boss-Base**.

Use the **Blind** type for Direction1.

Enter **.394"** for extrude Depth.

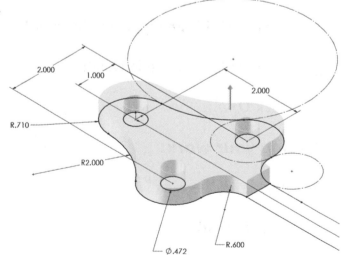

Click **OK**.

3. Creating the Tail-End sketch:

Select the <u>Front</u> reference plane from the Feature tree and open **new sketch**.

Sketch the **construction circles** and add the sketch geometry right over them.

Add the **Tangent** relation as noted.

Add the Smart Dimensions to fully define the sketch.

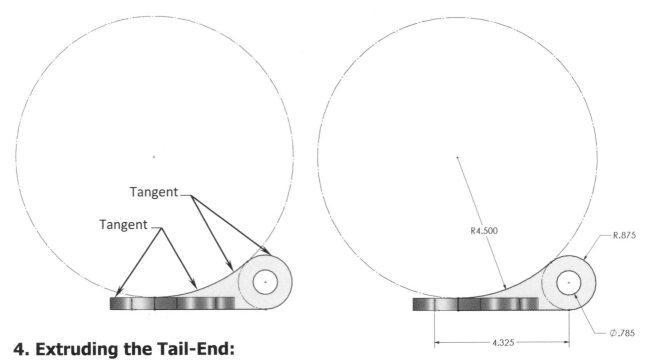

4. Extruding the Tail-End:

Click **Extruded Boss-Base**.

Use the **Mid Plane** type for Direction1.

Enter **.630"** for extrude Depth.

Click **OK**.

5. Saving your work:

Click **File, Save As**.

Enter **Extrudes_Exe2**.

Click **Save**.

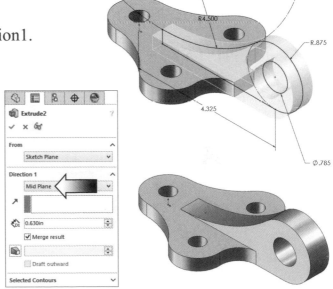

CHAPTER 5

Revolved Parts

Revolved Parts
Ball Joint Arm

The Revolve command rotates one or more sketch profiles about a centerline, up to 360° to create a <u>thin</u> or a <u>solid</u> feature.

Open profile
= Thin Feature

Closed profile
= Solid Feature

The revolved sketch should be a single continuous closed contour and it can be any shape or size without overlaps or gaps.

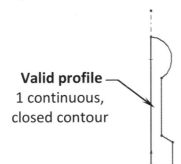 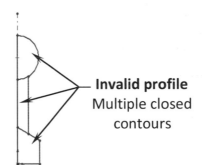

Valid profile
1 continuous,
closed contour

Invalid profile
Multiple closed
contours

If there are several centerlines in the same sketch, the center of rotation must be selected when creating the revolve feature.

The center of the revolve can either be a centerline, a line, an axis, or a linear model edge. The revolve feature can be a cut feature (which removes material) or a revolve boss feature (which adds material).

This chapter will guide you through the basics of creating the revolved parts.

Revolved Parts
Ball Joint Arm

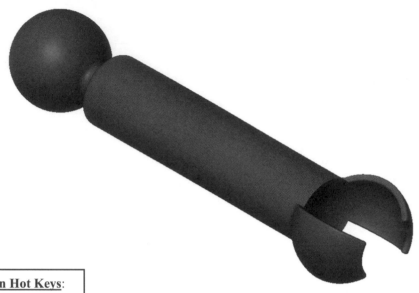

Dimensioning Standards: **ANSI**

Units: **INCHES** – 3 Decimals

Tools Needed:

 Insert Sketch

 Line

 Circle

 Rectangle

 Sketch Fillet

 Trim

 Add Geometric Relations

 Dimension

 Centerline

 Base/Boss Revolve

 Fillet/Round

 Mirror Features

1. Creating the Base Profile:

Select the <u>Front</u> plane from the FeatureManager tree.

Click 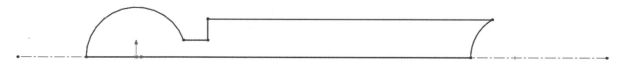 or select **Insert, Sketch**.

Sketch the profile using the **Line** and **Circle** commands.

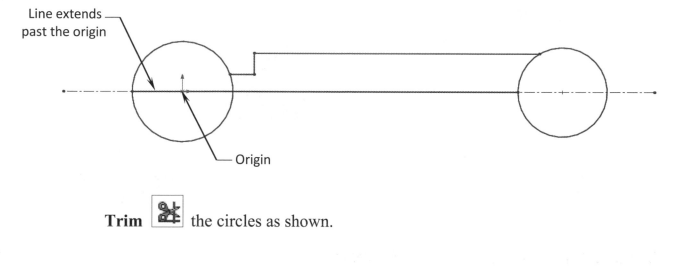

Line extends past the origin

Origin

Trim the circles as shown.

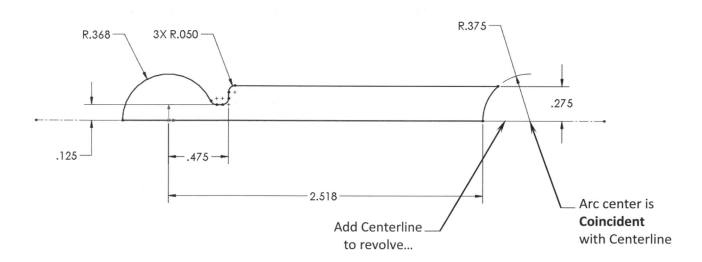

Add the **Dimensions** and **Relations** shown below to fully define the sketch. (It would be easier to add the R.050" fillets after the sketch is fully defined.)

R.368

3X R.050

R.375

.275

.125

.475

2.518

Add Centerline to revolve...

Arc center is **Coincident** with Centerline

2. Revolving the Base Feature:

Click **Revolve** on the Features tab or select: **Insert, Boss-Base, Revolve**.

Revolve Type: **Blind**.

Revolve Angle: **360 deg**.

Click **OK**.

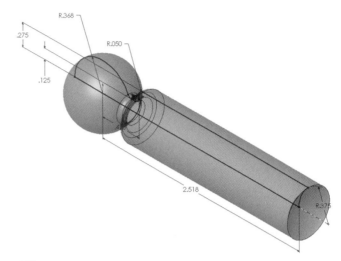

3. Sketching the Opened-End Profile:

Select the Top plane from the FeatureManager tree.

Click 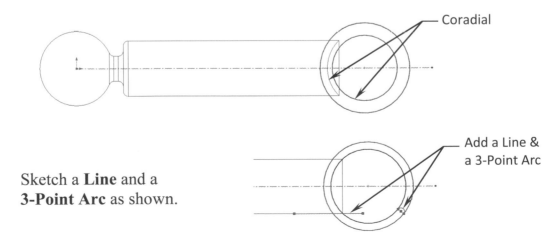 or select **Insert, Sketch**.

Switch to **Hidden Lines Visible** mode (under **View, Display**).

Sketch **2 circles** as shown.

Add a **Coradial** relation between the small circle and the hidden edge.

Coradial

Sketch a **Line** and a
3-Point Arc as shown.

Add a Line &
a 3-Point Arc

Add a **Tangent** relation between the small and the larger arcs. Add the same relation to the other arc.

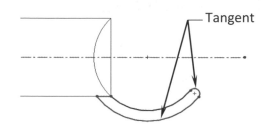

Trim 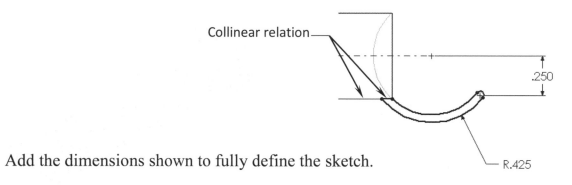 the line and the 2 circles.

Add a **Collinear** relation between the line and the bottom edge of the part.

Add the dimensions shown to fully define the sketch.

4. Revolving the Opened-End Feature:

Click **Revolve** 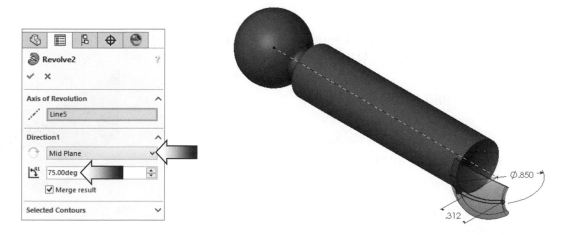 or select **Insert, Boss-Base, Revolve**.

Revolve Type: **Mid Plane**.

Revolve Angle: **75 deg**.

Click **OK**.

5. Mirroring the Revolved feature:

Click **Mirror** 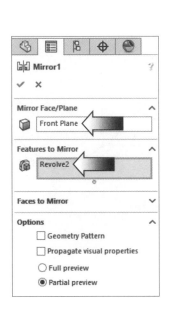 or select **Insert, Pattern Mirror, Mirror**.

For Mirror Plane, select the <u>Front</u> plane from the FeatureManager tree.

For Features to Mirror, select the **Revolve2** feature either from the graphics area or from the FeatureManager tree.

Click **OK**.

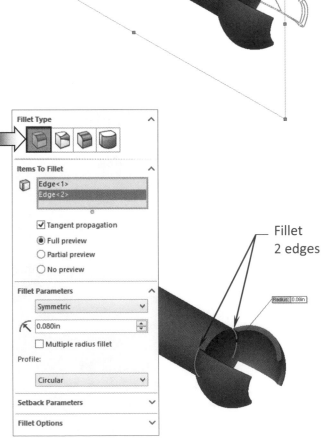

6. Adding the .080" Fillets:

Click **Fillet** or select **Insert / Features / Fillet / Round**.

Enter **.080** in. for Radius.

For Items to Fillet, select the <u>two edges</u> as shown.

Click **OK**.

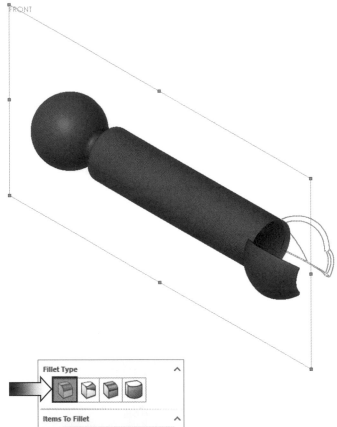

Fillet
2 edges

7. Adding the .015" Fillets:

Click **Fillet** or select **Insert, Features, Fillet/Round**.

Enter **.015** in. for Radius.

Select the <u>edges</u> of the 2 revolved features as shown below.

Click **OK**.

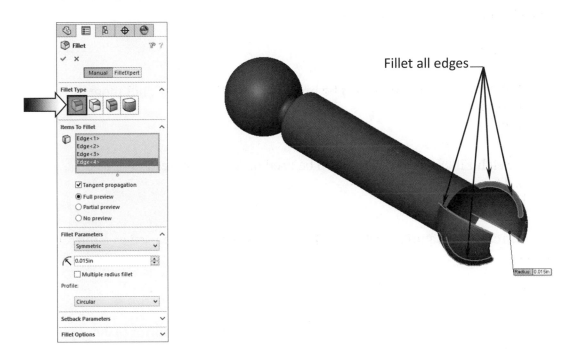

Fillet all edges

8. Saving Your Work:

Select **File, Save As, Ball-Joint-Arm, Save**.

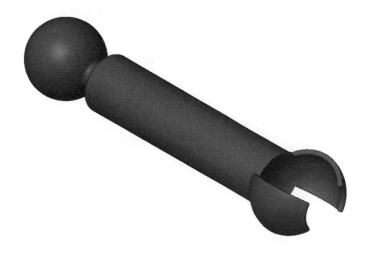

Questions for Review

1. A proper sketch profile for a revolved feature should be a single closed contour created on one side of the revolved centerline.
   a. True
   b. False

2. If there is more than one centerline in the same sketch, one centerline should be selected prior to revolving the sketch.
   a. True
   b. False

3. A revolve feature should **always** be revolved a complete 360°, not more or less.
   a. True
   b. False

4. The Sketch Fillet command can also be used on solid features.
   a. True
   b. False

5. To mirror a 3D feature, a plane or a planar surface should be used as a mirror plane.
   a. True
   b. False

6. To mirror a series of features, a centerline can be used as a mirror plane.
   a. True
   b. False

7. After a fillet feature is created, its parameters (selected faces, edges, fillet values, etc.) cannot be modified.
   a. True
   b. False

8. Either an axis, a model edge, or a sketch line can be used as the center of the revolve.
   a. True
   b. False

Exercise: Flat Head Screwdriver

1. Create the part using the drawing provided below.
2. Dimensions are in inches, 3 decimal places.
3. The part is symmetrical about the Top plane.
4. Unspecified radii to be R.050 max.
5. Use the instructions on the following pages, if needed.

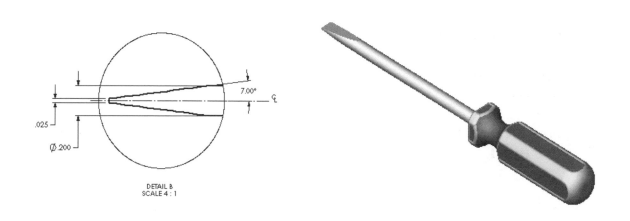

DETAIL B
SCALE 4 : 1

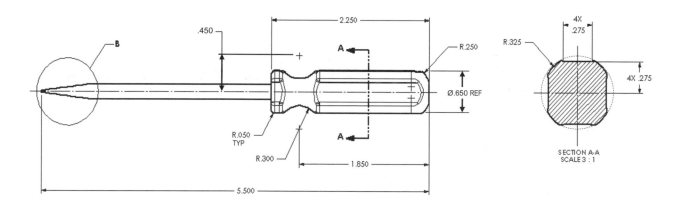

6. Save your work as **Flat Head Screwdriver**.

1. Creating the base sketch:

Select the <u>Front</u>
plane and open
a **new sketch**.

Sketch the profile
and add the dimensions
as shown.
(Note: Only add the Sketch Fillets after the sketch is fully defined.)

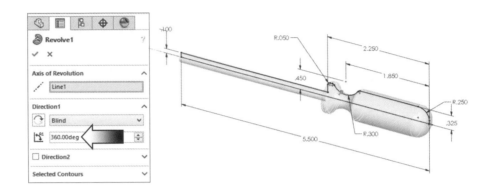

2. Revolving the base:

Click **Revolve,
Boss-Base**.

For Direction 1:
Use **Blind**.

Angle = **360°**

Click **OK**.

3. Creating the flat head:

Select the <u>Front</u>
plane and open
a **new sketch**.

Sketch the <u>3 lines</u> as
shown and add the
dimensions to fully
define the sketch.

Create an **extruded
cut** using **Through-
All - Both**.

(Since the sketch was
open, Through All Both
is the only extrude option
in this case).

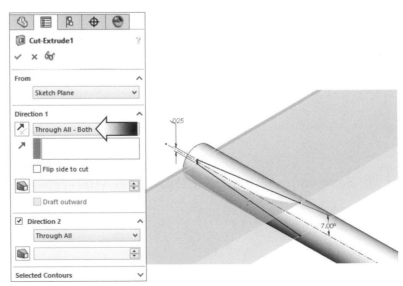

4. Creating the flat handle:

Select the <u>flat surface</u> on the right end of the handle and open a **new sketch**.

Sketch a rectangle and <u>mirror</u> it using the vertical and horizontal centerlines (the sketch can be left under defined for this example).

Sketch face

Equal

Create an **extruded cut** using the opposite end of the handle as the end condition for the Up-To-Surface option (Through-All can also be used to achieve the same result).

Up to Surface

5. Adding the .050" fillets:

Add a **.050"** fillet to the <u>2 faces</u> as indicated.

(It is quicker to select the faces rather than the edges to add this fillet. When a face is selected, SOLIDWORKS applies the fillet to all edges on that face).

Select 2 faces

6. Saving your work:

Save your work as **Flat Head Screwdriver**.

Derived Sketches
Center Ball Joint

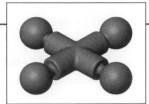

Derived Sketches
Center Ball Joint

The derived sketch option creates a copy of the original sketch and places it on the same or different plane but within the same part document.

* The derived sketch is the child feature of the original.

* The derived sketch (copy) cannot be edited. Its size and shape are dependent on the parent sketch.

* Derived sketches can be moved and related to different planes or faces with respect to the same model.

* Changes made to the parent sketch are automatically reflected in the derived sketches.

To break the link between the parent sketch and the derived copies, right-click on the derived sketch and select **Underived**.

After the link is broken, the derived sketches can be modified independently and will not update when the parent sketch is changed.

One major difference between the traditional copy / paste option and the derived sketch is:

* The copy/paste creates an **independent** copy. There is no link between the parent sketch and the derived sketch.

* The derived sketch creates a **dependent** copy. The parent sketch and the derived sketch are fully linked.

Derived Sketches
Center Ball Joint

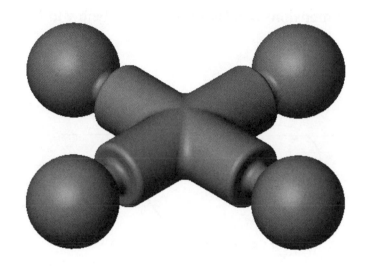

Dimensioning Standards: **ANSI**

Units: **INCHES** – 3 Decimals

Tools Needed:

⌐ Insert Sketch	✎ Line	✎ Centerline
⊙ Circle	⌐ Sketch Fillet	✂ Trim
⊥ Add Geometric Relations	⌂ Dimension	▢ Plane
Derived Sketch	▣ Fillet/Round	◉ Base/Boss Revolve

1. Creating the Base Profile:

Select the Front plane from the Feature Manager tree.

Click 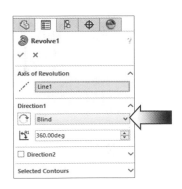 or select **Insert, Sketch**.

Sketch the profile using **Lines**, **Circles**, **Sketch Fillets,** and the **Trim Entities** tools (refer to step 1 on page 5-3 for reference).

Add the **Dimensions** and **Relations** shown below to fully define the sketch.

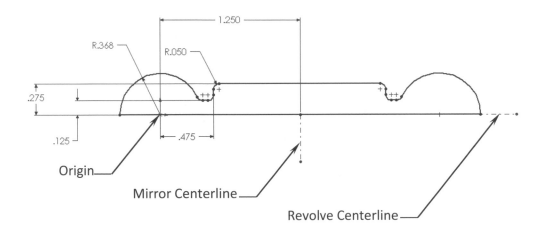

2. Revolving the Base Feature:

Click **Revolve** 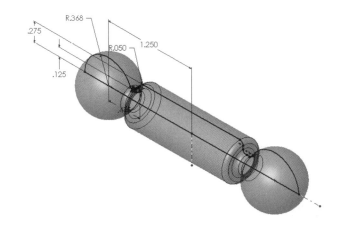 or select **Insert, Boss-Base, Revolve**.

Revolve Type: **Blind**.

Revolve Angle: **360 deg**.

Click **OK**.

3. Creating a new work plane:

Click **Plane** 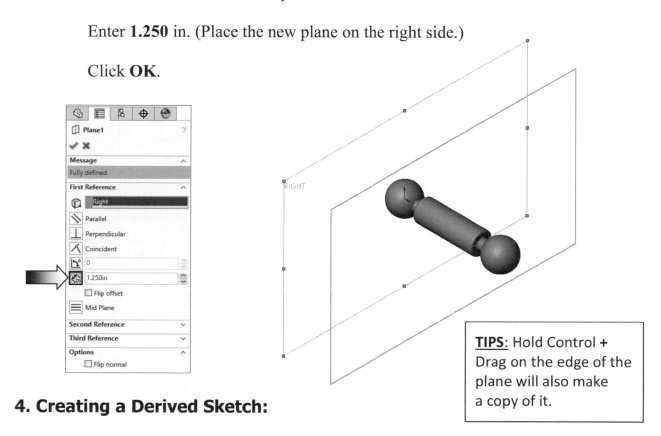 or select **Insert / Reference Geometry / Plane**.

Select the <u>Right</u> plane from the FeatureManager tree.

Click the **Offset Distance** option.

Enter **1.250** in. (Place the new plane on the right side.)

Click **OK**.

> **TIPS:** Hold Control +
> Drag on the edge of the
> plane will also make
> a copy of it.

4. Creating a Derived Sketch:

Hold the **Control** key, select the new plane (**Plane1**) and the **Sketch1** (under Base-Revolved1) from the FeatureManager tree.

Select **Derived Sketch** under the **Insert** menu.

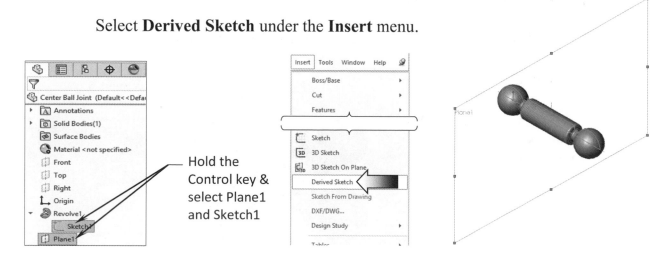

Hold the
Control key &
select Plane1
and Sketch1

A copy of Sketch1 is created and placed on Plane1, and the sketch is automatically activated for positioning.

5. Positioning the Derived Sketch:

Add a **Collinear** relation between the <u>Top</u> plane and the <u>Line</u> as indicated. (Select the plane from the FeatureManager tree.)

Add a **Collinear** relation between the <u>Front</u> plane and the <u>Centerline</u> as shown.

Click **OK**.

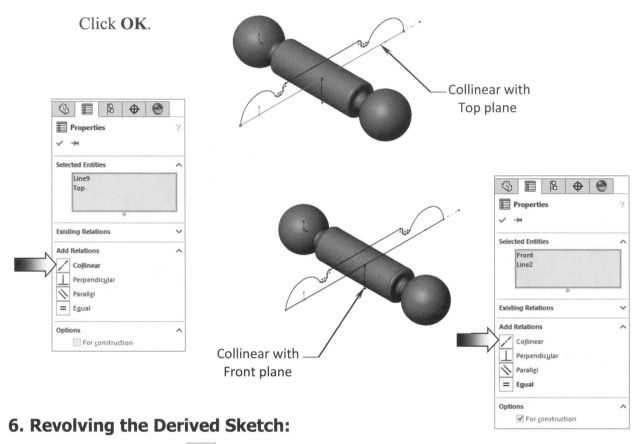

Collinear with
Top plane

Collinear with
Front plane

6. Revolving the Derived Sketch:

Click **Revolve** 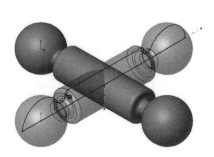 or select **Insert / Boss-Base / Revolve**.

Revolve Type: **Blind**.

Revolve Angle: **360 deg**.

Click **OK**.

7. Adding Fillets:

Click **Fillet** on the Features toolbar or select **Insert, Features, Fillet/Round**.

Enter **.100 in.** for radius.

Select the 2 <u>edges</u> shown.

Click **OK**.

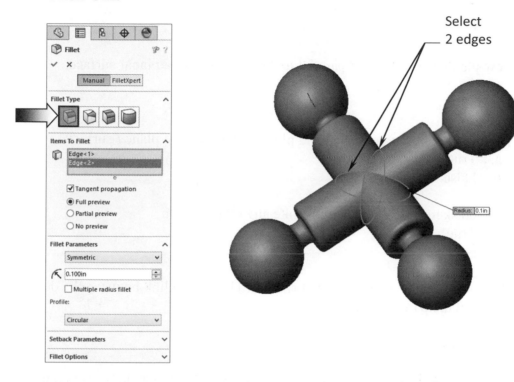

Select
2 edges

8. Saving Your Work:

Select **File, Save As, Center Ball Joint, Save**.

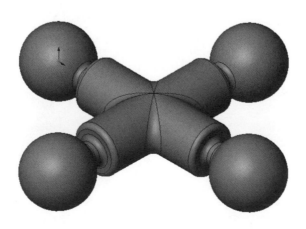

Questions for Review

1. The first feature in a part is the parent feature.
   a. True
   b. False

2. More than one centerline can be selected at the same time to revolve a sketch profile.
   a. True
   b. False

3. A new plane can be created parallel to another plane or a planar surface.
   a. True
   b. False

4. A derived sketch can be copied and placed on a different plane / surface.
   a. True
   b. False

5. A derived sketch can be edited just like any other sketch.
   a. True
   b. False

6. A derived sketch is an independent sketch. There is no link between the derived sketch and the parent sketch.
   a. True
   b. False

7. A derived sketch can only be positioned and related to other sketches / features.
   a. True
   b. False

8. When the parent sketch is changed, the derived sketches will be updated automatically.
   a. True
   b. False

Exercise: Revolved Parts - Wheel

1. Create the 3D model using the drawing provided below.
2. Dimensions are in inches, 3 decimal places.
3. The part is Symmetrical about the horizontal axis.
4. The 5 mounting holes should be created as a Circular Pattern.
5. Use the instructions on the following pages if needed.

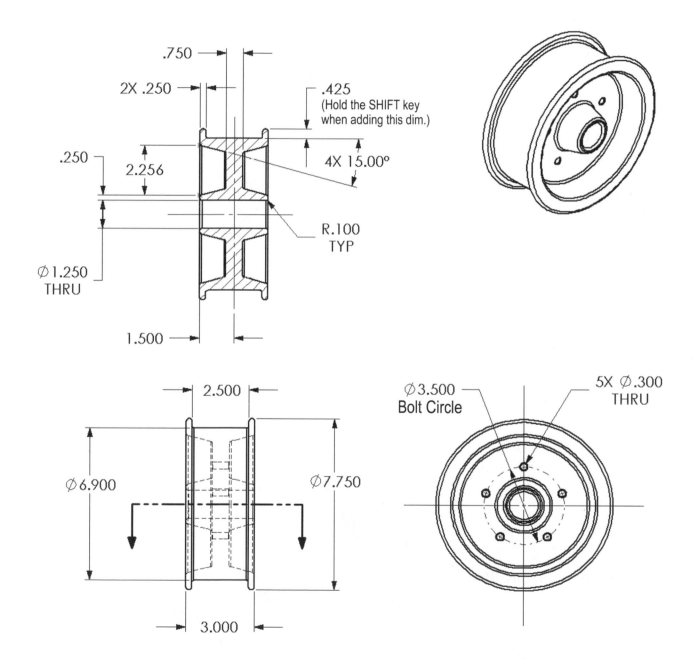

6. Save your work as **Wheel_Exe**.

1. Start with the Front plane.

Sketch **2 centerlines** and use the vertical centerline for Dynamic Mirror.

Sketch the profile as shown.

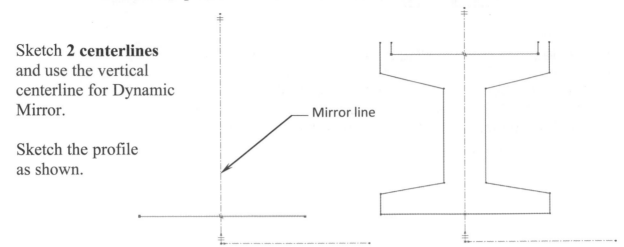

Mirror line

Add a **tangent arc** to close-off the upper portion of the sketch.

Add the dimensions and relations as indicated.

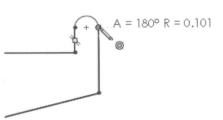

A = 180° R = 0.101

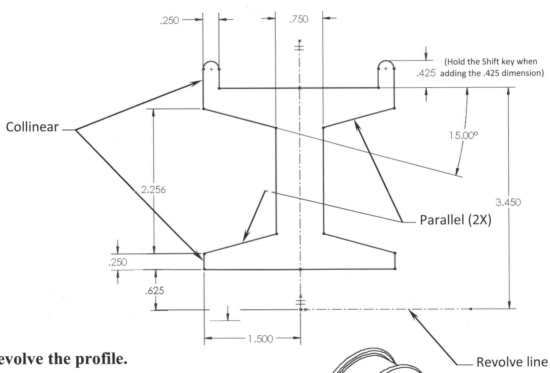

.250   .750

Collinear

2.256

.250

.625

1.500

(Hold the Shift key when adding the .425 dimension)

.425

15.00°

3.450

Parallel (2X)

Revolve line

2. Revolve the profile.

3. Add the 5 holes.

4. Save your work as Wheel_Exe.sldprt.

Exercise: **Plastic Bottle**
Extrude / Revolve / Sweep and Circular Pattern

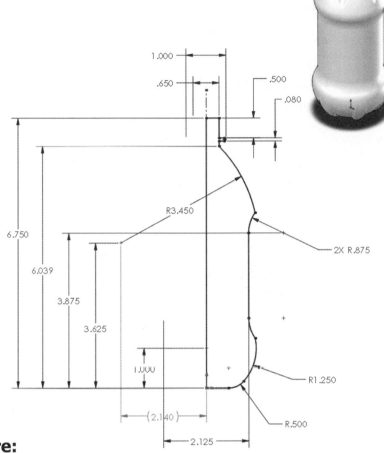

1. Copying the document:

<u>Go to:</u>
Training Files folder, and browse to the file named **Basic Solid Modeling_ Bottle EXE.sldprt**

Make a Copy of this file and **Open the copy**.

(To review how this part was made, open the sample part from the Training Files folder in the Built-Parts folder).

2. Revolving the Base feature:

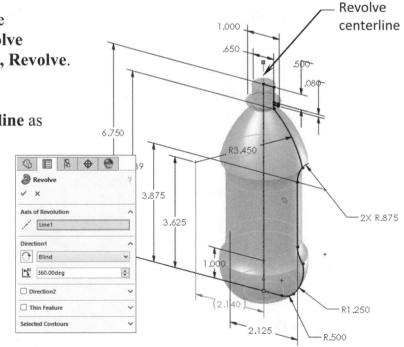

Select the **Sketch1** from the Feature tree and click **Revolve** or select **Insert, Boss-Base, Revolve**.

Select the **Vertical Centerline** as the Center of the rotation.

Direction: **Blind** (Default).

Revolve Angle: **360 deg**.

Click **OK**.

3. Adding fillets:

Click the **Fillet** command from the Features tab.

Use the default **Constant Size Fillet** option.

Enter **.500in**. for radius.

Select the **2 edges** indicated.

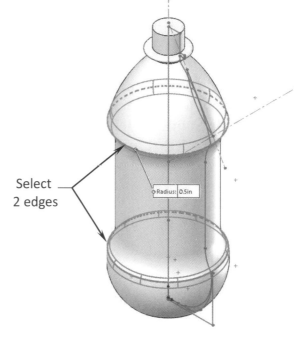

Select 2 edges

Click **OK**.

4. Creating the upper Cut:

Select the **Sketch2** and click **Revolved Cut** or select **Insert, Cut, Revolve**.

Select the **Angular Centerline (30°)** as the Center of the rotation.

Use the default **Blind** option.

Revolve Angle: **360 deg**.

Click **OK**.

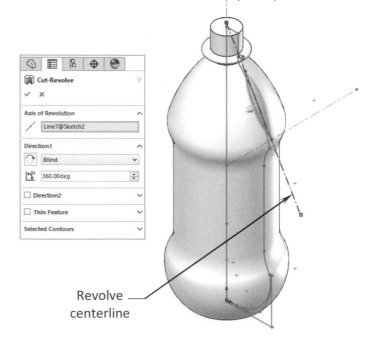

Revolve centerline

5. Creating the lower cut:

Select the **Sketch3** from the Feature-Manager tree and click **Revolved Cut** .

Revolve Direction: **Mid-Plane**

Revolve Angle: **40 Deg**.

Click **OK**.

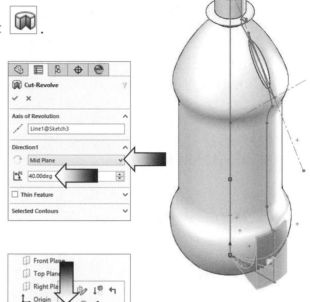

Expand each feature on the tree and hide all 3 sketches by clicking on each sketch name and selecting **Hide** (arrow).

6. Adding the .062" fillets:

Click the **Fillet** command from the Features toolbar.

Use the default **Constant Size Fillet** option.

Enter **.062in**. for radius.

Select the <u>4 edges</u> as noted.

Click **OK**.

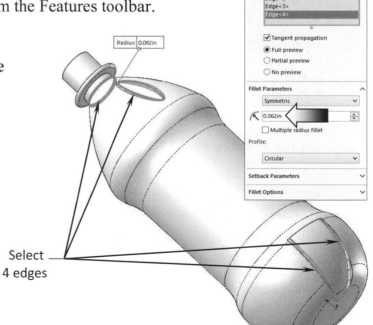

Select 4 edges

7. Adding the .125" Fillets:

Click the **Fillet** command.

Use the default
Constant Fillet Size
option once again.

Enter **.125in** for radius.

Select the <u>2 edges</u> shown.

Click **OK**.

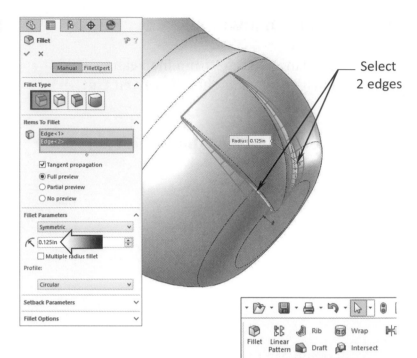

Select
2 edges

8. Creating a Circular Pattern:

Click **Circular Pattern** 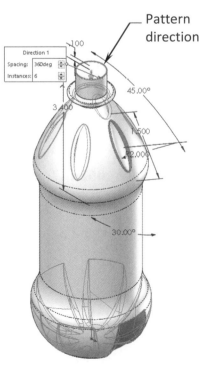 or select
Insert, Pattern-Mirror, Circular Pattern.

Select the **Circular Edge** to use
as the pattern direction.

Pattern Angle: **360deg**.

Number of Copies: **6**

Equal Spacing enabled.

Select the **Cut-Revolved1**, the
Cut-Revolved2, the **Fillet2**, and
Fillet3, either from the graphics
or from the feature tree.

Click **OK**.

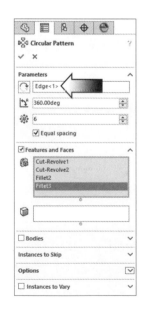

Pattern
direction

9. Shelling with Multi-Wall thicknesses:

Click the **Shell** command (arrow).

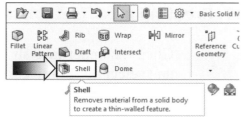

Enter **.025in** for the 1st wall thickness.

For Faces-to-Remove select the <u>upper face</u> as noted.

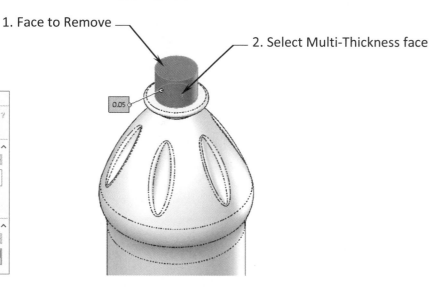

1. Face to Remove

2. Select Multi-Thickness face

For Multi-Thickness face, select the <u>side face</u> of the circular boss as indicated.

Change the 2nd wall thickness to **.050in.**

Click **OK**.

10. Creating a section view:

Click **Section View** from the View (Heads-Up) toolbar (arrow).

Use the default Front plane as the cutting plane.

Zoom in to verify the wall thicknesses. <u>Exit</u> the section command when done.

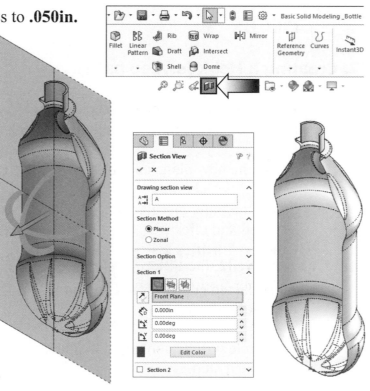

11. Creating an Offset-Distance plane:

Click the upper face shown and select the **Plane** command .

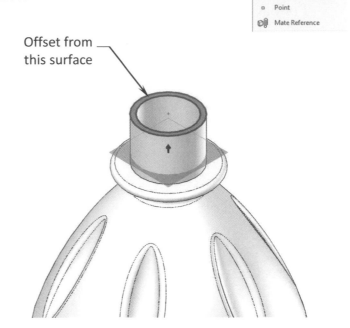

Enter **.400in**. for offset distance.

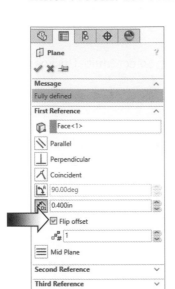

Enable the **Flip Offset** checkbox to place the plane below the upper surface.

Click **OK**.

12. Creating a Helix (sweep path):

A helix curve is normally used as a sweep path for a swept feature. A sketch profile is swept along the path to define the feature's shape and size.

Select the new plane and open a **new sketch** .

Select the outer circular edge as noted and click **Convert Entities** 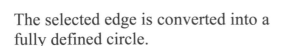.

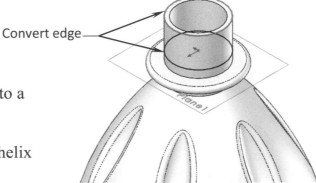

Convert edge

The selected edge is converted into a fully defined circle.

The circle will be converted to a helix in the next step.

Switch to the Features toolbar and click **Curves**, **Helix-Spiral** .

Use the default **Constant Pitch** option.

Change the Pitch to **.125in**.

For Revolutions, enter **2.5**.

Set Start Angle to **0.0deg**.

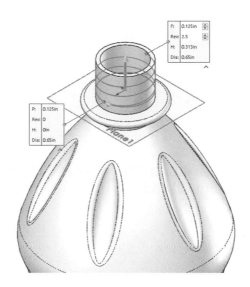

Set Direction to **Clockwise**.

Click **OK**.

13. Sketching the Thread contour (Sweep Profile):

Select the <u>Right</u> plane and open a **new sketch**.

Zoom in on the neck area, closer to the right end of the helix.

Sketch a **horizontal centerline** and **two normal lines** as shown.

Push **Esc** to turn off the Line command. Box-Select all three lines to highlight them (drag the cursor across the lines to select them).

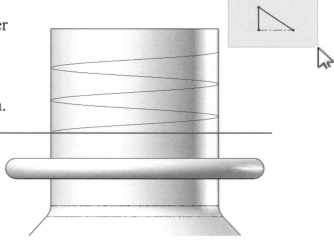

Click the **Mirror Entities** command 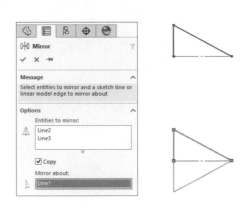 to mirror the selected lines.

Add the dimensions shown below.

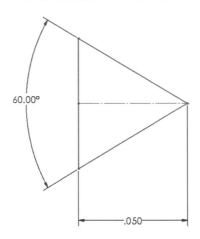

Add a <u>Sketch Fillet</u> of **.005in** to the right corner of the sketch.

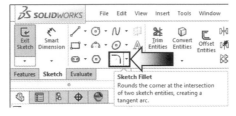

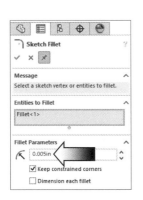

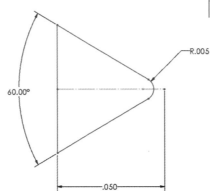

Add a **Pierce** relation between the <u>endpoint</u> on the left side of the centerline and the <u>helix</u>. The Pierce relation snaps the sketch endpoint to the closest end of the helix.

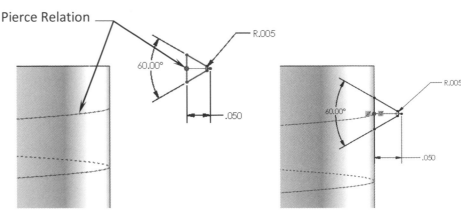

14. Adding the Threads:

Click **Swept Boss Base**  command on the Features toolbar.

For Sweep Profile, select the <u>triangular profile</u> either from the graphics area or from the Feature tree.

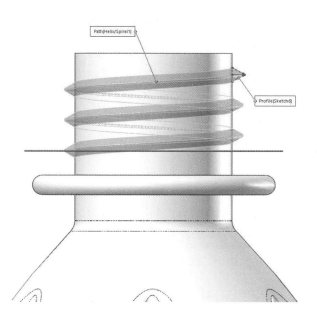

For Sweep Path, select the <u>Helix</u>.

The preview graphic shows the profile is swept along the path as shown above.

Click **OK**.

Hide the plane

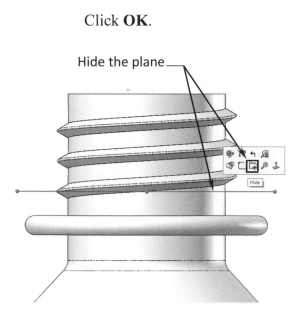

The 2 ends need to be rounded off

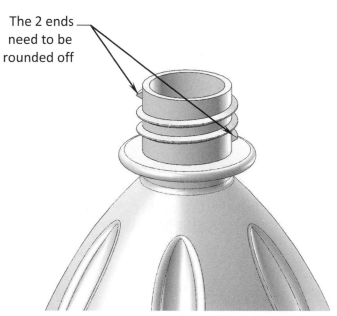

15. Closing the ends:

Select the <u>flat end</u> of the Swept feature and open a **new sketch**.

While the selected face is still highlighted, click **Convert-Entities**.

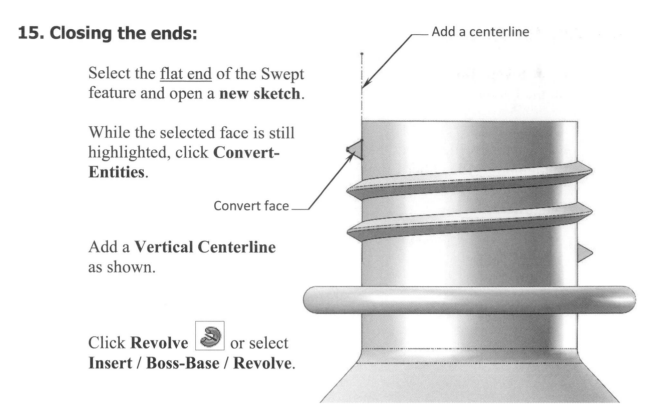

Add a centerline

Convert face

Add a **Vertical Centerline** as shown.

Click **Revolve** or select **Insert / Boss-Base / Revolve**.

Revolve **Blind** (Default) and click Reverse if needed.

Revolve Angle: **100 deg**.

Click **OK**.

Repeat Step 15 to round off the other end of the threads.

16. Saving your work:

Click **File / Save As /**

For the file name, enter **Basic Solid Modeling_Bottle Exe.**

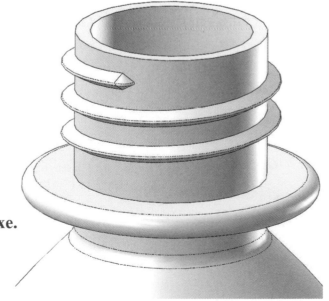

Click **Save.**

CHAPTER 6

Using the Rib & Shell Features

Rib  and Shell  Features
Plastic Tray

A **Rib** is a special type of extruded feature created from open or closed sketched contours. It adds material of a specified thickness in a specified direction between the contour and an existing part. A rib can be created using single or multiple sketches.

Rib features can have draft angles applied to them either inward or outward.

The **Detailed Preview**  Property Manager can be used with multi-body parts to enhance detail and select entities to display.

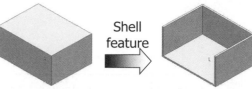
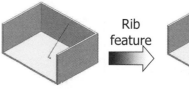

The **Shell** tool hollows out a part, leaves the selected faces open, and creates thin walled features on the remaining faces. If nothing (no face) is selected on the model, a solid part can be shelled, creating a closed hollow model.

Multiple thicknesses are also supported when shelling a solid model. In most cases, the model fillets should be applied before shelling a part.

One of the most common problems when the shell fails is when the wall thickness of the shell is smaller than one of the fillets in the model.

If errors appear when shelling a model, you can run the **Error Diagnostics**. The shell feature displays error messages and includes tools to help you identify why the shell feature failed. The diagnostic tool **Error Diagnostics** is available in the **Shell** Property Manager.

Using the Rib & Shell Features
Plastic Tray

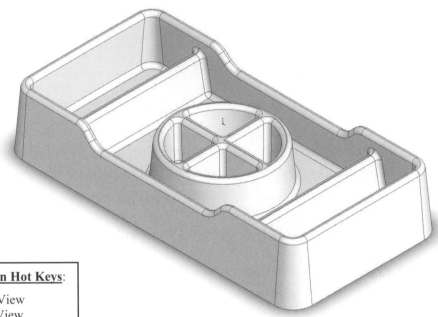

View Orientation Hot Keys:

Ctrl + 1 = Front View
Ctrl + 2 = Back View
Ctrl + 3 = Left View
Ctrl + 4 = Right View
Ctrl + 5 = Top View
Ctrl + 6 = Bottom View
Ctrl + 7 = Isometric View
Ctrl + 8 = Normal To
    Selection

Dimensioning Standards: **ANSI**

Units: **INCHES** – 3 Decimals

Tools Needed:

 Insert Sketch

 Line

 Circle

 Add Geometric Relations

 Dimension

 Fillet

 Boss/Base Extrude

 Rib

 Shell

1. Sketching the Base Profile:

Select the <u>Top</u> plane from the FeatureManager tree.

Click **Sketch** or select **Insert / Sketch**.

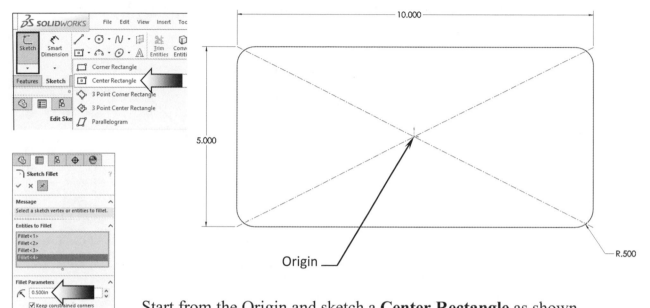

Start from the Origin and sketch a **Center Rectangle** as shown.

Add the **Sketch Fillets** and **dimensions** to fully define the sketch.

2. Extruding the Base feature:

Switch to the **Features** tab and click **Extruded Boss-Base**.

End condition: **Blind** and click **Reverse**.

Extrude depth: **2.00 in**.

Draft angle: **5.00 deg.**

Enable **Draft Outward**.

Click **OK**.

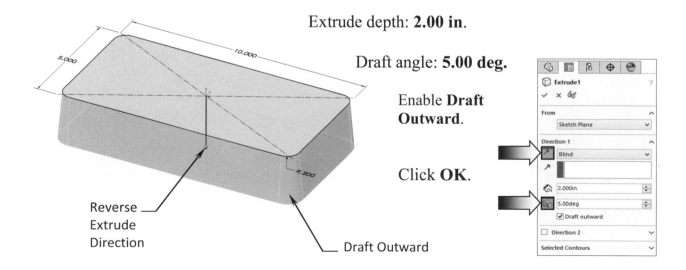

3. Adding the Side Cutouts:

Select the <u>Front</u> plane from the FeatureManager tree.

Click **Sketch** or select **Insert / Sketch.**

Sketch the profile below and add **Dimensions / Relations** needed to fully define the sketch.

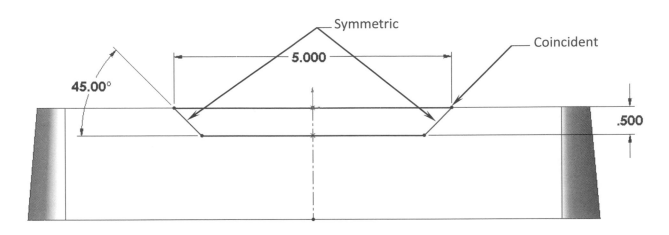

Click **Extruded Cut** .

Select **Through All-Both** for direction1.

Click **OK.**

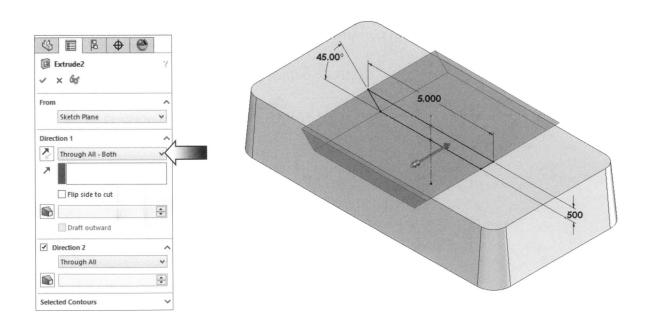

4. Removing more material:

Select the <u>upper surface</u> and open a **new sketch**.

Select the <u>outer edges</u> and create an **offset Entities** of **.250in** as indicated.

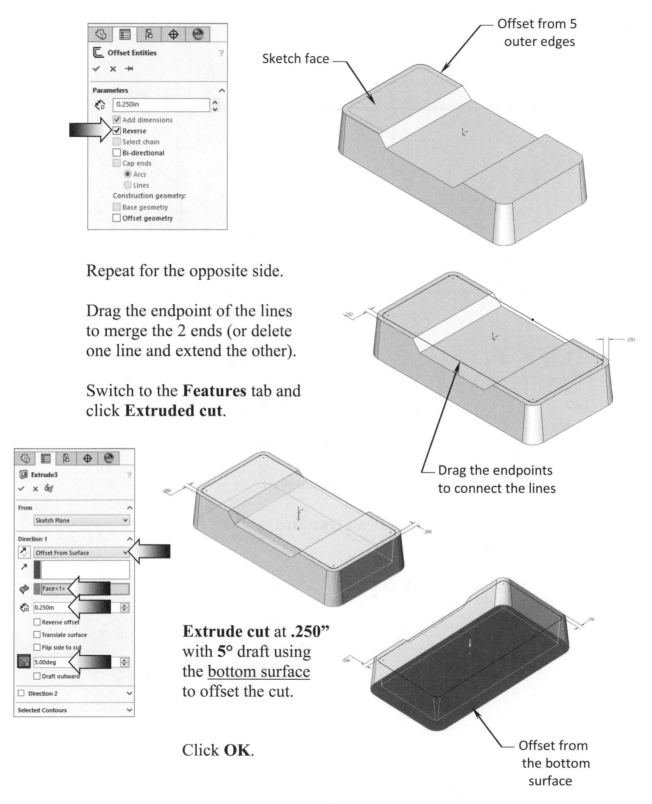

Sketch face

Offset from 5 outer edges

Repeat for the opposite side.

Drag the endpoint of the lines to merge the 2 ends (or delete one line and extend the other).

Switch to the **Features** tab and click **Extruded cut**.

Drag the endpoints to connect the lines

Extrude cut at **.250"** with **5°** draft using the <u>bottom surface</u> to offset the cut.

Offset from the bottom surface

Click **OK**.

5. Creating the Rib Profiles:

Select the <u>face</u> as indicated and open a **new sketch** .

Sketch the profile and add **dimensions / relations** needed to fully define the sketch.

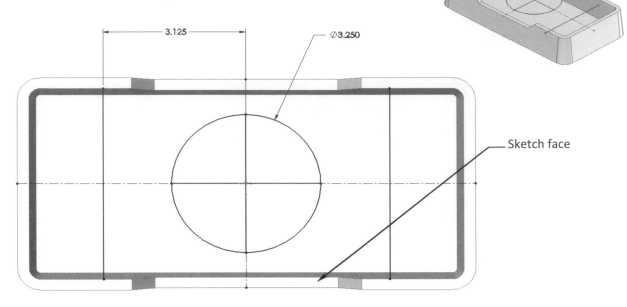

Sketch face

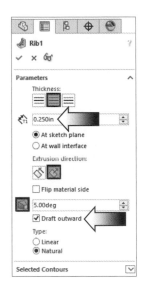

Switch to the **Features** tab and click **Rib** .

Select the option **Both-Sides** and enter **.250 in.** for thickness.

Select the **Normal To Sketch** option.

Click the **Draft** option and enter **5.00** deg.

Enable **Draft Outward** checkbox.

Click **OK**.

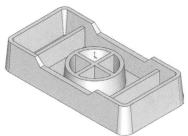

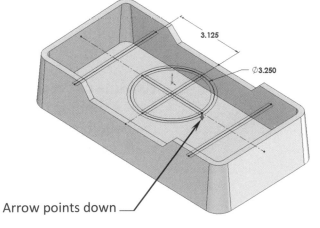

Arrow points down

6. Adding the .500" Fillets:

Click **Fillet** or select **Insert / Features / Fillet-Round**.

Enter **.500 in.** for radius and select the <u>8 Edges</u> as shown.

Click **OK**.

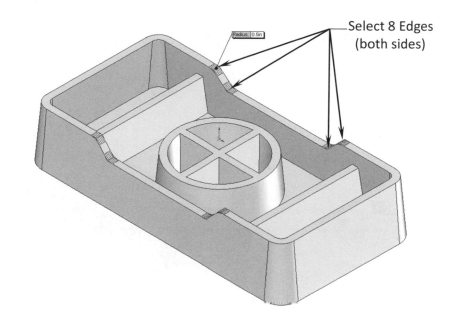

Select 8 Edges (both sides)

7. Adding the .125" Fillets:

Click **Fillet** or select **Insert / Features / Fillet-Round**.

Enter **.125 in.** for radius and select <u>all edges</u> **except** for the bottom edges.

Click **OK**.

Box-select

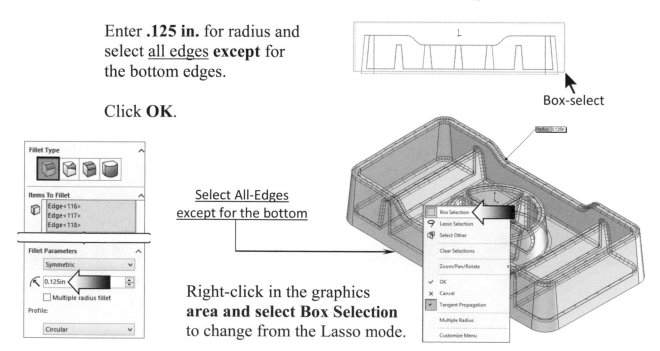

<u>Select All-Edges</u>
<u>except for the bottom</u>

Right-click in the graphics **area and select Box Selection** to change from the Lasso mode.

8. Shelling the lower portion:

The Shell command hollows out a part and leaves open the faces that you select.

Click **Shell** 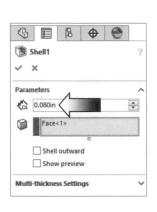 or select **Insert / Features / Shell**.

For Faces to Remove, select the <u>bottom face</u>.

Enter **.080** in. for Thickness.

Click **OK**.

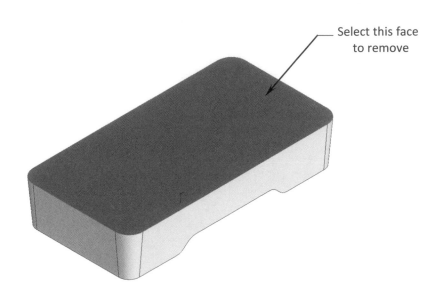

Select this face to remove

9. Saving Your Work:

Click **File / Save As**.

Enter **Rib and Shell** for file name.

Click **Save**.

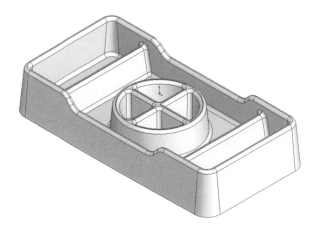

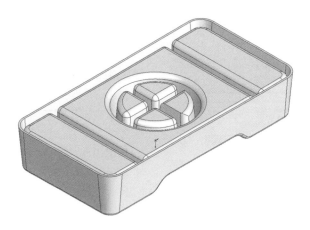

Questions for Review

1. A sketch can be repositioned after its relations/dimensions to the origin have been removed.
   a. True
   b. False

2. The Mid-Plane extrude option extrudes the profile to both directions and adds equal thickness on both sides.
   a. True
   b. False

3. Draft Outward is the only option available; the Draft Inward option is not available.
   a. True
   b. False

4. The Shell feature hollows out the part starting with the selected face.
   a. True
   b. False

5. If nothing (no faces) is selected, a solid model cannot be shelled.
   a. True
   b. False

6. A Rib is a special type of extruded feature; no drafts may be added to it.
   a. True
   b. False

7. A Rib can only be created with a closed-sketch profile; opened-sketch profiles may not be used.
   a. True
   b. False

8. The Rib features can be fully edited, just like any other feature in SOLIDWORKS.
   a. True
   b. False

7. FALSE   8. TRUE
5. FALSE   6. FALSE
3. FALSE   4. TRUE
1. TRUE   2. TRUE

Shell & Mirror Features

Shell & Mirror Features
Styrofoam Box

In most cases, creating a model as a solid and then shelling it out towards the end of the process would probably be easier than creating the model as a thin walled part from the beginning.

Although the model can be made as a thin walled part from the beginning of the design but this lesson will guide us through the process of creating the solid model first and the shell feature will be added towards the end.

The shell command hollows out a model, deletes the faces that you select, and creates a thin walled feature on the remaining faces. But if you do not select any faces of the model, it will still get shelled, and the model will become a closed hollow part. Multi-thickness can be created at the same time.

Mirror, on the other hand, offers a quick way to make a copy of one or more features where a plane or a planar surface is used to mirror about.

These are the options that are available for mirror:

* Mirror Features: used to mirror solid features such as Extruded Boss, Cuts, Fillets, and Chamfers.

* Mirror Faces: used when the model, or an imported part, has faces that make up the features but not the solid features themselves.

* Mirror Bodies: used to mirror solid bodies, surface bodies, or multibodies.

If the Mirror Bodies option is selected, the Merge Solids and the Knit Surfaces options appear, requiring you to select the appropriate checkboxes prior to completing the mirror function.

Shell & Mirror Features
Styrofoam Box

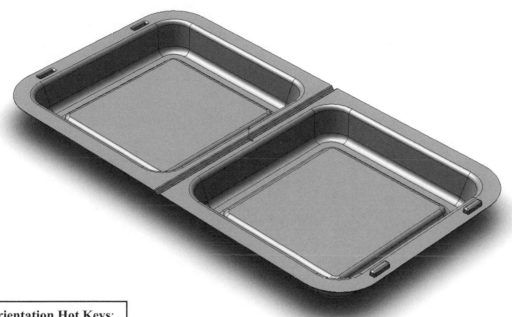

View Orientation Hot Keys:

Ctrl + 1 = Front View
Ctrl + 2 = Back View
Ctrl + 3 = Left View
Ctrl + 4 = Right View
Ctrl + 5 = Top View
Ctrl + 6 = Bottom View
Ctrl + 7 = Isometric View
Ctrl + 8 = Normal To
      Selection

Dimensioning Standards: **ANSI**

Units: **INCHES** – 3 Decimals

Tools Needed:

 Offset Entities   Extruded Boss-Base   Extruded Cut

 Mirror   Fillet/Round   Shell

1. Starting a new part:

Click **File / New / Part**. Set the Units to Inches and Number of Decimal to 3.

Select the <u>Top</u> plane and open a **new sketch** .

Sketch a **rectangle** a little bit to the right of the Origin as shown below.

Add a reference **centerline** from the origin and connect it to the midpoint of the line next to it.

Add the **dimensions** as shown to fully define this sketch.

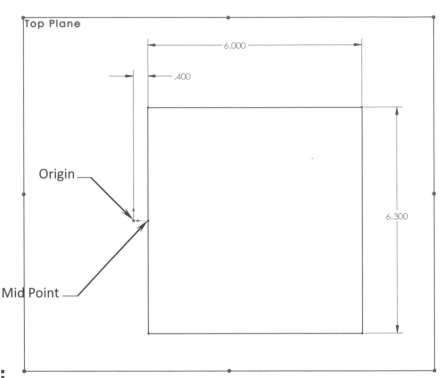

2. Extruding the base:

Switch to the **Features** tab and click **Extruded Boss-Base** .

Direction1: **Blind**

Click **Reverse** (arrow).

Depth: **1.000in**.

Draft: **10deg**.

Click **OK**.

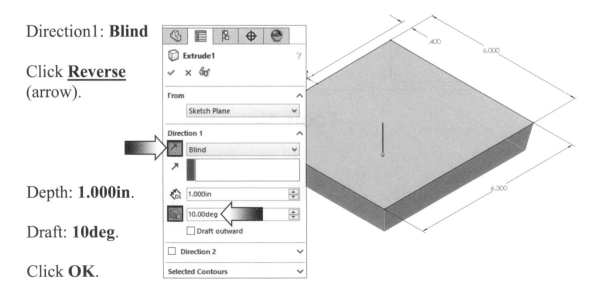

3. Adding the .750" fillets:

Click the **Fillet** 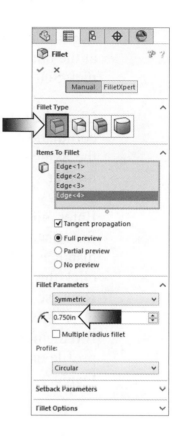 command from the **Features** tab.

Enter **.750in** for radius.

Select the **4 edges** as noted.

Click **OK**.

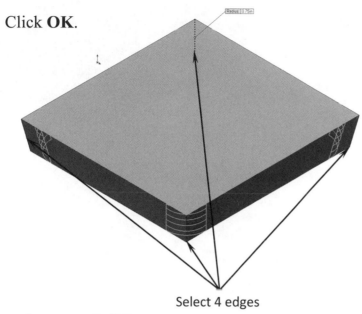

Select 4 edges

4. Adding the .250" fillets:

Press **Enter** to repeat the previous command or click the **Fillet** command again.

Enter **.250in** for radius.

Select one of the **edges** at the bottom of the part. The fillet should propagate automatically because the Tangent Propagation checkbox was enabled by default.

Select the
bottom edge

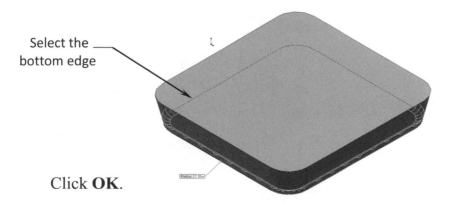

Click **OK**.

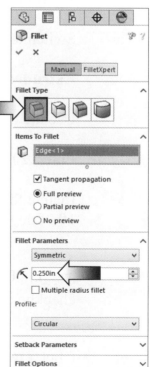

5. Creating an offset sketch:

Open a **new sketch**  on the <u>bottom face</u> as indicated.

Right-click one of the **edges** at the bottom as noted and pick **Select Tangency**.

Click the **Offset Entities** command on the Sketch toolbar.

Enter **.400in** for offset distance and click the **Reverse** checkbox to flip the new profile on the <u>inside</u>.

6. Creating a recess:

Switch to the **Features** tab and click **Extruded Cut** .

Use the default **Blind** type.

Enter **.200in** for extrude depth.

Enable the draft button and enter **10.00deg**. for Angle (arrow).

<u>Clear</u> the Draft Outward checkbox.

Click **OK**.

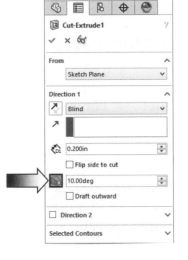

7. Adding the .125" fillets:

From the **Features** tab, click the **Fillet** command .

Enter **.125in** for radius.

Select the rectangular <u>face</u> of the recess. The fillet propagates along all edges of the selected face.

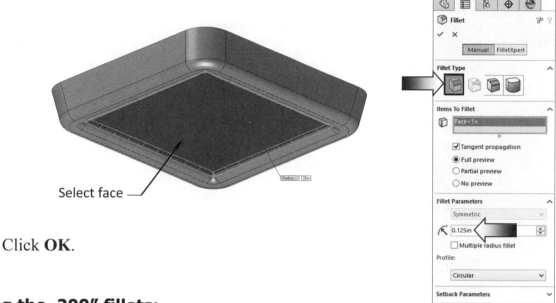

Select face —

Click **OK**.

8. Adding the .200" fillets:

Press **Enter** to repeat the last command (or click the Fillet command again).

Enter **.200in** for radius.

Select <u>one</u> of the <u>upper edges</u> of the recess. The fillet propagates around the recess feature.

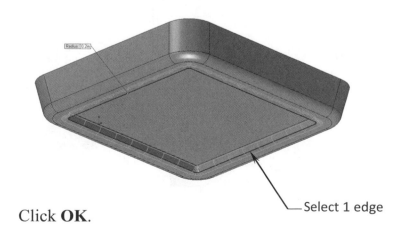

Select 1 edge

Click **OK**.

9. Creating the rim:

Select the <u>upper face</u> of the part and open a **new sketch** 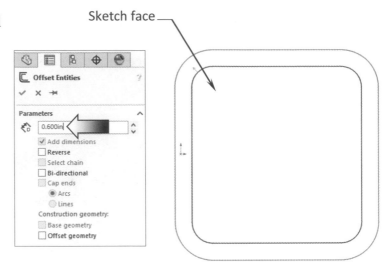.

The selected face should still be highlighted.

Click the **Offset Entities** command on the Sketch tab.

Enter **.600in** for offset distance.

Click Reverse if needed to flip the offset to the <u>outside</u>.

Click **OK**.

Delete the 2 corner radiuses and extend the lines by dragging and dropping the endpoints, or by using the Merge relation to close-off both ends of the profile.

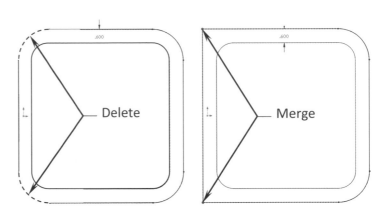

Click **Extruded Boss-Base**.

Use the default **Blind** type.

Enter **.400in** for extrude depth.

Enter **10deg** for draft angle and enable the **Draft Outward** checkbox.

Click **OK**.

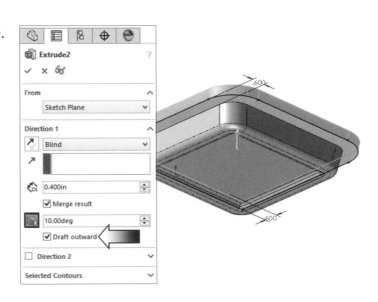

10. Adding another .125" fillet:

Click the **Fillet** command 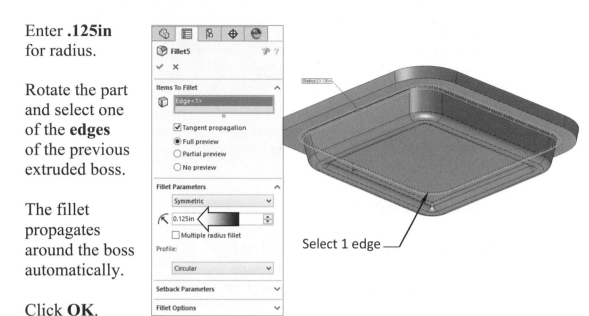 from the Features toolbar.

Enter **.125in** for radius.

Rotate the part and select one of the **edges** of the previous extruded boss.

The fillet propagates around the boss automatically.

Click **OK**.

Select 1 edge

11. Creating the fold feature:

Select the Front plane and open a **new sketch**.
Press **Ctrl + 8** to rotate the view normal to the sketch.

Sketch a **triangle** using the **Line** tool. Add the dimensions shown.

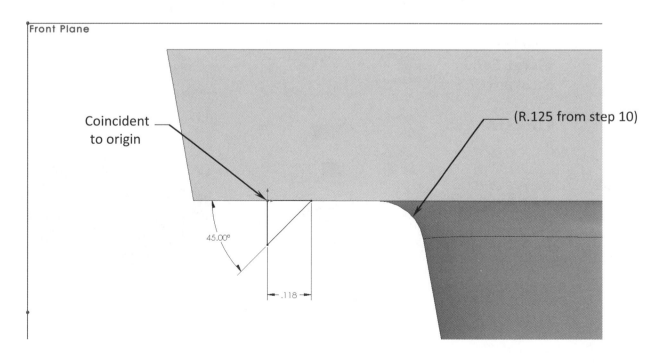

Front Plane

Coincident to origin

(R.125 from step 10)

45.00°

.118

Switch to the **Features** tab and click **Extruded Boss-Base**.

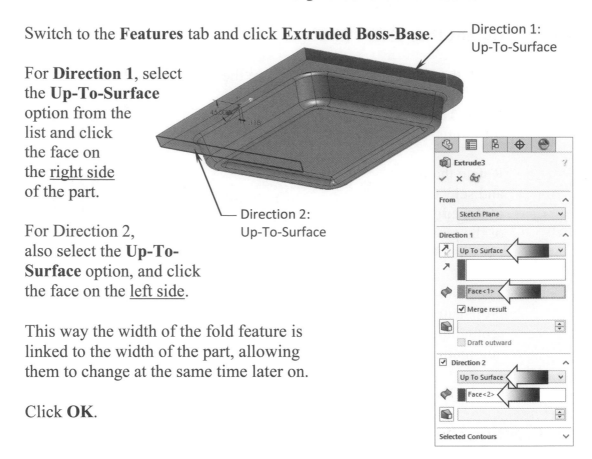

Direction 1:
Up-To-Surface

For **Direction 1**, select
the **Up-To-Surface**
option from the
list and click
the face on
the right side
of the part.

Direction 2:
Up-To-Surface

For Direction 2,
also select the **Up-To-
Surface** option, and click
the face on the left side.

This way the width of the fold feature is
linked to the width of the part, allowing
them to change at the same time later on.

Click **OK**.

12. Mirroring a solid body:

Select the Right plane from the Feature tree and click the **Mirror** button.

Expand the **Bodies to Mirror** section and select the part in the graphics area.

Enable the
Merge Solid
checkbox.

Click **OK**.

Notes: To mirror one or
more *features* of a
part, use the
Features-To-Mirror
option.

To mirror the entire
part use the Bodies-to-Mirror option.

13. Creating the lock feature:

Rotate the part, select the bottom face of the rim and open a **new sketch** .

Sketch a **centerline** from the origin; it will be used as the mirror line.

Sketch a small **rectangle** above the centerline.

Use the mirror function to mirror the rectangle below the centerline.

Add the dimensions as shown to fully define the sketch.

Sketch face

Switch to the **Features** tab and click **Extruded Cut** .

Use the default **Blind** type.

Enter **.150in** for depth.

Enable the **Draft** checkbox and enter **3deg** for taper angle.

Click **OK**.

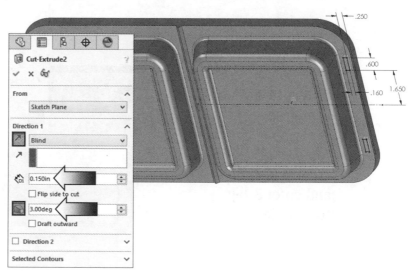

14. Creating the lock cavity:

Select the bottom face of the rim & open a **new sketch** .

Similar to the last step, sketch a **centerline** from the origin and a small **rectangle** above it.

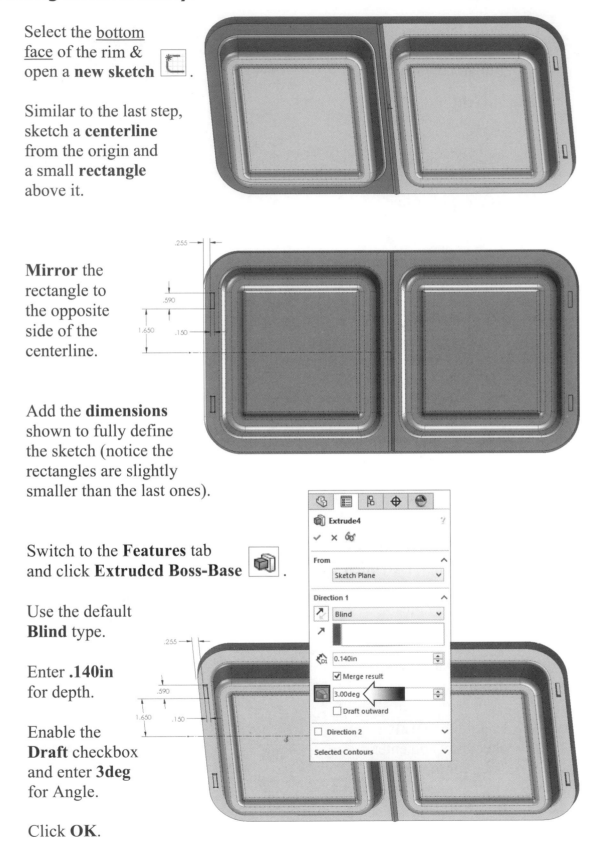

Mirror the rectangle to the opposite side of the centerline.

Add the **dimensions** shown to fully define the sketch (notice the rectangles are slightly smaller than the last ones).

Switch to the **Features** tab and click **Extruded Boss-Base** .

Use the default **Blind** type.

Enter **.140in** for depth.

Enable the **Draft** checkbox and enter **3deg** for Angle.

Click **OK**.

15. Adding the .032" fillets:

Click the **Fillet** command from the Features toolbar.

Enter **.032in** for radius.

Expand the FeatureManager tree (push the arrow symbol) and select the last 2 features, the **Cut-Extrude2** and the **Extrude4**.

This method is a little bit quicker than selecting the edges individually.

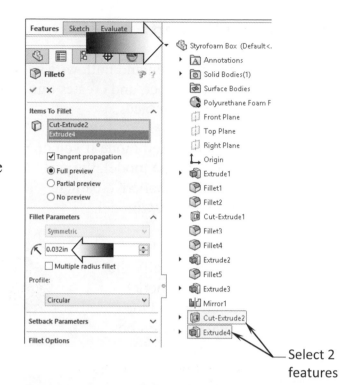

All edges of the selected features are filleted using the same radius value. You should create the fillets separately if you want the features to have different radius values.

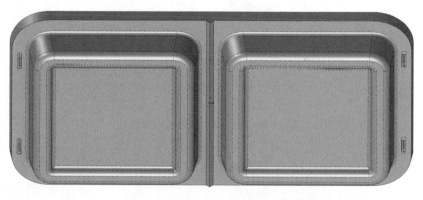

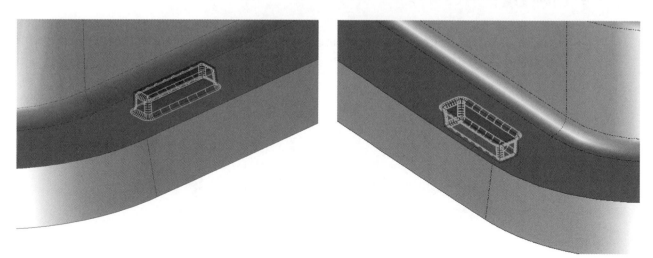

Rotate and zoom closer to verify the fillets on both features.

16. Shelling the part:

The shell command hollows a solid model, deletes the faces you select, and creates thin walled features on the remaining faces.

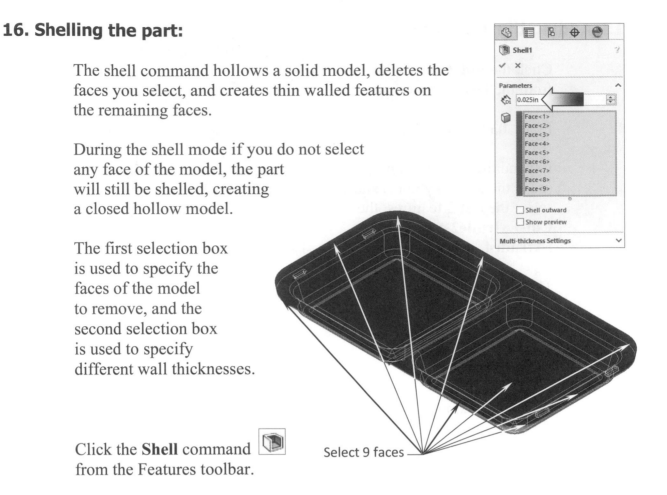

During the shell mode if you do not select any face of the model, the part will still be shelled, creating a closed hollow model.

The first selection box is used to specify the faces of the model to remove, and the second selection box is used to specify different wall thicknesses.

Click the **Shell** command from the Features toolbar.

Select 9 faces

Enter **.025in** for wall thickness.

Select the <u>upper face</u> of the model and the <u>8 faces</u> of the rim to remove.

Click **OK**.

Inspect your model against the image shown below. Edit the shell to correct it if needed.

17. Adding the .050" fillets:

Zoom in on the fold section in the middle of the part.

Click **Fillet**

Enter **.050in** for radius.

Select the <u>3 edges</u> as noted (2 on the top and 1 on the bottom).

Click **OK**.

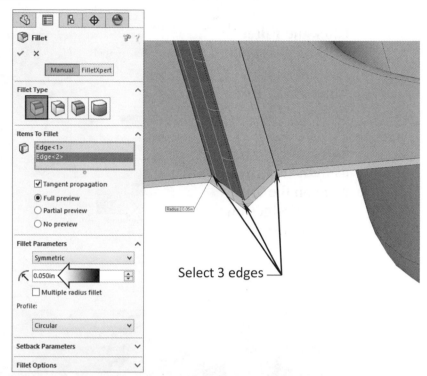

Select 3 edges

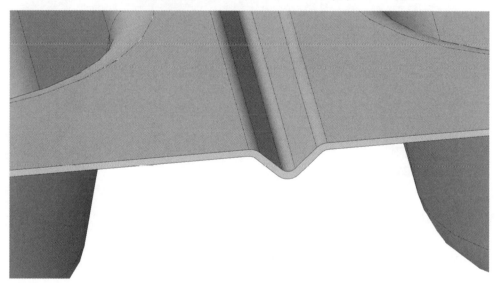

Verify the resulting fillets.

We could have used the Multiple Radius option to create both fillets, the .050" and the .032" (in the next step), but the preview graphics will get very busy, making it hard to identify which edges should be used to apply the correct fillets.

Instead, we are going to create the two fillets separately; that will also give you a chance to practice using the fillet command once again.

18. Adding the .032" fillets:

Click the **Fillet** command again.

Enter **.032"** for radius.

Select the 3 edges as indicated (one edge on the top and two edges on the bottom).

Click **OK**.

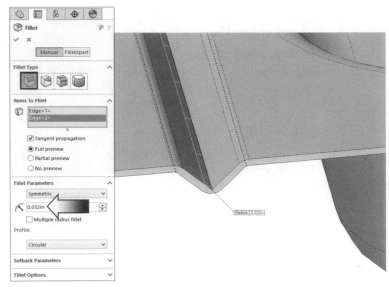

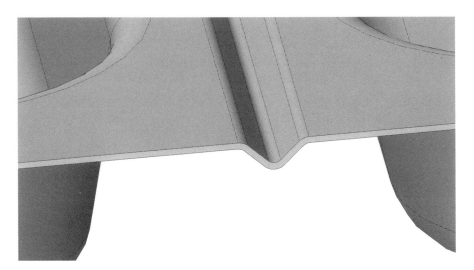

19. Saving your work:

Save your work as **Using Shell Mirror**.

Questions for Review

1. It is easier to create a thin walled part by starting as a solid and then shell it out towards the end.
    a. True
    b. False

2. The shell command hollows out a solid model and removes the faces that you select.
    a. True
    b. False

3. If you do not select any faces of the model, the part will still be shelled into a closed volume.
    a. True
    b. False

4. Only one wall thickness can be created with the shell command.
    a. True
    b. False

5. To mirror a feature, you must use a centerline to mirror about.
    a. True
    b. False

6. To mirror one or more features, a planar face or a plane is used as the center of the mirror.
    a. True
    b. False

7. To mirror the entire part, you must use mirror-features and select all features from the tree.
    a. True
    b. False

8. To mirror the entire part, you must use mirror-body and select the part from the graphics area or from the Solid Bodies folder.
    a. True
    b. False

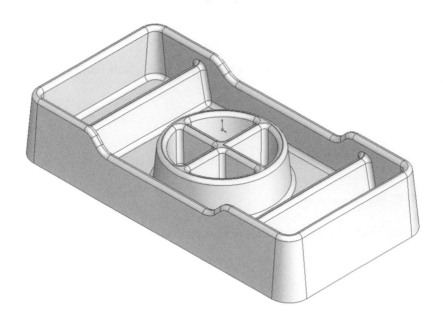

CHAPTER 7

Patterns

Linear Patterns
Test Tray

The Linear Pattern is used to arrange multiple instances of selected features along one or two linear paths.

The following four references are needed to create a linear pattern:

* Direction of the pattern (linear edge of the model or an axis).

* Distance between the pattern instances.

* Number of pattern instances in each direction.

* Feature(s) to the pattern.

If the seed feature(s) is changed, all instances within the pattern will be updated automatically.

The following options are available in Linear Patterns:

* Pattern instances to skip. This option is used to hide / skip some of the instances in a pattern.

* Pattern seed only. This option is used when only the original feature gets repeated, not its instances.

This chapter will teach us the use of the pattern commands such as Linear, Circular, and Curve Driven Patterns.

Linear Patterns
Test Tray

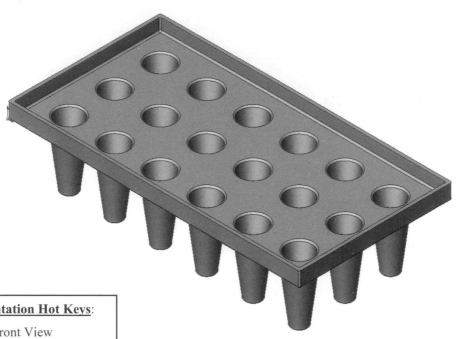

View Orientation Hot Keys:

Ctrl + 1 = Front View
Ctrl + 2 = Back View
Ctrl + 3 = Left View
Ctrl + 4 = Right View
Ctrl + 5 = Top View
Ctrl + 6 = Bottom View
Ctrl + 7 = Isometric View
Ctrl + 8 = Normal To
        Selection

Dimensioning Standards: **ANSI**

Units: **INCHES** – 3 Decimals

Tools Needed:

Insert Sketch	Rectangle	Circle
Dimension	Add Geometric Relations	Base/Boss Extrude
Shell	Linear Pattern	Fillet/Round

1. Sketching the Base Profile:

Select the Top plane from the FeatureManager tree.

Click **Sketch** or select **Insert, Sketch**.

Sketch a **Corner Rectangle** starting at the Origin.

Add the **Dimensions** shown below.

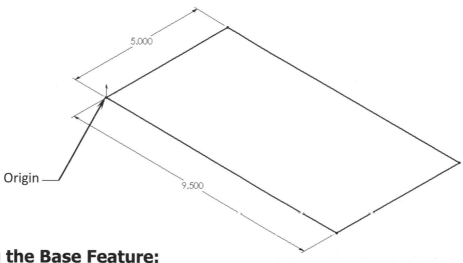

2. Extruding the Base Feature:

Switch to the **Features** tab and click or select **Insert, Boss-Base, Extrude**.

End Condition: **Blind.**

Extrude Depth: **.500in**.

Click **OK**.

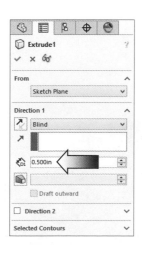

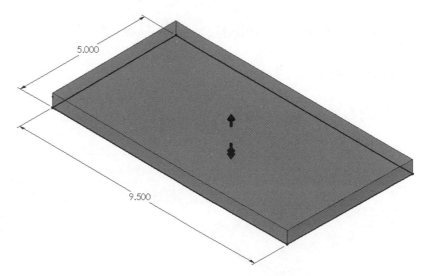

3. Sketching the seed feature:

Select the <u>face</u> indicated as the sketch plane.

Click **Sketch**  or select **Insert, Sketch**.

Sketch a **Circle** and add the **Dimensions** shown.

Sketch Face

Ø1.000

1.000

1.000

4. Extruding a seed feature:

Switch to the **Features** tab and click 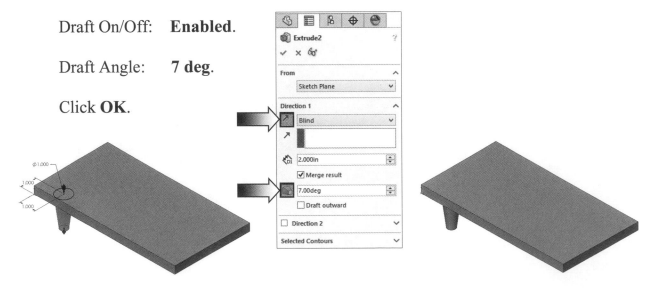 or select **Insert, Boss-Base, Extrude**.

End Condition: **Blind, Reverse Direction** enabled.

Extrude Depth: **2.00 in**.

Draft On/Off: **Enabled**.

Draft Angle: **7 deg**.

Click **OK**.

5. Creating a Linear Pattern:

Click 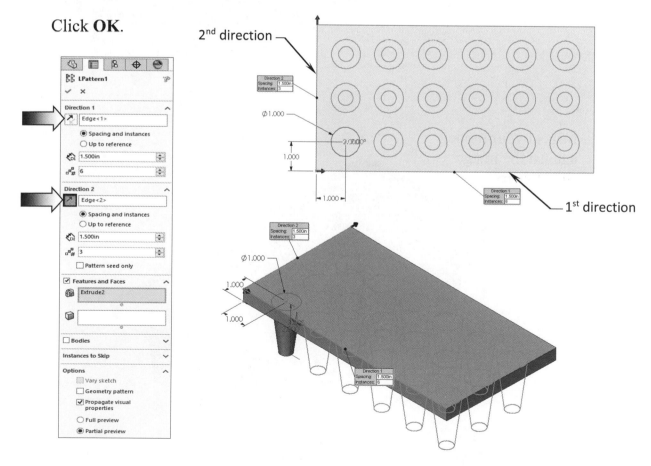 or select **Insert, Pattern Mirror, Linear Pattern**.

For **Direction 1**:

Select the bottom **horizontal edge** as Pattern Direction1.

Enter **1.500** in. for Spacing.

Enter **6** for Number of Instances.

For **Direction 2**:

Select the **vertical edge** as Pattern Direction2.

Enter **1.500** in. for Spacing.

Enter **3** for Number of Instances.

For Features to Pattern select **Extrude2** from the FeatureManager tree.

Click **OK**.

> **Linear Patterns**
>
> The Linear Pattern option creates multiple instances of one or more features uniformly along one or two directions.

6. Shelling the Base feature:

Select the upper face as shown.

Click or select **Insert, Features, Shell**.

Enter **.100** in. for Thickness.

Click **OK**.

> ### Shell
>
> The Shell command hollows out the part, starting with the selected face. Constant or multi-wall thickness can be done in the same operation.

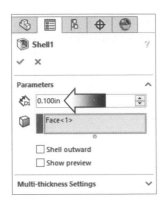

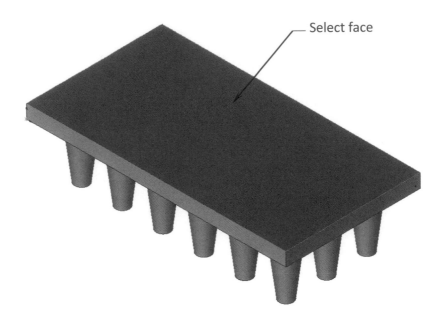

— Select face

Inspect your model against the image shown here.

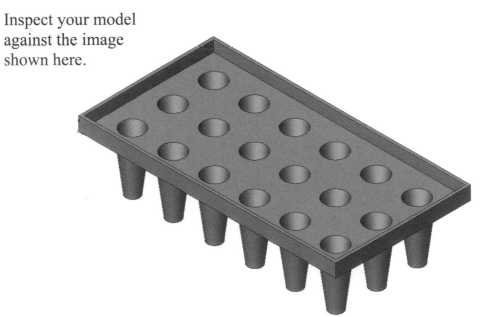

7. Adding Fillets:

Click **Fillet** or select **Insert, Features, Fillet / Round.**

Enter **.050** in. for radius.

Select all the <u>edges</u>. (If the Box-Select is not the default option, right-click in the graphics area and select Box-Selection. Control+A also works well.)

Click **OK**.

💡 **Fillets**
A combination of faces and edges can be selected within the same fillet operation. (Box-select the entire part to select all edges.)

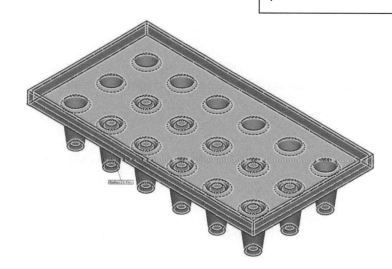

8. Saving your work:

Select **File, Save As, Test Tray, Save**.

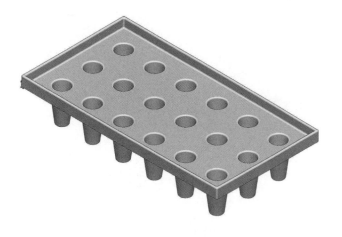

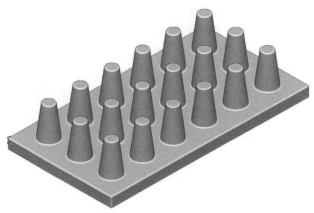

Questions for Review

1. SOLIDWORKS only allows you to pattern one feature at a time. Patterning multiple features is not supported.
   a. True
   b. False

2. Only the spacing and number of copies are required to create a linear pattern.
   a. True
   b. False

3. SOLIDWORKS does not require you to specify the 2nd direction when using both directions option.
   a. True
   b. False

4. The Shell feature hollows out the part and creates a wall thickness specified by the user.
   a. True
   b. False

5. After the Shell feature is created, its wall thickness cannot be changed.
   a. True
   b. False

6. Both faces and edges of a model can be used to create a fillet.
   a. True
   b. False

7. The value of the fillet can be changed at any time.
   a. True
   b. False

8. "Pattern a pattern" is not supported in SOLIDWORKS.
   a. True
   b. False

7. TRUE   8. FALSE
5. FALSE   6. TRUE
3. FALSE   4. TRUE
1. FALSE   2. FALSE

Circular Patterns
Spur Gear

Circular Patterns
Spur Gear

One or more instances can be copied in a circular fashion or around an Axis.

The center of the pattern can be defined by a circular edge, an axis, a temporary axis, a cylindrical face, or an angular dimension.

Below are the references required to create a circular pattern:

* Center of rotation.

* Spacing between the instances.

* Number of copies.

* Feature(s) to copy.

Only the original feature can be edited; changes made to the original are automatically updated within the pattern.

The features to the pattern can be selected directly from the graphics area or from the FeatureManager tree.

The instances in a pattern can be skipped. Skip is similar to suppress, the instances can be skipped or un-skipped during or after the pattern is made.

The Geometry Pattern option creates the pattern using only the geometry (faces and edges) of the features, rather than patterning and solving each instance of the feature.

Circular Patterns
Spur Gear

View Orientation Hot Keys:

Ctrl + 1 = Front View
Ctrl + 2 = Back View
Ctrl + 3 = Left View
Ctrl + 4 = Right View
Ctrl + 5 = Top View
Ctrl + 6 = Bottom View
Ctrl + 7 = Isometric View
Ctrl + 8 = Normal To
          Selection

Dimensioning Standards: **ANSI**

Units: **INCHES** – 3 Decimals

Tools Needed:

Insert Sketch	Line	Center Line
Dynamic Mirror	Add Geometric Relations	Dimension
Convert Entities	Trim Entities	Base/Boss Revolve
Circular Pattern	Base/Boss Extrude	Extruded Cut

1. Sketching the Body profile:

Select the <u>Front</u> plane from the FeatureManager Tree.

Click **Sketch** or select **Insert, Sketch**.

Sketch a **Centerline** starting at the Origin.

Select the **Dynamic Mirror** tool and click the centerline to activate the Dynamic Mirror mode (or select **Tools / Sketch Tools / Dynamic Mirror**).

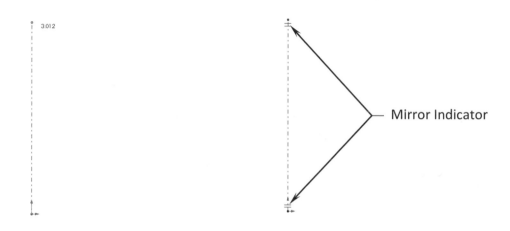

Mirror Indicator

Sketch the profile below using the **Line** tool.

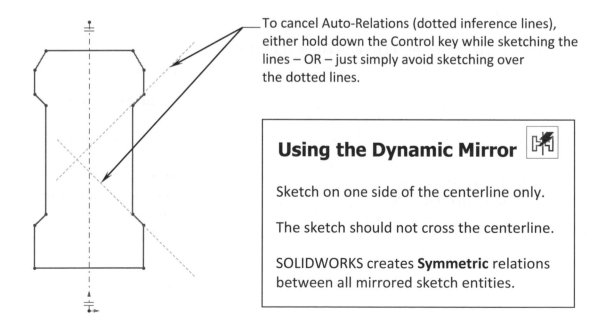

To cancel Auto-Relations (dotted inference lines), either hold down the Control key while sketching the lines – OR – just simply avoid sketching over the dotted lines.

Using the Dynamic Mirror

Sketch on one side of the centerline only.

The sketch should not cross the centerline.

SOLIDWORKS creates **Symmetric** relations between all mirrored sketch entities.

Add the following **Geometric Relations** to the entities indicated below:

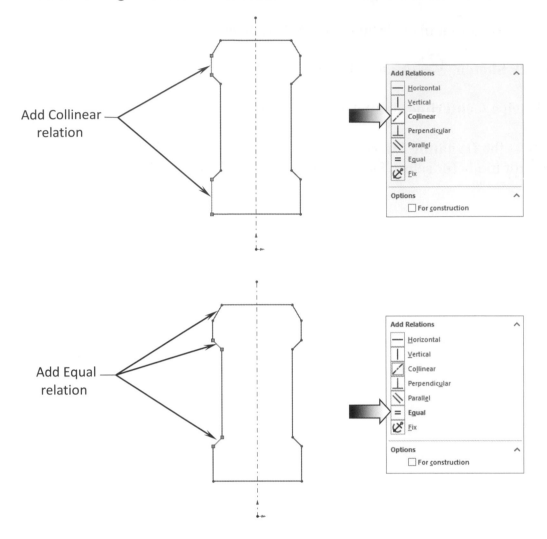

Add Collinear
relation

Add Equal
relation

Sketch a horizontal **Centerline** starting at the Origin.

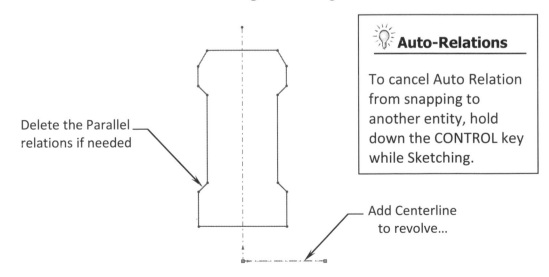

Delete the Parallel
relations if needed

🔆 **Auto-Relations**

To cancel Auto Relation
from snapping to
another entity, hold
down the CONTROL key
while Sketching.

Add Centerline
to revolve...

Add the **dimensions** shown below:

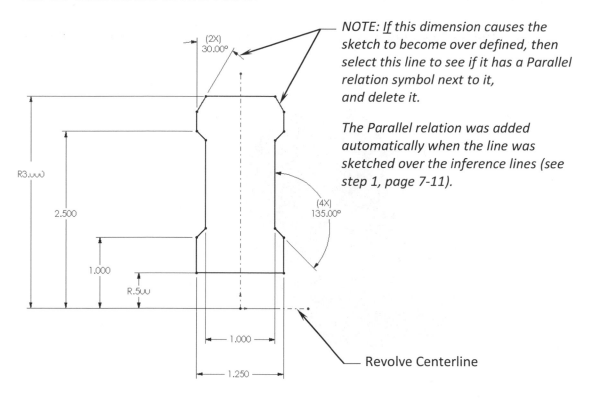

NOTE: If this dimension causes the sketch to become over defined, then select this line to see if it has a Parallel relation symbol next to it, and delete it.

The Parallel relation was added automatically when the line was sketched over the inference lines (see step 1, page 7-11).

Revolve Centerline

2. Revolving the Base Body:

Select the <u>horizontal centerline</u> as indicated above.

Click  or select **Insert, Boss-Base, Revolve**.

Revolve direction: **One Direction**

Revolve Angle: **360°**

Click **OK**.

3. Sketching the Thread Profile:

Select the <u>face</u> as indicated and click **Sketch** or select **Insert, Sketch**.

Sketch Face

Click **Normal To** from the Standard View Toolbar (or press Control+8).

Sketch a vertical **Centerline**, starting at the origin.

Select the **Mirror** tool and click the centerline to activate the Dynamic-Mirror mode.

Dynamic Mirror symbol

NOTE: The Dynamic Mirror command can also be selected under: Tools, Sketch Tools, Dynamic Mirror.

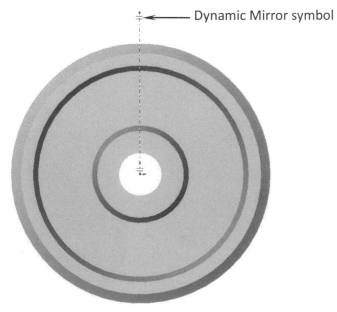

Sketch the profile using the **Line** and the **3-Point-Arc** tools.

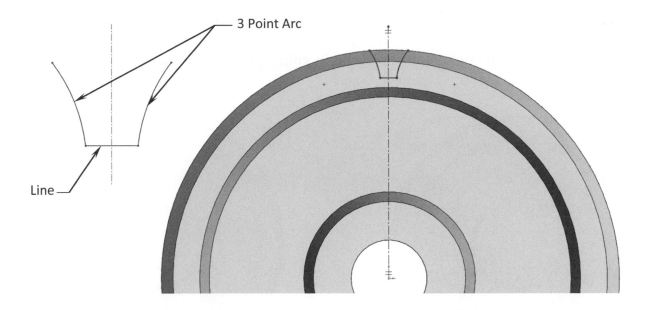

4. Converting the entities:

Select the **outer edge** of the part and click **Convert Entities** 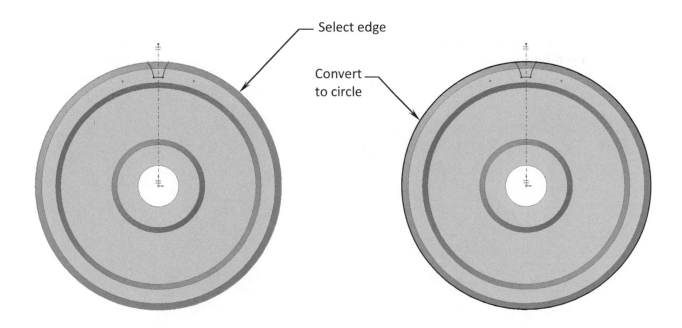 from the Sketch tab.

The selected edge is converted to a circle.

5. Trimming the Sketch Entities:

Select the **Trim** tool 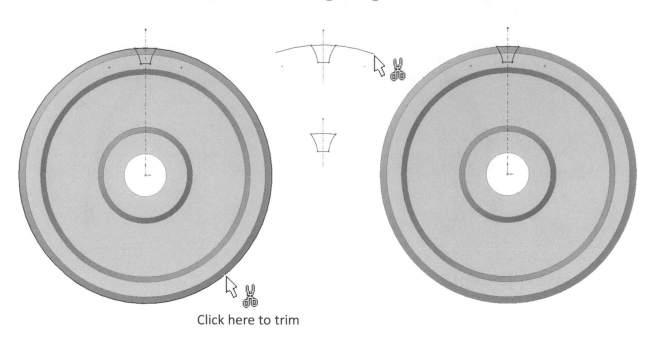 from the Sketch tab; select the **Trim-to-Closest**

option ▮▐ , and click on the **lower right edge** of the circle to trim.

Click here to trim

6. Adding Dimensions:

Select the **Smart Dimension** tool from the Sketch tab and add the dimensions shown below.

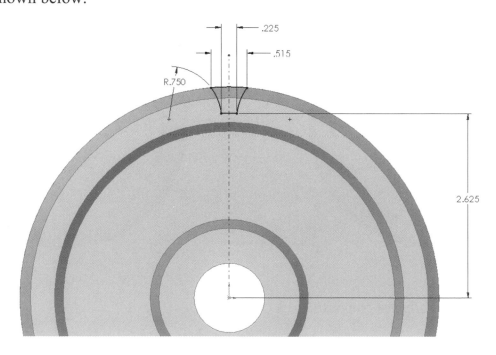

Switch to the Isometric View or press **Ctrl + 7**.

7. Cutting the First Tooth:

Switch to the **Features** tab and click **Extruded Cut** .

End Condition: **Through All**.

Click **OK**.

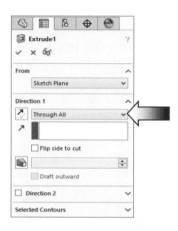

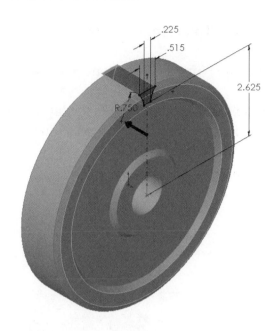

8. Circular Patterning the Tooth:

Click the drop-down arrow under the **Hide/Show Items** (arrow) and select the **Temporary Axis** command (arrow). A temporary axis in the center of the hole is created automatically; this axis will be used as the center of the pattern.

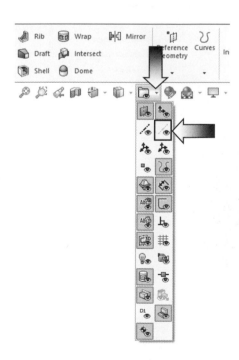

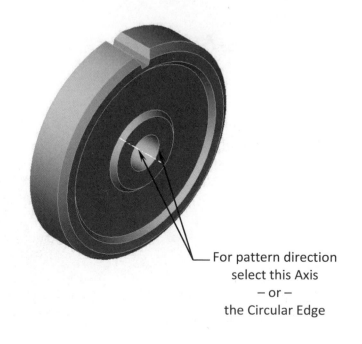

For pattern direction select this Axis
– or –
the Circular Edge

Click **Circular Pattern** 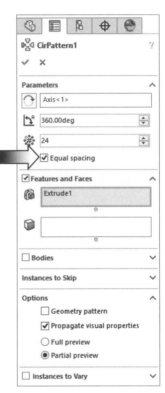 <u>below</u> the Linear Pattern drop-down.

Click on the center **axis** as Direction 1.

Click the **Equal Spacing** check box.

Set the Total Angle to **360°**.

Set the Number of Instances to **24**.

Click inside the Features to Pattern box and select one of the faces of the cut feature (or select the previous **Extruded-Cut** feature from the tree).

Click **OK**.

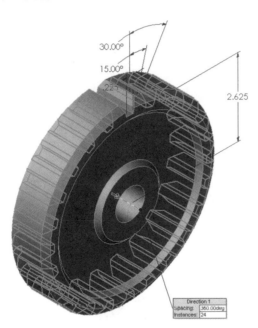

The cut feature is patterned 24 times around the center axis of the model.

Change to different orientations to inspect the patterned feature.

9. Adding the Keyway:

Select the <u>face</u> indicated and open a **new sketch**  or select **Insert, Sketch**.

Sketch Face

Click **Normal To**  from the Standard Views toolbar (Ctrl + 8).

Sketch a vertical **Centerline** starting at the Origin.

Select the **Mirror** tool [icon] and click the centerline to activate the Dynamic-Mirror mode.

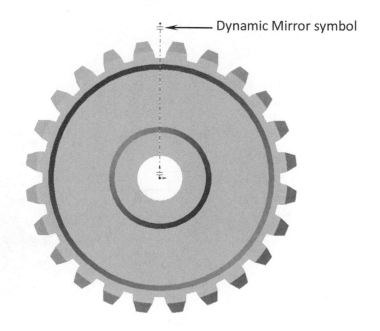

Dynamic Mirror symbol

💡 Sketch Mirror

Use the Mirror option in a sketch to make a symmetrical profile.

When a sketch entity is changed, the mirrored image will also change.

Sketch the profile of the keyway and add dimension shown below:
Note: It would be quicker to convert the edge of the hole and trim.

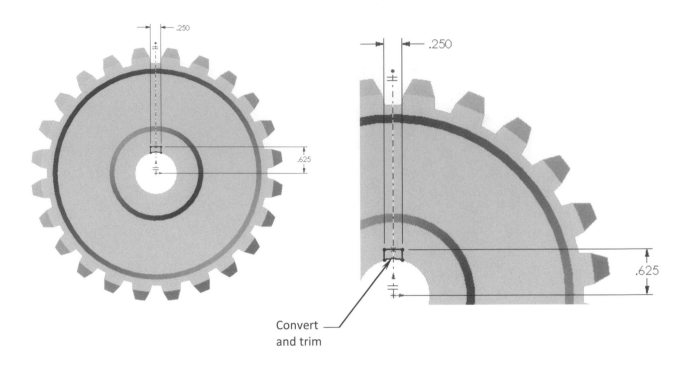

Convert
and trim

10. Extruding a Cut:

Switch to the **Features** tab and click **Extruded Cut** .

End Condition: **Through All**

Click **OK**.

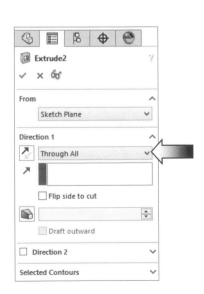

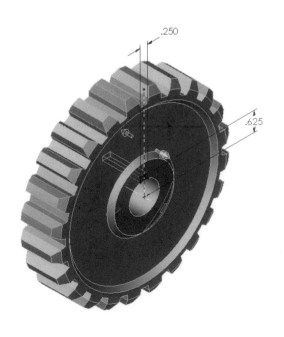

11. Saving Your Work:

Select **File / Save As**.

Enter **Spur Gear** for the file name.

Click **Save**.

Questions for Review

1. All entities in a revolve sketch should not cross over the revolved centerline.
    a. True
    b. False

2. The system creates **symmetric** relations to all mirrored sketch entities.
    a. True
    b. False

3. An Equal relation makes the entities equal in size.
    a. True
    b. False

4. The system creates an On-Edge relation to all converted entities.
    a. True
    b. False

5. The Trim Entities tool is used to trim 2D sketch entities.
    a. True
    b. False

6. The center of the circular pattern can be defined by an axis, a linear edge, or an angular dimension.
    a. True
    b. False

7. The center of rotation, spacing, number of copies, and features to copy are required when creating a circular pattern.
    a. True
    b. False

8. When the original feature is changed, all instances in the pattern will also change.
    a. True
    b. False

7. TRUE   8. TRUE
5. TRUE   6. TRUE
3. TRUE   4. TRUE
1. TRUE   2. TRUE

Circular Patterns
Circular Base Mount

Circular Patterns
Circular Base Mount

As mentioned in the first half of this chapter, the Circular Pattern command creates an array of feature(s) around an axis.

The 4 references below are required to create circular patterns:

* Center axis (Temporary Axis, Axis, an Edge, etc.)

* Spacing between each instance

* Number of instances in the pattern

* Feature(s) to the pattern

Only the original feature may be edited, and changes made to the original feature will automatically be passed onto the instances within the pattern.

The features to the pattern can be selected directly from the graphics area or from the Feature Manager tree.

Instances in a pattern can be skipped. The skipped instances can be edited during or after the pattern is complete.

The Temporary axis can be toggled on or off (View/Temporary Axis).

This 2nd half of the chapter will teach us the use of the Circular Pattern command as well as the Curve Driven Pattern command.

Circular Patterns
Circular Base Mount

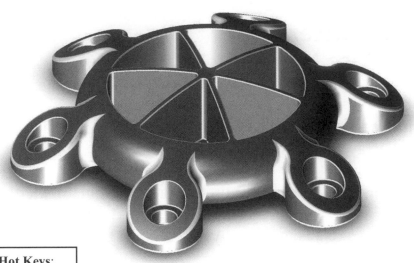

Dimensioning Standards: **ANSI**

Units: **INCHES** – 3 Decimals

Tools Needed:

Insert Sketch	Line	Mirror Dynamic
Add Geometric Relations	Dimension	Base/Boss Revolve
Revolve Cut	Circular Pattern	Fillet/Round

1. Creating the Base Sketch:

Select the <u>Front</u> plane, start a **new sketch** .

Sketch the profile on the right side of the revolve centerline as shown below.

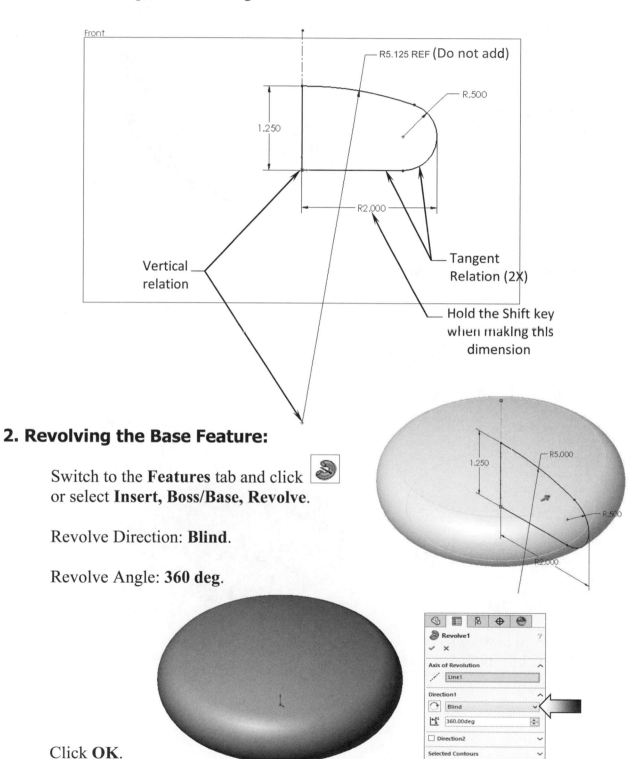

2. Revolving the Base Feature:

Switch to the **Features** tab and click or select **Insert, Boss/Base, Revolve**.

Revolve Direction: **Blind**.

Revolve Angle: **360 deg**.

Click **OK**.

3. Creating the first Side-Tab sketch:

Select the Top
plane and open
a **new sketch**.

Sketch the profile
on the right; add
the dimensions and
relations as shown
in the image to fully
define the sketch.

Add the **R.250**
after the sketch
is fully defined.

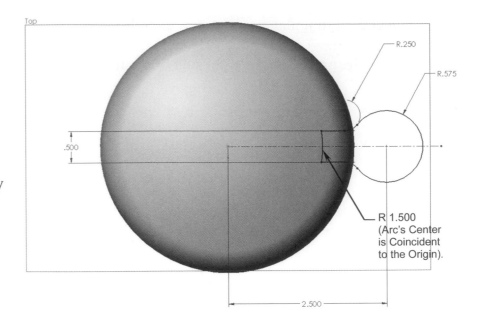

4. Extruding the Side-Tab:

Switch to the **Features** tab and click ⬚ or select **Insert / Boss-Base / Extrude**.

Direction 1, select **Up To Surface**.

For End Condition, select the upper surface.

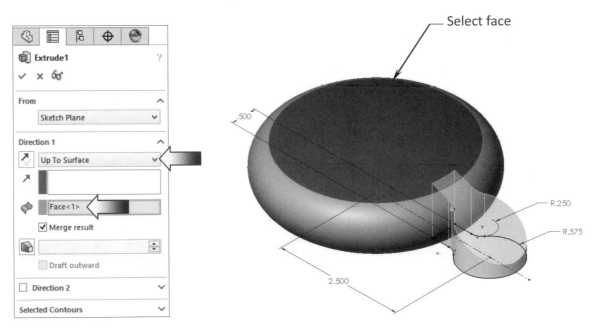

Click **OK**.

5. Adding a Counterbore Hole:

Sketch the profile of the Counterbore* on the <u>Front</u> plane.

*(To create a **Virtual Diameter** dimension 1st click the centerline then select any other entity and place the dimension on the other side of the centerline.)*

Add the 2 Diameter and the 2 Depth dimensions.

Add the relations needed to fully define the sketch.

* *The Hole Wizard command can also be used to create the counterbore hole.*

Ø.625

.800

Ø.375

Virtual Diameter: Add dimension from the center-line to the end point of the line on the left of the profile.

Ø.625

.500

.800

- Ø.375 — Mid-Point

6. Cutting the C'Bore:

Click 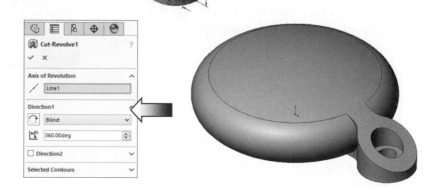 or select **Insert, Cut, Revolve**.

Revolve Direction: **Blind**.

Revolve Angle: **360 deg**.

Click **OK**.

7. Creating the circular pattern:

Click 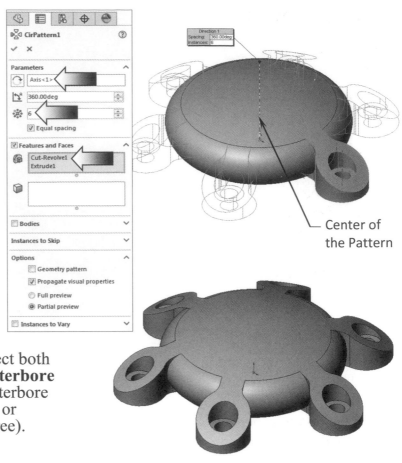 **Circular Pattern** (below Linear).

Click **View, Hide/Show** select the **Temporary-Axis** option.

Select the **Axis** in the center of the part as noted.

Set Pattern Angle to **360°**.

Enter **6** for the Number of Instances.

For Features to Pattern, select both the **Side Tab** and the **Counterbore** hole (either select the Counterbore hole from the graphics area or from the FeatureManager tree).

Click **OK**.

Center of the Pattern

8. Creating a new Plane:

Click  or select **Insert, Reference Geometry, Plane**.

Select **Offset Distance** option.

For 1st References elect the **TOP** plane from the Feature-Manager tree.

Enter **1.300 in.** for Distance.

Click **OK**.

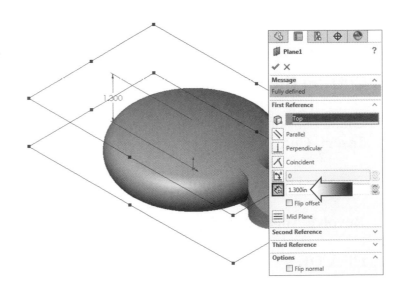

9. Creating the Pocket Sketch:

Select the <u>new plane</u> (Plane1) and open a **new sketch**.

Use the **Dynamic Mirror** and sketch the profile as shown.

Add the dimensions and relations as needed to fully define the sketch.

Use **Circular Sketch Pattern** to make a total of 6 instances of the pocket.

10. Cutting the Pockets:

Switch to the **Features** tab and click **Extruded Cut**.

For direction 1, select **Offset From Surface**.

Select the **bottom surface** to offset from.

Enter **.125 in**. for Depth.

Click **OK**.

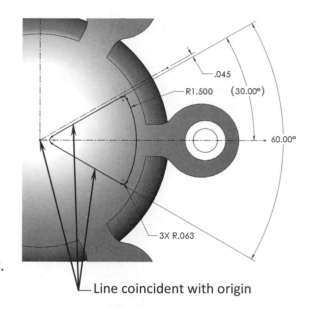

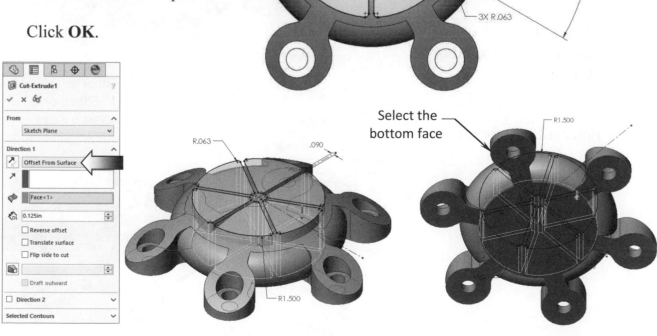

11. Adding the .0625" Fillets:

Click **Fillet**  or select **Insert / Features / Fillet-Round**.

Enter **.062 in**. for radius.

Select all upper and lower edges of the <u>6 tabs</u>.

Click **OK**.

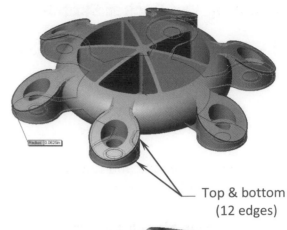

Top & bottom
(12 edges)

12. Adding the .125" Fillets:

Click **Fillet** 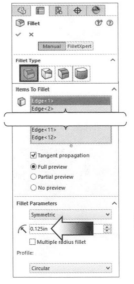 or select **Insert, Features, Fillet-Round.**

Enter **.125 in**. for radius.

Select all edges on the sides of the <u>6 tabs</u> or select the 6 faces as noted.

Click **OK**.

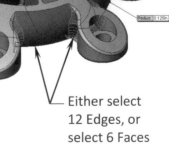

Either select
12 Edges, or
select 6 Faces

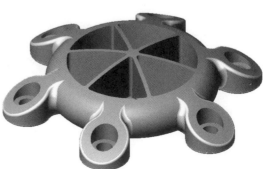

13. Adding the .015" Fillets:

Click **Fillet**  or select
Insert, Features, Fillet-Round.

Enter **.015 in.** for
radius.

Select <u>all edges</u>
of the 6 Pockets
and the 6
Counterbores.

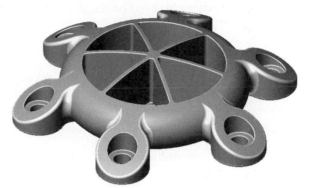

Click **OK**.

14. Saving your work:

Click **File, Save As.**

Enter **Circular Base Mount** for file name.

Click **Save**.

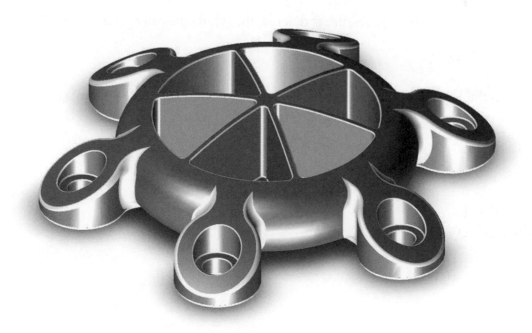

Questions for Review

1. The Circular Patterns command can also be selected from Insert / Pattern Mirror / Circular Pattern.
   a. True
   b. False

2. A Temporary Axis can be used as the center of the pattern.
   a. True
   b. False

3. A linear edge can also be used as the center of the circular pattern.
   a. True
   b. False

4. The Temporary Axis can be toggled ON / OFF under View / Temporary Axis.
   a. True
   b. False

5. The instances in the circular pattern can be skipped during and after the pattern is created.
   a. True
   b. False

6. If an *instance* of the patterned feature is deleted, the whole pattern will also be deleted.
   a. True
   b. False

7. If the *original* patterned feature is deleted, the whole pattern will also be deleted.
   a. True
   b. False

8. When the Equal spacing check box is enabled, the total angle (360°) must be used.
   a. True
   b. False

7. TRUE   8. FALSE
5. TRUE   6. FALSE
3. TRUE   4. TRUE
1. TRUE   2. TRUE

Exercise - Circular Patterns

1. Opening a part document:

Browse to the Training Files folder and open the part document named: **Circular Pattern_Exe.sldprt**

There is a rib feature already created in the model. This rib will get trimmed to its final shape and size, and then circular patterned 10 times.

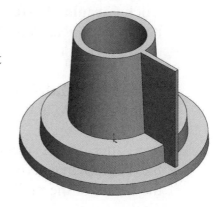

2. Sketching the trimmed profile:

Open a <u>new sketch</u> on the **Front** plane.

Sketch the profile using <u>3 lines</u> and <u>2 arcs</u>.

Add the dimensions and relations shown to fully define the sketch.

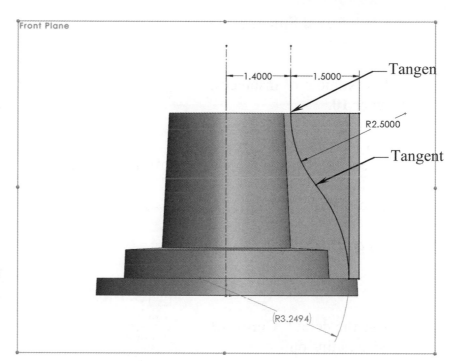

Front Plane

1.4000 1.5000 Tangen

R2.5000

Tangent

(R3.2494)

3. Making a revolved cut:

Switch to the **Features** tab and click **Revolved Cut**.

Use the default **Blind** type and **360°** angle.

Click **OK**.

Cut-Revolve1

Axis of Revolution
Line1

Direction1
Blind
360.00deg

Direction2

Selected Contours

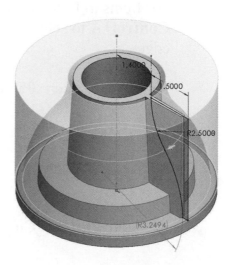

4. Creating a circular pattern:

Click **Circular Pattern** under the Linear Pattern drop-down list.

Click **View**, **Hide/Show**, **Temporary Axis**.

For Pattern Axis, select the **Temporary Axis** in the center of the model.

Click **Equal Spacing** and enter **360°**.

For Number of Instances, enter **10**.

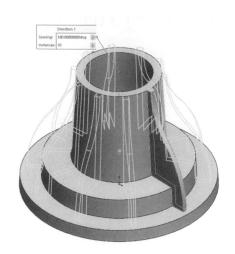

For Features and Faces, select the **Rib** and the **Cut-Revolved** from the feature tree.

Click **OK**.

5. Adding fillets:

Click **Fillet**.

Use the default **Constant Size** radius option.

For Items to Fillet, press **Control+A** to select all edges.

For Radius, enter: **.050in**.

Click **OK**.

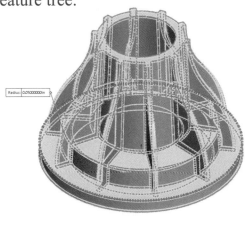

6. Saving your work:

Save your work as: **Circular Pattern_Completed.**

Curve Driven Pattern and Hole Wizard

Curve Driven Patterns
Universal Bracket

The **Curve Driven Pattern** PropertyManager appears when you create a new curve driven pattern feature or when you edit an existing curve driven pattern feature.

The PropertyManager controls the following properties:

Pattern Direction: Select a curve, edge, sketch entity, or select a sketch from the Feature-Manager to use as the path for the pattern. If necessary, click Reverse Direction to change the direction of the pattern.

Number of Instances: Set a value for the number of instances of the seed feature in the pattern.

Equal spacing: Sets equal spacing between each pattern instance. The separation between instances depend on the curve selected for Pattern Direction and on the Curve method.

Spacing: (Available if you do not select Equal spacing.) Set a value for the distance between pattern instances along the curve. The distance between the curve and the Features to Pattern is measured normal to the curve.

Curve method: Defines the direction of the pattern by transforming how you use the curve selected for Pattern Direction. Select one of the following:

* **Transform curve**. The delta X and delta Y distances from the origin of the selected curve to the seed feature are maintained for each instance.

* **Offset curve**: The normal distance from the origin of the selected curve to the seed feature is maintained for each instance.

Alignment method: Select one of the following:
* **Tangent to curve**: Aligns each instance tangent to the curve selected for Pattern direction.

* **Align to seed**: Aligns each instance to match the original alignment of the seed feature of **Curve method** and **Alignment method** selections.

* **Face normal:** (For 3D curves only.) Select the face on which the 3D curve lies to create the curve driven pattern.

Curve Driven Pattern and Hole Wizard
Universal Bracket

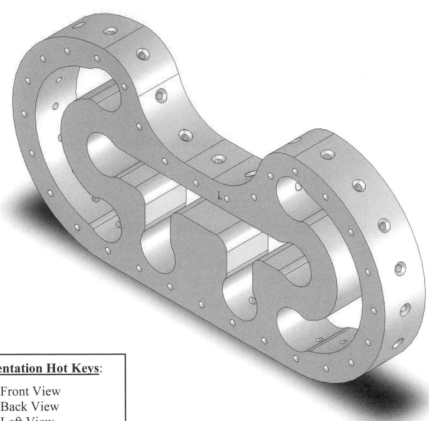

Dimensioning Standards: **ANSI**

Units: **INCHES** – 3 Decimals

Tools Needed:

 Insert Sketch

 Convert Entities

 Offset Entities

 Boss Base Extrude

 Hole Wizard

 Curve Driven Pattern

1. Opening the existing file:

Go to: The Training Files folder
<u>Open</u> a copy of the part document named: **Curve Driven Pattern.sldprt**

<u>Edit</u> the **Sketch1.**

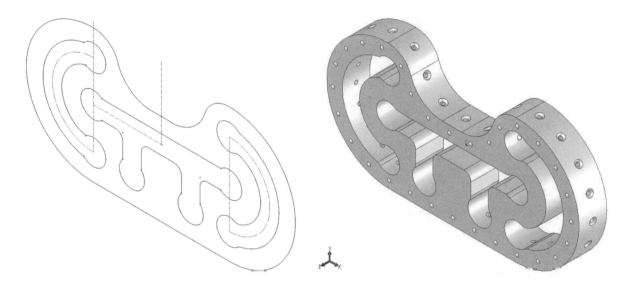

Make sure the sketch is fully defined before extruding it.

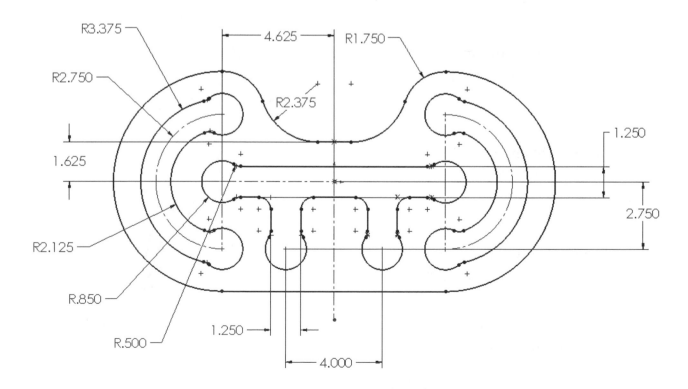

2. Extruding the Base:

Switch to the **Features** tab and click 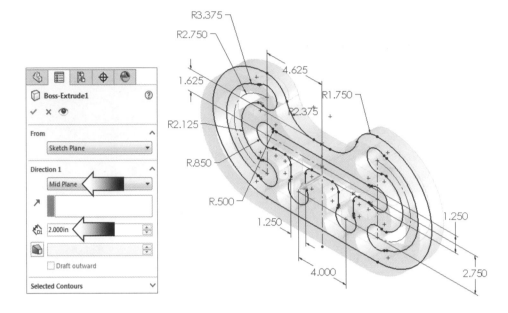 or select:
Insert / Features / Boss Base Extrude.

Use **Mid Plane**
for Direction1.

Enter **2.00 in**. for
Depth.

Click **OK**.

3. Creating the sketch of the 1ˢᵗ hole:

Select the <u>face</u> as indicated and open a **new sketch** .

Sketch a **Centerline** at the
Mid-Point of the two arcs.

Add a **Circle** on the Mid-
Point of the centerline.

Add a **Ø.250 in**. dimension.

The sketch should be fully
defined at this point.

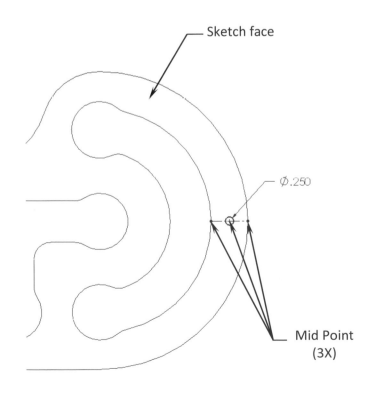

Sketch face

Ø.250

Mid Point
(3X)

4. Cutting the hole:

Switch to the **Features** tab and click 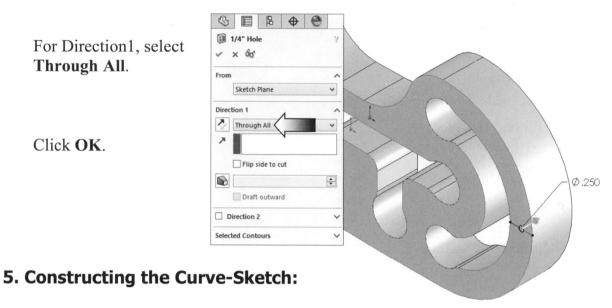 or select **Insert, Cut, Extrude**.

For Direction1, select **Through All**.

Click **OK**.

5. Constructing the Curve-Sketch:

Select the <u>face</u> as noted* and open a **new sketch**.

Select all **Outer-Edges** of the part (Right-click one of the edges & Select Tangency).

Click 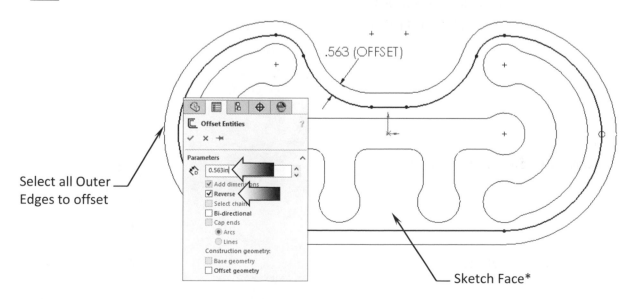 or select **Tools / Sketch Entities / Offset Entities**.

Enter **.563 in**. for Offset Distance.

Click Reverse if necessary, to place the new profile on the **inside**.

<u>Exit</u> the sketch and rename the sketch to **CURVE1**.

Select all Outer Edges to offset

Sketch Face*

6. Creating the 1st Curve-Driven Pattern:

Click or select **Insert, Pattern Mirror, Curve Driven Pattern**.

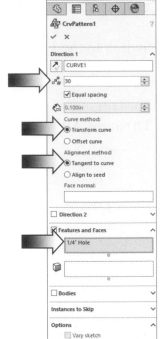

For Direction1, select the **Curve1** from the Feature-Manager tree.

For Number of Instances, enter **30**.

Enable the **Equal-Spacing** check box.

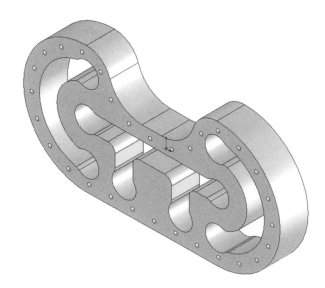

For Features to Pattern, select the **Cut-Extrude1** (the hole) from the FeatureManager tree.

Click **OK**.

Hide the Curve1.

7. Constructing the 2ⁿᵈ Curve:

Select the <u>Front</u> plane and open a **new sketch** .

Select all **Outer-Edges** of the

part and click or select **Tools, Sketch Entities, Convert Entities**.

<u>Exit</u> the sketch and rename the sketch to **CURVE2**.

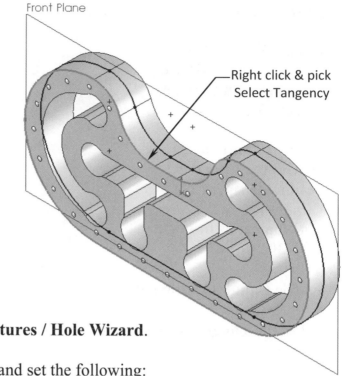

Front Plane

—Right click & pick Select Tangency

8. Adding the Hole Wizard:

Click or select **Insert / Features / Hole Wizard**.

Select the **Counter-Sink** option and set the following:

* Standard: **ANSI Inch**
* Type: **Flat Head Screw** (100)
* Size: **1/4**
* Fit: **Normal**
* End Condition: **Blind**
* Depth: **1.250 in**.

Select the **Position** tab (arrow) and click the **3D Sketch** button 3D Sketch .
This option allows the holes to be placed on non-planar surfaces as well.

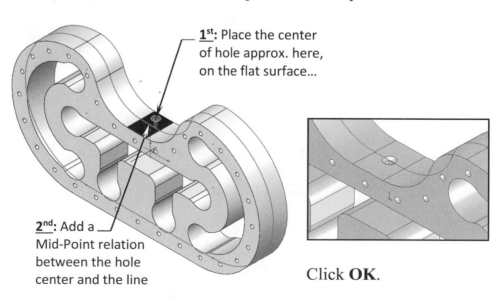

1ˢᵗ: Place the center of hole approx. here, on the flat surface...

2ⁿᵈ: Add a Mid-Point relation between the hole center and the line

Click **OK**.

9. Creating the 2ⁿᵈ Curve Driven Pattern:

Click 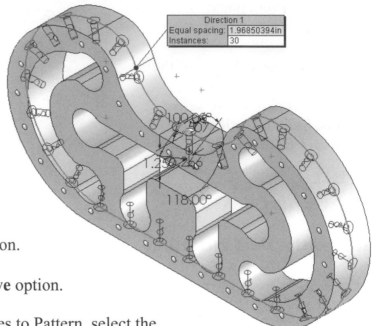 or select **Insert, Pattern/Mirror, Curve Driven Pattern** (under Linear Pattern).

For Direction1, select the **Curve2** from the FeatureManager tree.

For Number of Instances, enter **30**.

Enable **Equal Spacing** option.

Select the **Tangent to Curve** option.

For Features to Pattern, select the **CSK-Hole** from the FeatureManager tree.

Click **OK**.

10. Saving a copy of your work:

Click **File / Save As**.

Enter **Curve Driven Pattern** for file name.

Click **Save**.

CHAPTER 8

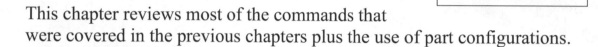

Part Configurations

Part Configurations
Machined Block

This chapter reviews most of the commands that
were covered in the previous chapters plus the use of part configurations.

Upon successful completion of this lesson, you will have a better
understanding of how and when to:

* Sketch on planes and planar surfaces.

* Sketch fillets and model fillets.

* Dimensions and Geometric Relations.

* Extruded Cuts and Bosses.

* Linear Patterns.

* Using the Hole-Wizard option.

* Create new Planes.

* Mirror features.

* Create new Part Configurations, an option that allows the user
  to create and manage families of parts and assemblies.

Configuration options are available in Part and Assembly environments.

After the model is completed, it will be used again in a drawing chapter to
further discuss the details of creating an Engineering drawing.

Part Configurations
Machined Block

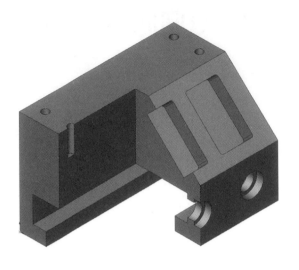

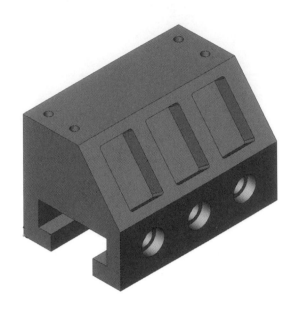

Dimensioning Standards: **ANSI**
Units: **INCHES** – 3 Decimals

Tools Needed:

Insert Sketch	Line	Rectangle
Circle	Sketch Fillet	Dimension
Add Geometric Relations	Extruded Boss/Base	Extruded Cut
Hole Wizard	Linear Pattern	Fillet

1. Sketching the base profile:

Select the <u>Front</u> plane from the FeatureManager tree.

Click **Sketch** or select **Insert, Sketch**.

Sketch the profile below using the **Line** tool.

Add the ordinate dimensions shown.

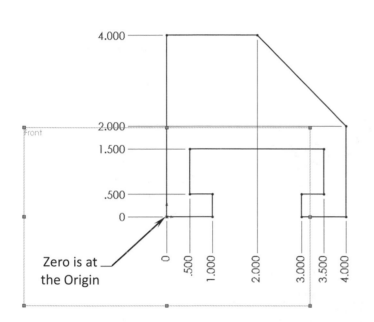

Ordinate Dimensions

To create Ordinate dimensions:

1. Click the small drop-down arrow below the Smart-Dimension command and select either Vertical or Horizontal Ordinate option.

2. First click the Origin point to start the Zero dimension, then click the next entity to create the next dimension, and so on.

2. Extruding the base feature:

Switch to the **Features** tab and click or select: **Insert, Boss/Base, Extrude**.

End Condition: **Blind**.

Reverse Direction: **Enabled**.

Extrude Depth: **6.00 in**.

Click **OK**.

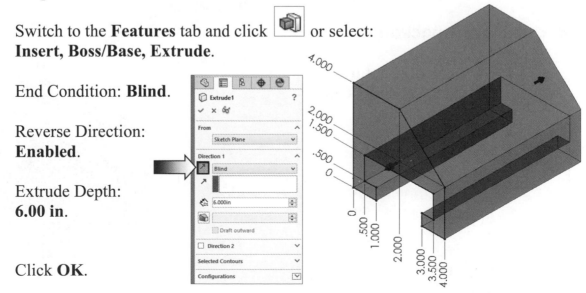

3. Creating the pocket profiles:

Select the <u>face</u> as indicated below for sketch plane.

Click **Sketch** or select **Insert, Sketch**.

Sketch the profile below using the **Corner Rectangle** and **Sketch Fillet** tools.

Add the dimensions or relations shown below to fully define the sketch*.

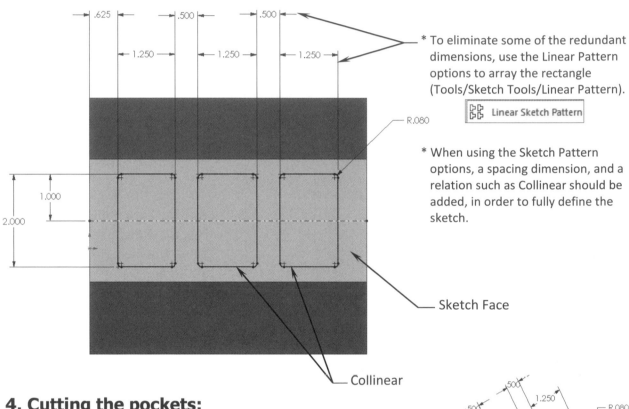

* To eliminate some of the redundant dimensions, use the Linear Pattern options to array the rectangle (Tools/Sketch Tools/Linear Pattern).

Linear Sketch Pattern

* When using the Sketch Pattern options, a spacing dimension, and a relation such as Collinear should be added, in order to fully define the sketch.

Sketch Face

Collinear

4. Cutting the pockets:

Switch to the **Features** tab and click **Extruded Cut**.

End Condition: **Blind**.

Extrude Depth: **.500 in**.

Click **OK**.

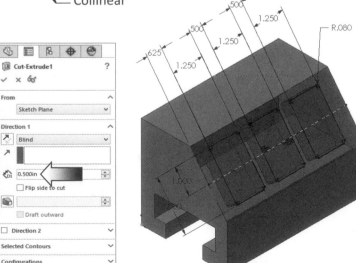

5. Adding a Counterbore:

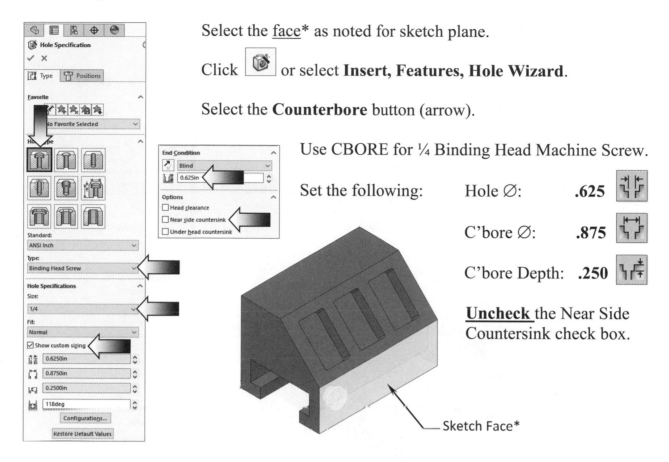

Select the <u>face</u>* as noted for sketch plane.

Click or select **Insert, Features, Hole Wizard**.

Select the **Counterbore** button (arrow).

Use CBORE for ¼ Binding Head Machine Screw.

Set the following:

Hole ∅:	**.625**
C'bore ∅:	**.875**
C'bore Depth:	**.250**

Uncheck the Near Side Countersink check box.

—— Sketch Face*

Click the **Positions Tab** (arrow).

Note: We need to review the pattern and the mirror commands, so only one hole will be created and then we will pattern it to create the other instances.

Add the 2 Dimensions shown below to position the C'bore.

Click **OK**.

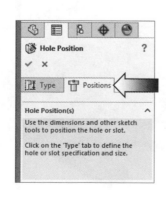

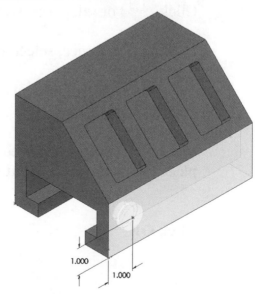

6. Patterning the Counterbore:

Click  or select **Insert, Pattern Mirror, Linear Pattern**.

Select the **bottom edge** for direction.

Enter: **2.00 in**. for Spacing.

For Number of Instances, enter: **3**

For Features to Pattern, select the **Counterbore**.

Click **OK**.

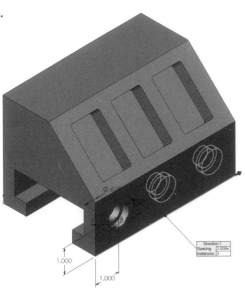

7. Creating the Mirror-Plane:

Click 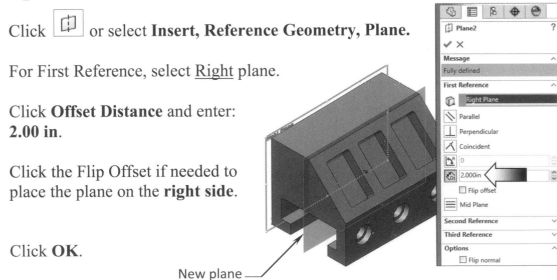 or select **Insert, Reference Geometry, Plane**.

For First Reference, select <u>Right</u> plane.

Click **Offset Distance** and enter: **2.00 in**.

Click the Flip Offset if needed to place the plane on the **right side**.

Click **OK**.

New plane

8. Mirroring the Counterbores:

Select **Mirror** under **Insert, Pattern/Mirror** drop-down list.

For Mirror Face/Plane, select the <u>new plane</u> (Plane1).

For Features to Mirror, select the **Counterbore** and the **LPattern1**.

Click **OK**.

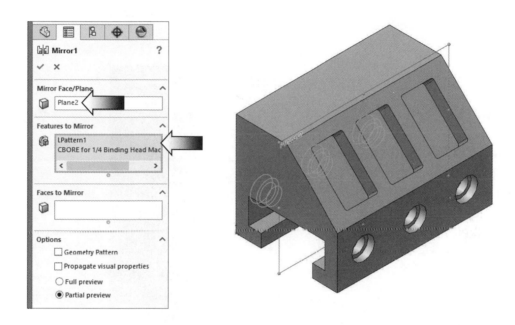

Rotate 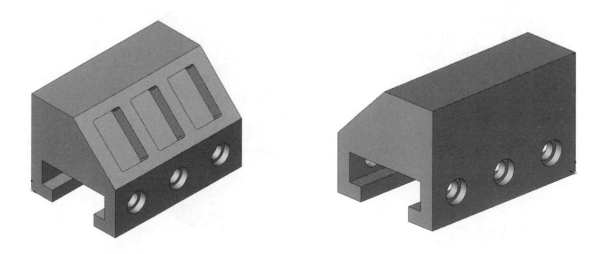 the model and inspect the mirrored feature from the back side.

9. Creating the blind holes:

Select the <u>top face</u>* of the part as the new sketch plane.

Click **Sketch** or select **Insert, Sketch**.

Sketch **4 Circles** as shown.

Add the Dimensions and relations below, to fully define the sketch.

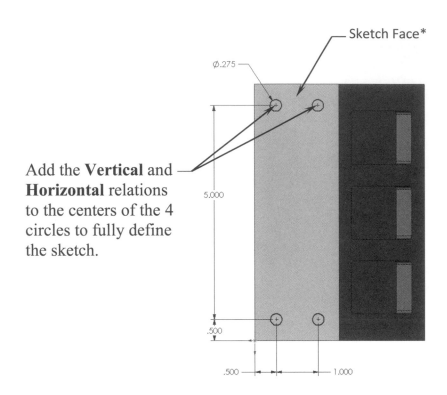

Add the **Vertical** and **Horizontal** relations to the centers of the 4 circles to fully define the sketch.

10. Cutting the 4 holes:

Click or select **Insert, Cut, Extrude**.

End Condition: **Blind**

Extrude Depth: **1.00 i**

Click **OK**.

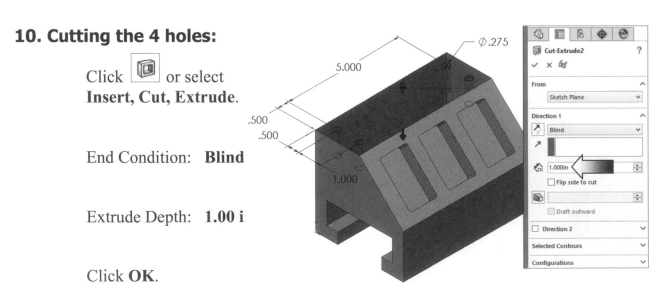

11. Creating a Cutaway section: (in a new configuration).

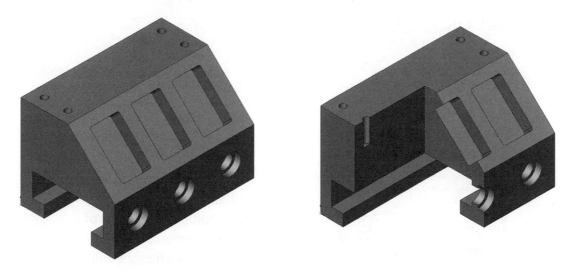

Default Configuration   **New Cutaway Configuration**

At the top of the FeatureManager tree, select the **Configuration tab**.

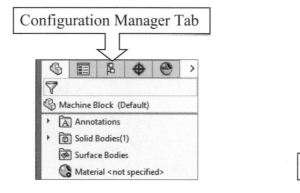

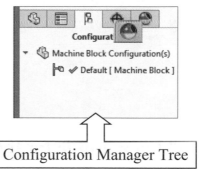

Right-click the name of the part and select **Add Configuration**.

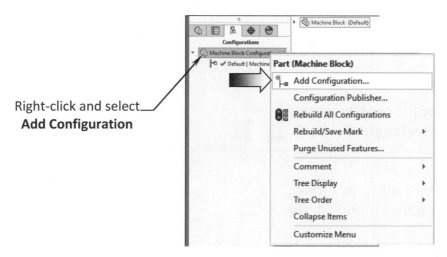

Under Configuration Name, enter **Cutaway View.**

Click **OK** [OK] .

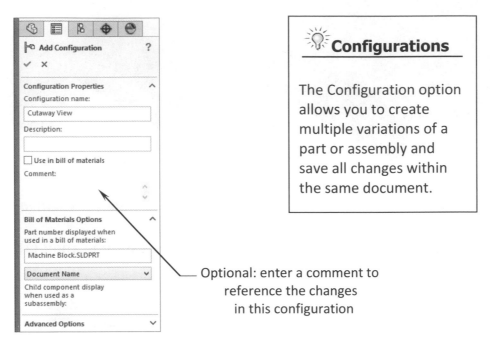

Configurations

The Configuration option allows you to create multiple variations of a part or assembly and save all changes within the same document.

Optional: enter a comment to reference the changes in this configuration

Sketch face

12. Sketching a profile for the cut:

Select the <u>face</u> indicated as the new sketch plane.

Click [icon] or select **Insert, Sketch**.

Sketch the profile as shown using the **Corner Rectangle** tool.

Add the dimensions or relations shown in the image to fully define the sketch.

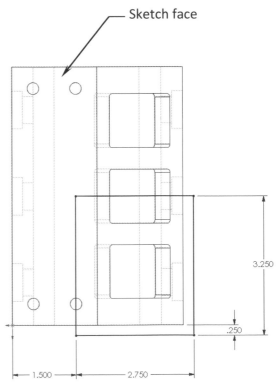

13. Creating the section cut:

Click 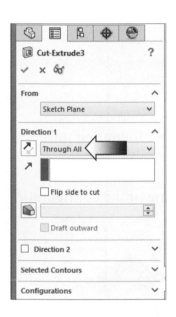 or select **Insert, Cut, Extrude**.

End Condition: **Through All.**

Click **OK**.

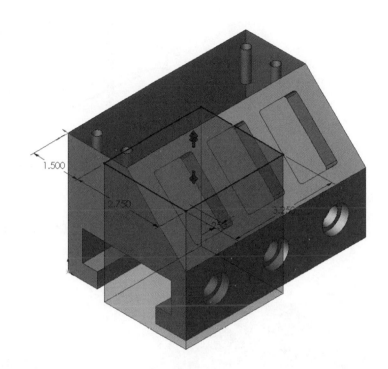

Inspect your Cutaway section view against the image shown below.

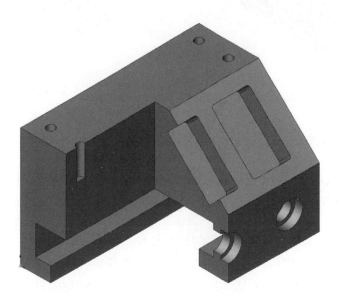

14. Switching between the Configurations:

Double-click the **Default** configuration to switch back to the original part.

Double-click the **Cutaway View** configuration to see the cut feature.

Double-click to toggle
between Configurations*

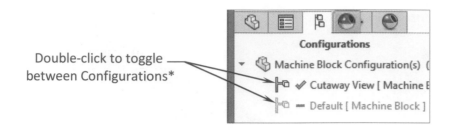

* The **Green Checkmark** icon next to the name of the configuration means it is **active**.

* The **Grey** Checkmark (or Dashed) icon next to the name of the configuration means it is **inactive**.

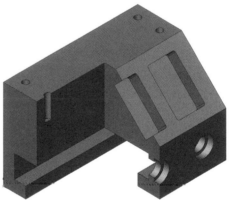

Default Configuration

Cutaway Configuration

The Cutaway configuration will be used again in one of the drawing chapters.

There is no limit on how many configurations can be created and saved in a part document.

If a part has many configurations, it might be better to use a design table to help manage them (refer to chapter 20 in this textbook for more information on how to create multiple configurations using the Design Table option).

15. Splitting the FeatureManager pane:

Locate the **Split Handle** (the round dot) on top of the FeatureManager tree and drag it down about halfway.

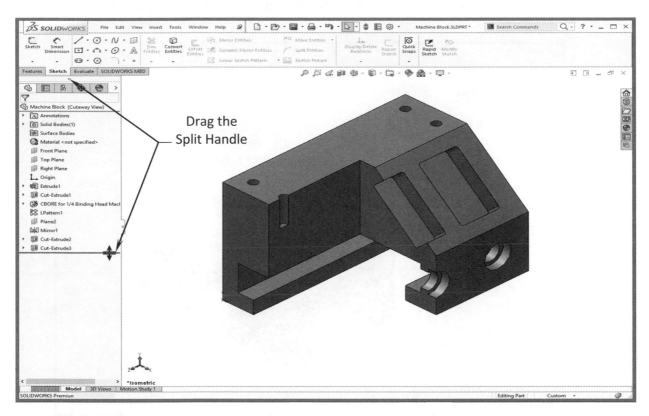

Drag the
Split Handle

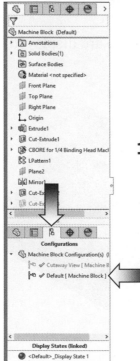

Click the ConfigurationManager tab (arrow) to change the lower half to ConfigurationManager.

Double-click the **Default** configuration to activate it.

16. Creating a new configuration:

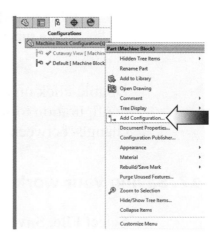

Right-click the name of the part and select **Add Configuration** (arrow).

For the name of the new configuration, enter **Machine Features Suppressed**.

Click **OK** to close the config. dialog (see next page)...

Hold the Control key and select the following cut features from the tree: **Cut-Extrude1, C'Bore for 1/4 Binding Head, LPattern1, Mirror1,** and **Cut-Extrude2**.

Release the Control key and click the **Suppress** button from the pop-up window.

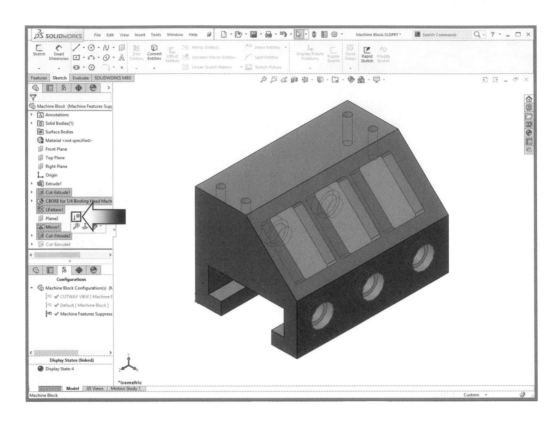

The machine features are now suppressed and their feature icons are displayed in grey color on the feature tree.

The machine features
are suppressed

Double-click on each
configuration to see the
changes between each one.

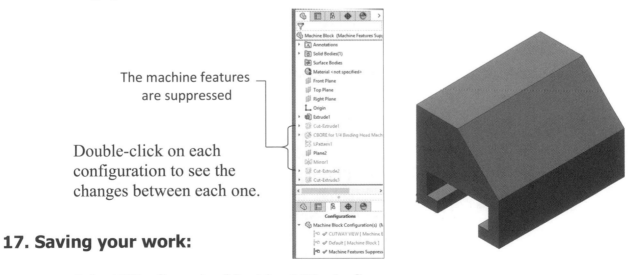

17. Saving your work:

Select **File, Save As, Machined Block, Save**.

18. Optional:

Create 3 additional configurations and change the dimensions indicated below: (Select **This Configuration** option from the pull-down list after each change).

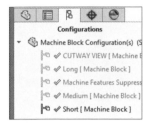

Short Configuration:

* Overall length: **4.00in**
* Number of Pockets: **2**
* Number of holes: **2**

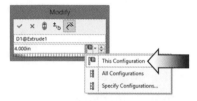

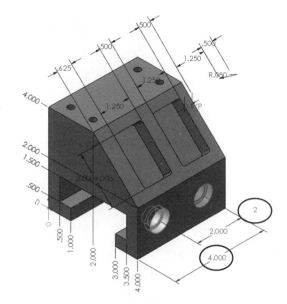

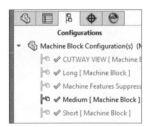

Medium Configuration:
(Change from Default)
* Overall length: **6.00in**
* Number of Pockets: **3**
* Number of holes: **3**

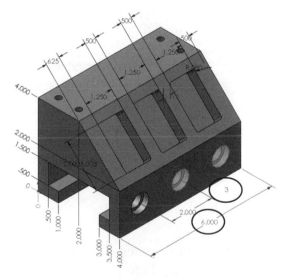

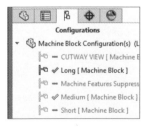

Long Configuration:

* Overall length: **8.00in**
* Number of Pockets: **3**
* Number of holes: **4**

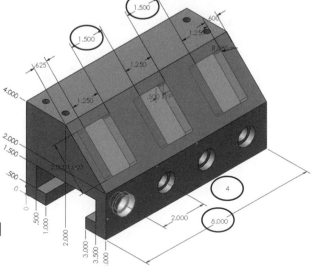

Questions for Review

1. The five basic steps to create an extruded feature are:
   - Select a sketch Plane OR a planar surface
   - Activate Sketch button (Insert / Sketch)
   - Sketch the profile
   - Define the profile (Dimensions / Relations)
   - Extrude the profile
     a. True
     b. False

2. When the extrude type is set to Blind, a depth dimension must be specified as the End-Condition.
     a. True
     b. False

3. More than one closed sketch profile on the same surface can be extruded at the same time to the same depth.
     a. True
     b. False

4. In the Hole Wizard definition, the Counterbore parameters such as bore diameter, hole depth, etc. are fixed; they cannot be changed.
     a. True
     b. False

5. Holes created using the Hole Wizard cannot be patterned.
     a. True
     b. False

6. Configurations in a part can be toggled ON / OFF by double clicking on their icons.
     a. True
     b. False

7. Every part document can only have *one* configuration in it.
     a. True
     b. False

7. FALSE

5. FALSE   6. TRUE

3. TRUE   4. FALSE

1. TRUE   2. TRUE

Exercise 1: Using Vary-Sketch

1. Opening an existing part:

Browse to the training folder and open a part document named **Vary Sketch.sldprt**.

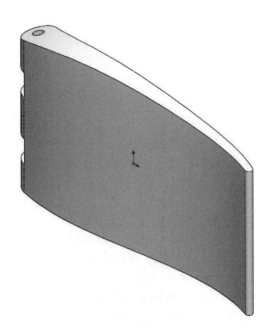

2. Creating a new configuration:

Switch to the ConfigurationManager tree.

Create a new configuration called **With Cutouts**.

3. Creating the cutouts:

Open a **new sketch** on the Front plane.

Create 2 separate **Offset-Entities** at **.275in each**.

Place the new entities on the inside.

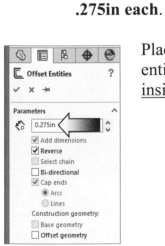

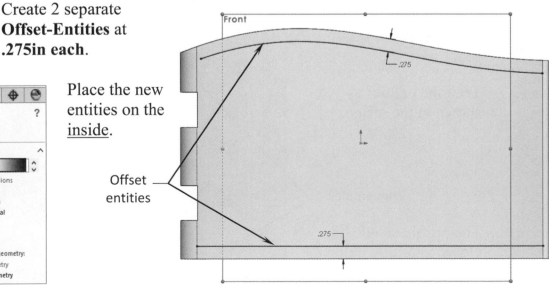

Offset entities

Drag the endpoints outward to extend the spline and the line.

Drag endpoints outward

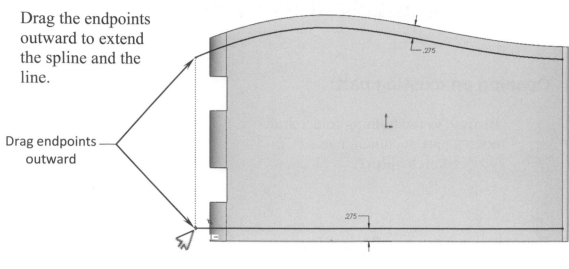

Every converted entity will have an On-Edge relation added to it to reference the model edge that it was created from.

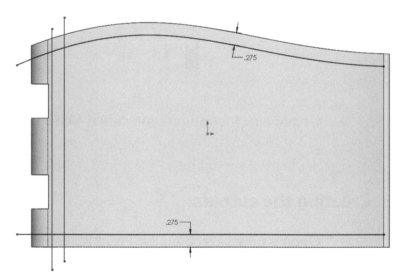

Add 2 additional lines. Make the lines a little longer so that trimming them will be easier.

Trim the entities as shown in the image.

Trim entities

The profile should be closed and fully defined.

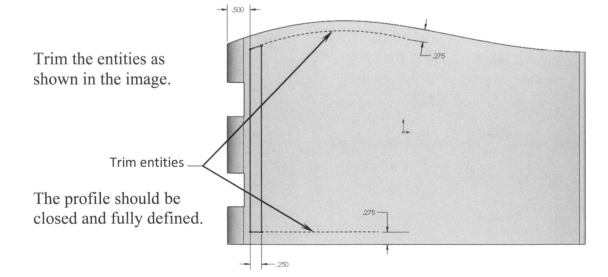

Click **Extruded Cut**.

Use **Through All-Both** to cut through both sides.

Click **OK**.

4. Creating a linear pattern:

Select the **Linear Pattern** command from the Features tab.

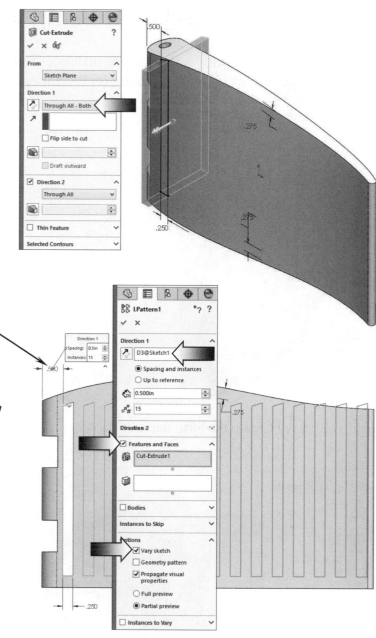

Pattern direction

For Pattern Direction select the dimension **.500"**

Enter **.500in** for spacing.

Enter **15** for number of instances.

For Features and Faces to pattern select: **Cut-Extrude1**, which is the rectangular cut from the previous step.

Enable the **Vary Sketch** checkbox.

Click **OK**. Toggle between the 2 configurations to see the differences.

Save your work as **Vary Sketch_Exe1**.

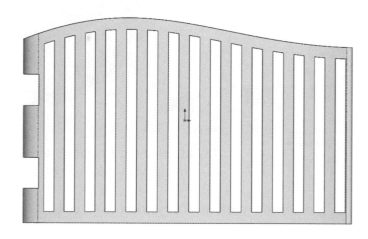

Exercise 2: Using Vary-Sketch

1. The Vary Sketch allows the pattern instances to change dimensions as they repeat.
2. Create the part as shown, focusing on the Linear Pattern & Vary-Sketch option.

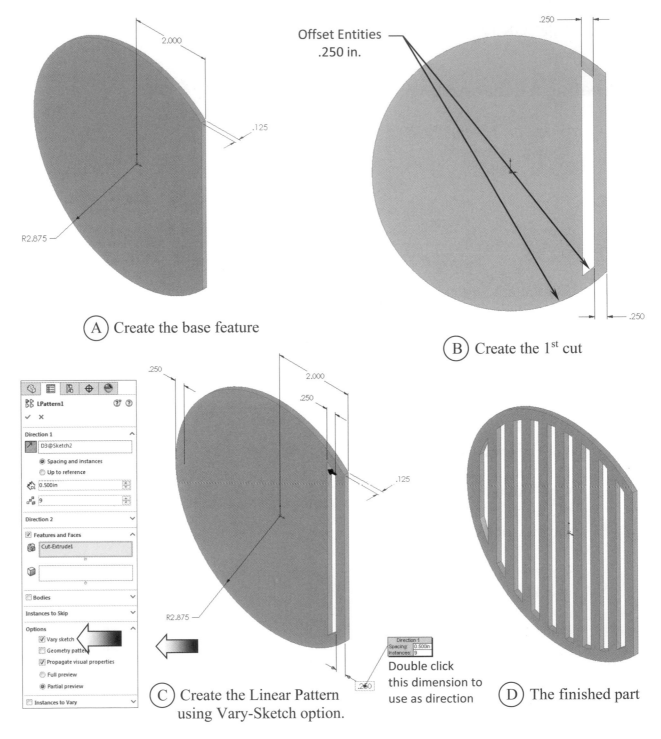

(A) Create the base feature

(B) Create the 1st cut

Offset Entities .250 in.

(C) Create the Linear Pattern using Vary-Sketch option.

Double click this dimension to use as direction

(D) The finished part

3. Save your work as **Vary Sketch_Exe2.**

CHAPTER 9

Modeling Threads

Modeling Thread - External
Threaded Insert

In most cases, threads are not modeled in the part. Instead, they are represented with dashed lines and callouts in the drawings. But for non-standard threads, they should be modeled in the part so that they can be used in some applications such as stereolithography for 3D printing or finite element analysis for simulation studies, etc.

Two sketches are required to model the threads, using the traditional method:

* A sweep path (a helix that controls the pitch, revolutions, starting angle, and left- or right-hand threads).

* A sweep profile (shape and size of the threads).

The swept cut command is used when creating the threads. However, in some cases, the swept boss command can also be used to add material to the sweep path when making the external threads.

The new Thread feature creates the thread profile and the helical path based on the user's inputs. The on-screen parameters allow you to define the start thread location, specify an offset, set end conditions, specify the type, size, diameter, pitch and rotation angle, and choose options such as right-hand or left-hand thread.

This chapter and its exercises will guide you through some special techniques on how internal and external threads are created in SOLIDWORKS.

Modeling Threads – External
Threaded Insert

View Orientation Hot Keys:

Ctrl + 1 = Front View
Ctrl + 2 = Back View
Ctrl + 3 = Left View
Ctrl + 4 = Right View
Ctrl + 5 = Top View
Ctrl + 6 = Bottom View
Ctrl + 7 = Isometric View
Ctrl + 8 = Normal To
        Selection

Dimensioning Standards: **ANSI**

Units: **INCHES** – 3 Decimals

Tools Needed:

Insert Sketch	Line	Thread Feature
Dimension	Add Geometric Relations	Mirror

1. Sketching the base profile:

Select the <u>Front</u> plane from the FeatureManager tree.

Click **Sketch** 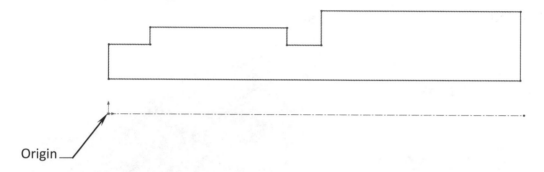 or select **Insert, Sketch**.

Sketch the profile as shown below using the **Line** tool.

Origin

Add the **Ordinate** dimensions shown below to fully define the sketch.

(Click the Smart Dimension drop-down arrow to access the Ordinate Dimension options.)

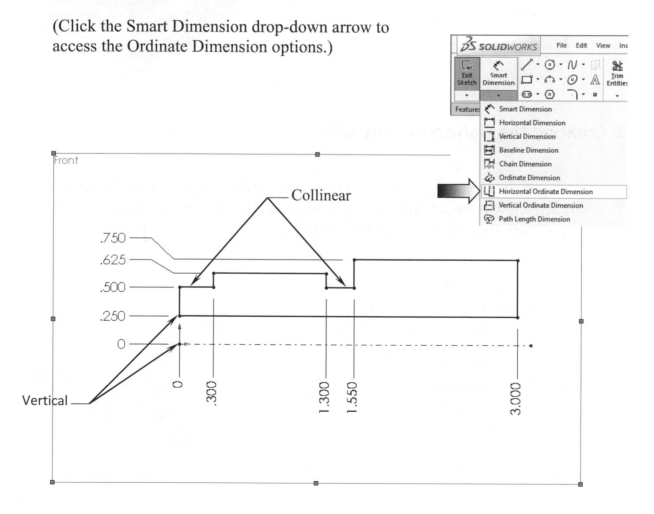

2. Revolving the base feature:

Click **Revolve**  on the Features toolbar or select: **Insert, Boss/Base, Revolve**.

Revolve Direction: **Blind**.

Revolve Angle: **360°**

Click **OK**.

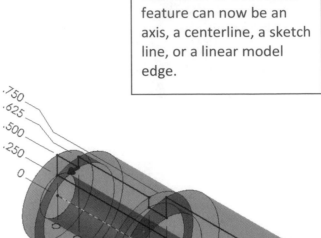

3. Creating the right-hand threads:

Using the traditional method, you can create helical threads on cylindrical faces by sweeping a thread profile along a helical path. This option requires a little bit more work than the new Thread feature.

The new thread feature creates the thread profile and the helical path based on the user's inputs. The on-screen parameters allow you to define the start thread location, specify an offset, set end conditions, specify the type, size, diameter, pitch and rotation angle, and choose options such as right-hand or left-hand thread.

Click **Threads** ⬚ (under Hole Wizard) or select **Insert, Features, Threads**.

For Edge of Cylinder, select the **circular edge** as noted.

For Start Location, select the **face** shown.

Enable the **Offset** checkbox.

For Offset Distance, enter **.0625in**.

Click the **Reverse Direction** button.

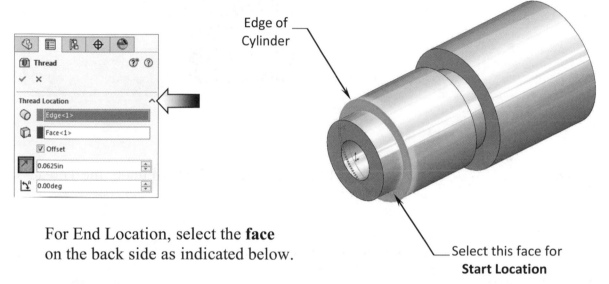

For End Location, select the **face** on the back side as indicated below.

Enable the **Offset** checkbox and enter **.0625in.** for Offset Distance.

Click the **Reverse Direction** button.

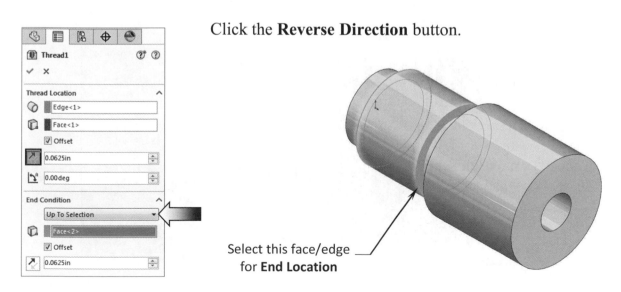

Under the Specification section, keep the Type at its default **Inch Die**.

For Size, select **1.2500-12** from the drop-down list.

For Thread Method, select the **Cut Thread** option.

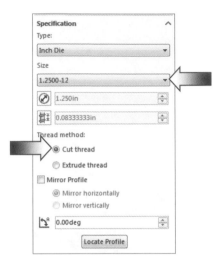

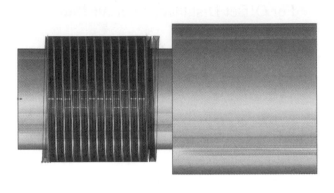

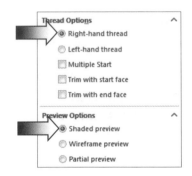

Select the **Right-Hand Thread** under the Thread Options.

Select **Shaded Preview** under the Preview Options.

Click **OK**.

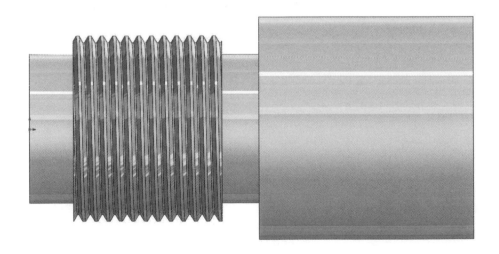

4. Using the Mirror Bodies option:

Select **Mirror** ⊞ from the Features tab or
click: **Insert, Pattern/Mirror, Mirror**.

For Mirror Face/Plane, select the **face** indicated.

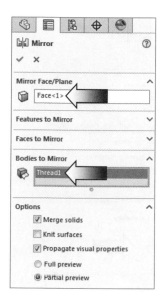

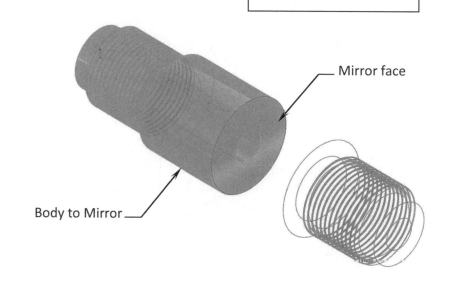

Mirror face

Body to Mirror

> **Mirror Bodies**
>
> To mirror all features in a body, either a planar face or a plane should be used as a contact surface.

Expand the **Bodies to Mirror** section and select the **model** in the graphics area as noted.

Enable the **Merge Solids** checkbox. This option will join or combine the two bodies into a single solid.

Click **OK**.

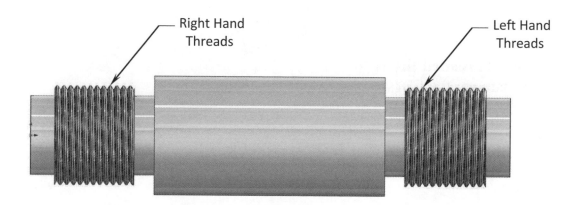

Right Hand
Threads

Left Hand
Threads

5. Adding chamfers:

Click **Chamfer** 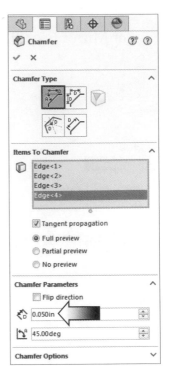 or **Insert, Features, Chamfers**.

Select the **4 edges** as indicated.

Enter **.050 in**. for Depth.

Enter **45°** for Angle.

Click **OK**.

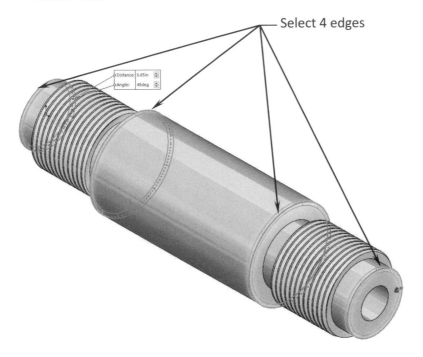

Select 4 edges

6. Saving the finished part:

Select **File / Save as**.

Enter **Threaded Insert** for the name of the file.

Click **Save**.

Questions for Review

1. It is proper to select the sketch plane first before activating the sketch command.
   a. True
   b. False

2. The ordinate dimension options can be found under the smart dimension drop-down menu.
   a. True
   b. False

3. The center of a revolve feature can now be an axis, a centerline, a line, or a linear model edge.
   a. True
   b. False

4. Threads cannot be modeled; they can only be represented with dashed lines and callouts.
   a. True
   b. False

5. Either a planar surface or a plane can be used to perform a mirror-bodies operation.
   a. True
   b. False

6. Several model edges can be chamfered at the same time if their values are the same.
   a. True
   b. False

7. The mirrored half is an independent feature; its geometry can be changed, and the original half will get updated accordingly.
   a. True
   b. False

8. The mirror function will create the left-hand threads from the right-hand threads.
   a. True
   b. False

7. FALSE   8. TRUE
5. TRUE   6. TRUE
3. TRUE   4. FALSE
1. TRUE   2. TRUE

Exercise: Modeling Threads - Internal

1. Dimensions provided are for modeling practice purposes.
2. Dimensions are in Inches, 3 decimal places.
3. Use the instructions on the following pages, if needed.
4. Save your work as **Nut_Internal Threads**.

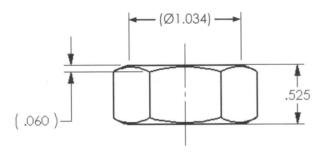

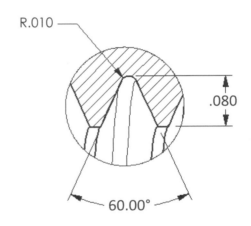

DETAIL B

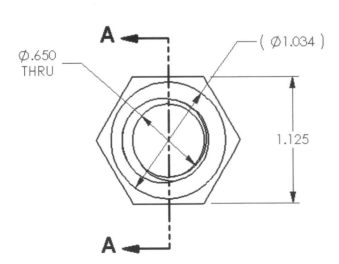

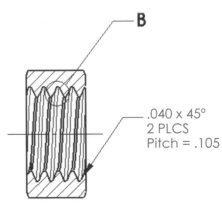

SECTION A-A

1. Starting with the base sketch:

Select the <u>Front</u> plane and open a **new sketch**.

Sketch a **6-sided Polygon** and add the dimensions and relation shown.

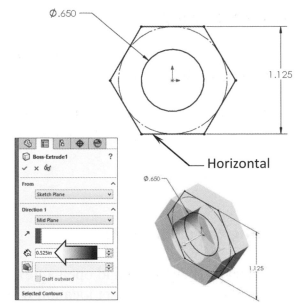

2. Extruding the base:

Extrude the sketch using **Mid Plane** and **.525"** thick.

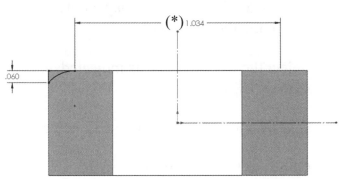

3. Removing the Sharp edges:

Select the <u>Top</u> plane and open a **new sketch**.

Sketch the profile shown.
* See the note on page 9-14 about Virtual Diameter dim.

Add the dimensions and the Tangent relation as noted.

Mirror the profile using the **horizontal centerline**.

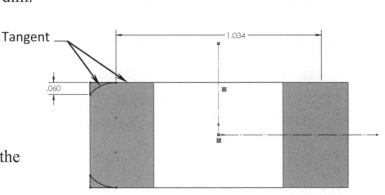

Revolve Cut with **Blind** and **360°** angle.

Click **OK**.

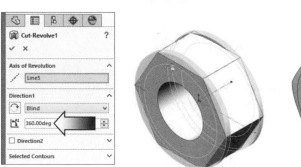

4. Creating a new plane:

Select the <u>front face</u> of the part and click **Insert, Reference Geometry, Plane**.

The **Offset Distance** should be selected automatically, enter **.105"** for distance.

Click **OK**.

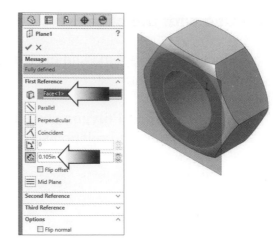

5. Creating the sweep path:

Select the <u>new plane</u> and open a **new sketch**. (Starting from this offset location will prevent an undercut from happening where the thread starts.)

Convert this edge

Select the circular edge as indicated and click **Convert Entities**.

The selected edge turns into a sketch circle.

Click the **Helix** command or select: **Insert / Curve / Helix-Spiral**.

Enter the following:

* Pitch = **.105"**
* Reverse Direction: **Enabled**.
* Revolutions: **7**
* Start Angle: **0.00 deg**.
* **Clockwise**

Click **OK**.

6. Creating the sweep profile:

Select the <u>Top</u> plane, open a **new sketch** and sketch the profile using Dynamic Mirror.

Add the dimensions as shown to fully define the sketch before adding the fillet.

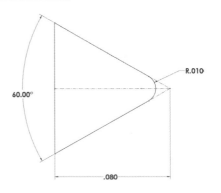

Add a **.010"** sketch fillet to the tip.
Add a **Pierce** relation between the endpoint of the centerline and the 1st revolution of the helix.

Exit the sketch.

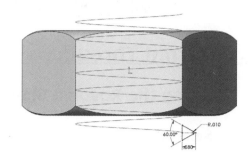

Pierce

7. Creating the swept cut:

Using the traditional method, we will sweep-cut the thread profile along the helix.

Switch to the **Features** toolbar.

Click **Swept Cut**.

Select the triangular sketch for Profile.

Select the Helix for Path.

The Profile is swept along the Path.

Click **OK**.

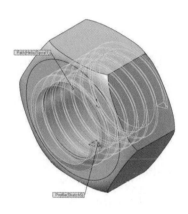

8. Saving your work:

Select **File, Save as**.

Enter **Nut_Internal Threads** for the file name.

Click **Save**.

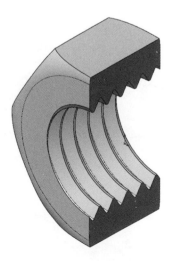

Exercise: Internal & External Thread

A. Internal Threads:

1. Creating the Revolve Sketch:

Select the <u>Front</u> plane and open a **new sketch** plane as shown.

Sketch the profile and add relations & dimensions as shown.

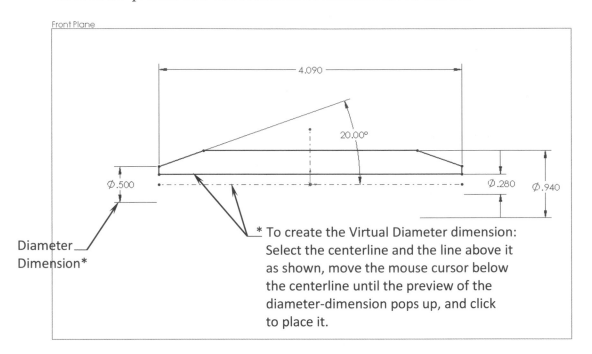

Front Plane

4.090

20.00°

Ø.500

Ø.280

Ø.940

Diameter Dimension*

* To create the Virtual Diameter dimension: Select the centerline and the line above it as shown, move the mouse cursor below the centerline until the preview of the diameter-dimension pops up, and click to place it.

2. Revolving the Body:

Revolve the sketch a full **360deg**. and click **OK**.

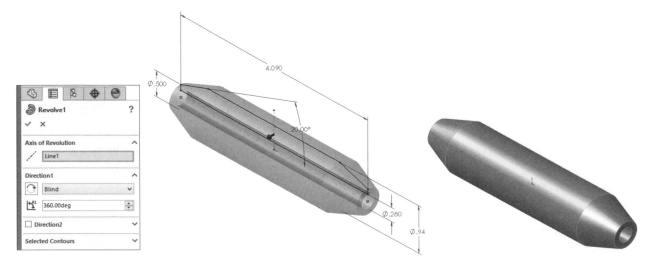

3. Creating the cutout features:

Using the <u>Front</u> plane, sketch the profiles of the cutouts as shown below.

Use the Mirror function where applicable. Add Relations & Dimensions to fully define the sketch.

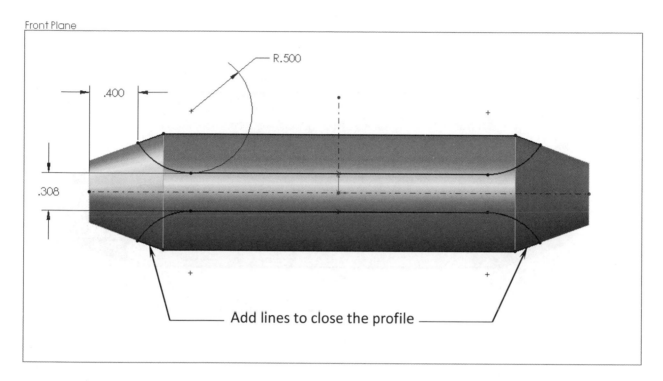

Add lines to close the profile

4. Extruding the Cutouts:

Click **Extruded Cut** and select **Through All-Both**.

Click **OK**.

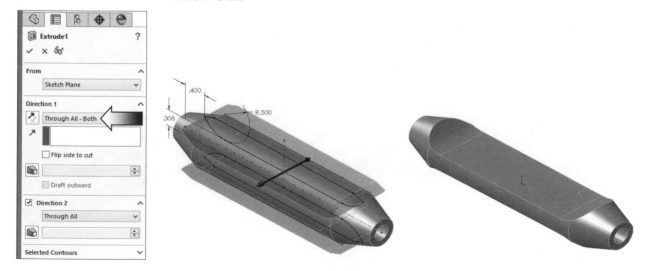

5. Sketching the Slot Profile:

Using the <u>Top</u> plane, sketch the profile of the slot (use the Straight Slot command).

Add Relations & Dimensions to fully define the sketch.

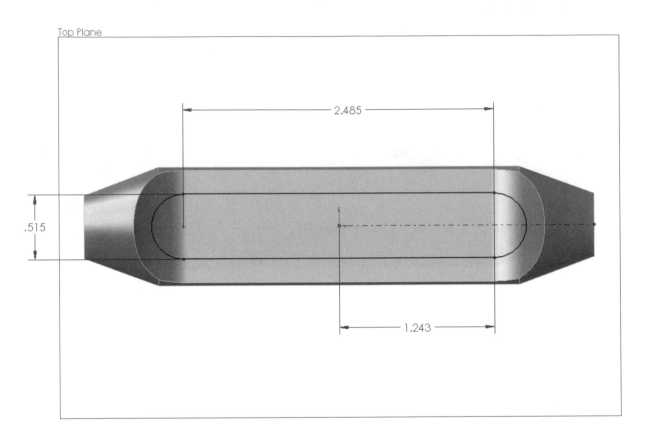

6. Cutting the Slot:

Click **Extruded Cut** and select **Through All-Both**.

Click **OK**.

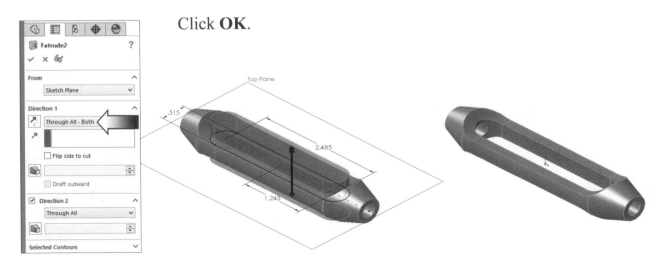

7. Adding Chamfers:

Add a **chamfer** to both ends of the holes.

Chamfer Depth = **.050 in**.   Chamfer Angle = **45 deg**.

Click **OK**.

Select 2
edges

8. Adding Fillets:

Add **fillets** to the **4 edges** as noted.

Radius = **.250 in**.

Click **OK**.

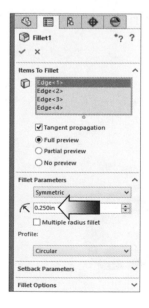

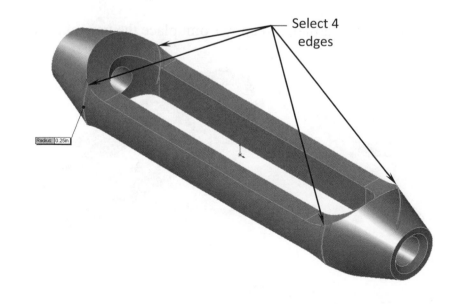

Select 4
edges

9. Filleting all edges:

Box-Select the entire part and add a fillet of **.032in**. to all edges (or use the shortcut Control+A to select all edges of the model).

Click **OK**.

10. Creating an Offset-Distance Plane:

Click **Plane** or select **Insert, Reference Geometry, Plane**.

Click **Offset-Distance** option and enter **2.063 in**.

Select the <u>Right</u> plane to copy from; place the new plane on the <u>right side</u>.

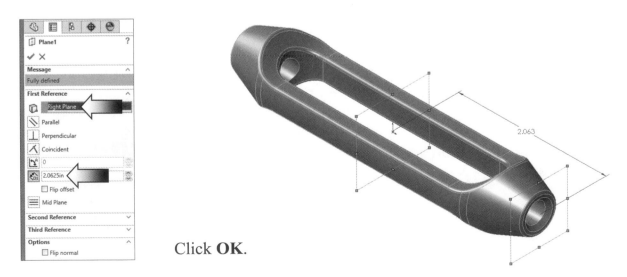

Click **OK**.

11. Creating the Helix (the sweep path):

Select the <u>new plane</u> to open a **new sketch**.

Convert the **Inner Circular Edge** into a Circle.

(Using an offset of "1-pitch" can help prevent the under-cut from showing where the thread starts.)

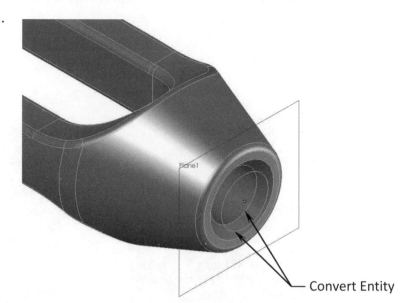

— Convert Entity

Click **Helix** or select **Insert, Curve, Helix-Spiral**.

Pitch: **.055 in**.   Revolutions: **12**

Start Angle: **0 deg**.   Direction: **Clockwise**

Click **OK**.

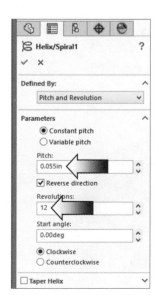

12. Sketching the Thread Profile:

Select the <u>Top</u> reference plane and open a **new sketch**.

Sketch the thread profile as shown. Add the relation and dimensions needed to define the profile.

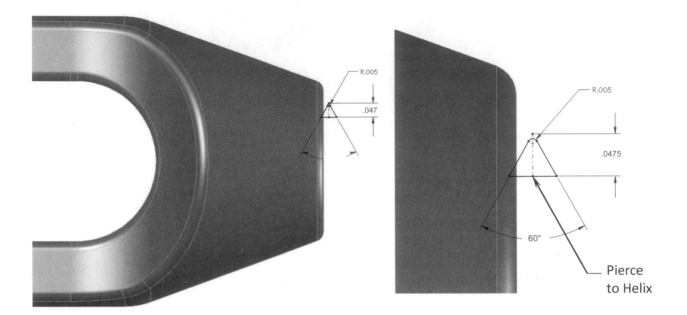

<u>Exit</u> the Sketch.

13. Sweeping the Cut:

Click **Cut-Sweep** or select **Insert, Cut, Sweep**.

Select the triangular sketch as Profile and select the helix as Sweep Path.

Click **OK**.

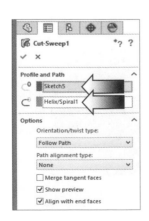

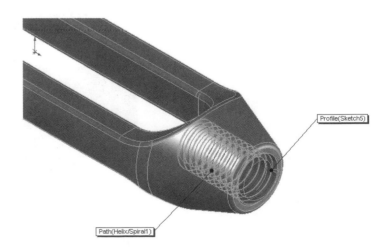

14. Mirroring the Threads:

Click **Mirror** or select **Insert, Pattern Mirror, Mirror**.

Select the **Right** plane as the Mirror Plane.

For Features to Mirror select the **Cut-Sweep1** feature.

Click **OK**.

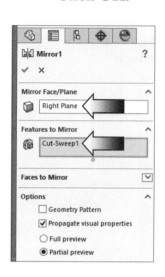

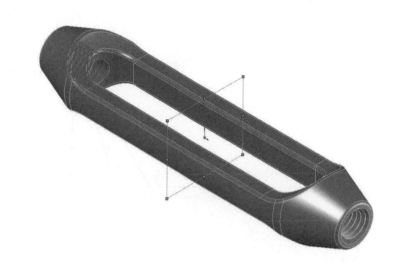

15. Creating the Cross-Section to view the Threads:

Select the <u>Front</u> plane from the FeatureManager tree.

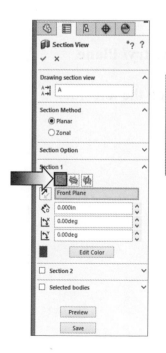

Click **Section View** or select **View / Display / Section**.

Zoom in on the threaded areas and examine the thread details.

Click **Cancel** when you are done viewing.

Save the part as **Internal Threads**.

B. External Threads:

1. Sketching the Sweep Path:

Select the <u>Front</u> plane and open a **new sketch**.

Sketch the profile shown; add relations & dimensions needed to fully define the sketch.

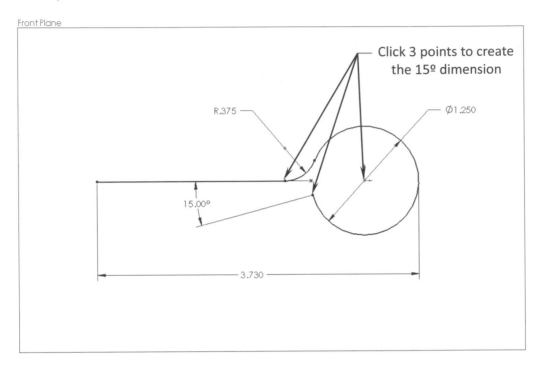

Front Plane

Click 3 points to create the 15º dimension

R.375

Ø1.250

15.00°

3.730

2. Creating a plane Perpendicular to a line:

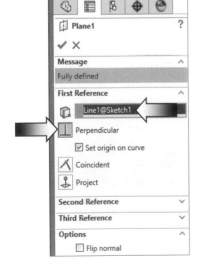

Click **Plane** or select **Insert / Reference Geometry/ Plane**.

Select the **Horizontal Line** and the **Endpoint** on the left side.

A new plane is created normal to the line.

Click **OK**.

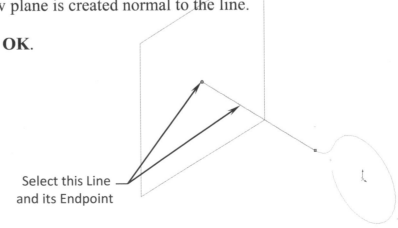

Select this Line and its Endpoint

3. Sketching the Sweep Profile:

Select the <u>new plane</u> and open a **new sketch**.

Sketch a **Circle** as shown and add a diameter dimension.

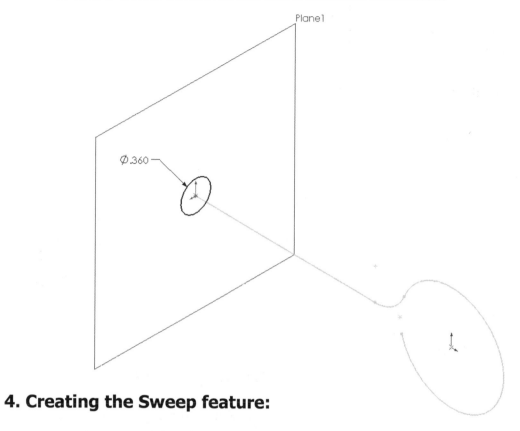

4. Creating the Sweep feature:

Click **Swept Boss-Base** or select **Insert / Features / Sweep**.

Select the **Circle** as Sweep Profile and select the **Sketch1** as Sweep Path.

Click **OK**.

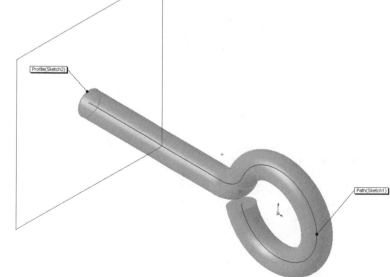

5. Adding Chamfers:

Click **Chamfer** or select
**Insert, Features,
Chamfer**.

Select the **2 Circular
Edges** at the 2 ends.

Enter **.050 in**. for Depth.

Enter **45 deg**. for Angle.

Click **OK**.

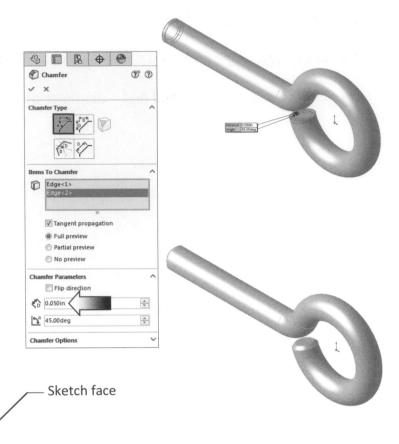

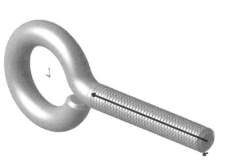

Sketch face

6. Creating the Helix:

Select the <u>face</u> as indicated
and open a **new sketch**.

Select the **circular edge** at the
end as shown and click Convert
Entities or select **Tools, Sketch
Tools/Convert Entities**.

Click **Helix** or select **Insert,
Curve, Helix-Spiral**.

Enter **.055** in. for Pitch.

Enter **39** for Revolutions.

Enter **0 deg**. for Start Angle.

Click **OK**.

Convert
Entity

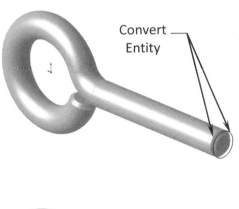

7. Creating the Thread Profile:

Either copy the previous thread profile or recreate it.

Add a **Pierce** relation to position the thread profile at the end of the helix.

Click **OK**.

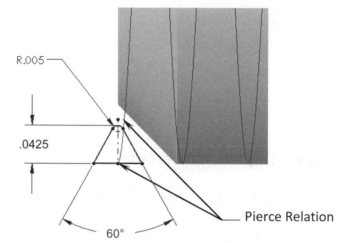

NOTE:
There should be a clearance between the 2 threaded parts so that they can be moved back and forth easily.

8. Sweeping Cut the Threads:

Click **Swept Cut** or select **Insert, Cut, Sweep.**

Select the **triangular sketch** as the Sweep Profile.

Select the **Helix** as the Sweep path.

Click **OK**.

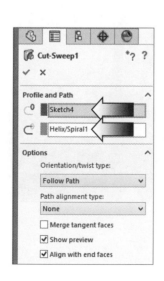

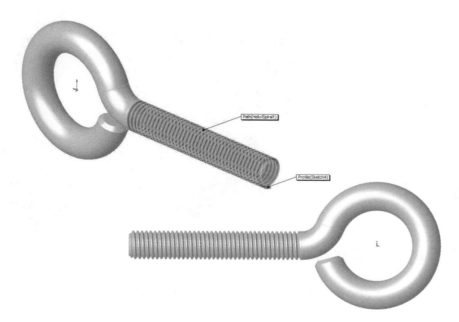

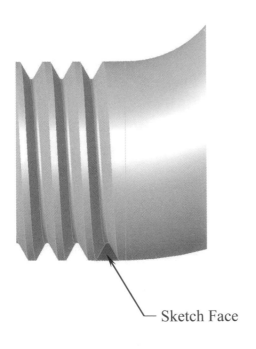

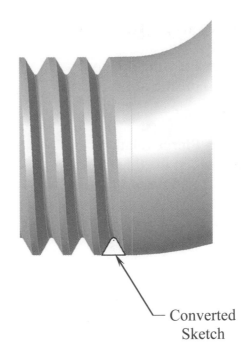

Sketch Face

Converted
Sketch

9. Removing the Undercut:

Select the <u>face</u> as noted and open a **new sketch**.

Convert the selected **face** into a new triangular sketch.

Click **Extruded-Cut** or select **Insert, Cut, Extrude**.

Select **Through All** for End Condition.

Click **OK**.

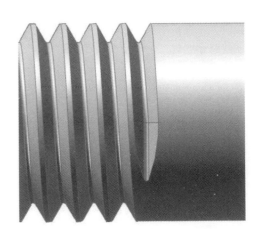

10. Saving your work:

Save a copy of your work as **External Threads**.

11. Optional:

(This step can also be done after completing the Bottom Up Assembly chapter.)

Start a New Assembly document and assemble the 2 components.

Create an Assembly Exploded View as shown below.

Save the assembly as **TurnBuckle.sldasm**.

<u>Exercise:</u> Creating Threads
Automatic vs. Manual

SOLIDWORKS offers a couple of methods to create threads: The **Automatic** option accelerates creating an extruded or cut thread on a hole or shaft; or the **Manual** option where you can define your own thread profile and sweep it along a helical path.

When you define threads in inches, pitches are defined in terms of revolutions per unit of measure (for example, threads per inch).

When you define threads in metric units, pitches are defined in terms of units of measure per revolution (for example, millimeters per thread).

Automatic 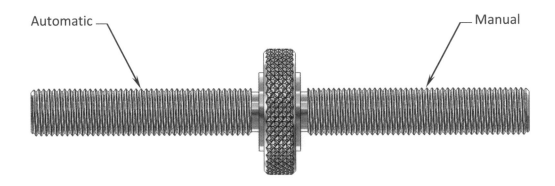 Manual

1. Opening a part document:

Select **File, Open**.

Browse to the training folder and open the part document named: **Thread Examples.sldprt**.

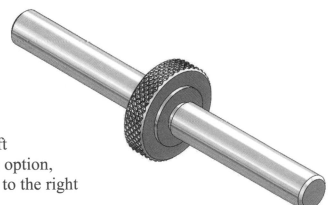

We will create the threads on the left side of the part using the Automatic option, and the Manual option will be done to the right side.

After the threads are created on both sides of the part, we will compare the results to see if one method is better or easier than the other, despite the number of steps it may take to create the threads.

2. Using the Automatic thread option:

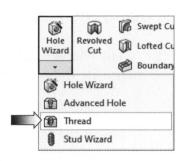

This option uses the PropertyManager to fill out the information regarding the threads that are being made.

Switch to the **Features** tab, expand the **Hole Wizard** drop-down menu and select **Thread**.

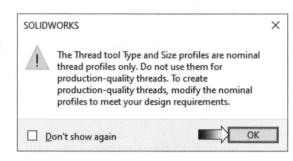

Click **OK** to close the warning dialog.

For Edge of Cylinder, select the **edge** ① as indicated (not the chamfer edge).

For Optional Start Location, select the **face** ② on the far left hand side.

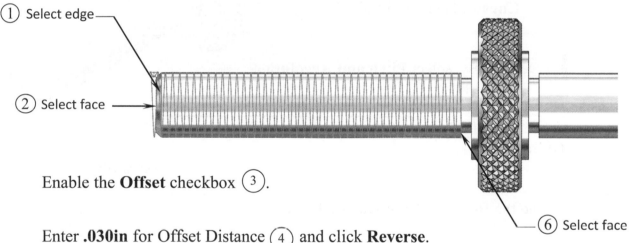

Enable the **Offset** checkbox ③.

Enter **.030in** for Offset Distance ④ and click **Reverse**.

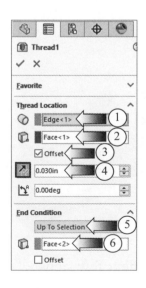
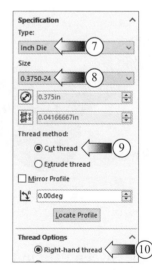

For End Condition, select **Up To Selection** ⑤.

Select the **face** on the far right hand side of the same cylinder ⑥.

For Type, select **Inch Die** from the list ⑦.

For Size, select **0.3750-24** from the list ⑧.

For Thread Method, click **Cut Thread** ⑨.

For Thread Option, click Right Hand Thread ⑩.

Click **OK**.

3. Using the Manual thread option:

The Manual option requires a thread profile to be swept along a helical path.

Select the <u>face</u> on the far right side of the 2nd cylinder and open a **new sketch**.

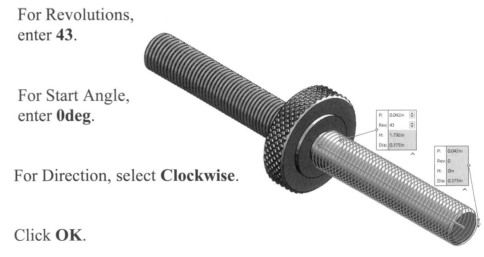

Select face ⎯

Select the **circular edge** of the cylinder (not the edge of the chamfer) and click: **Convert Entities**.

Switch to the **Features** tab and select: **Curves, Helix and Spiral**.

For Defined By, select **Pitch and Revolution**.

Use the default **Constant Pitch** option and enter **.042in** for Pitch.

Select edge ⎯

Enable the **Reverse Direction** checkbox.

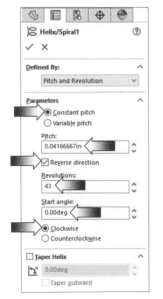

For Revolutions, enter **43**.

For Start Angle, enter **0deg**.

For Direction, select **Clockwise**.

Click **OK**.

This Helix will be used as the Sweep Path in the next few steps.

4. Copying the Thread Profile:

Expand the Thread1 feature on the FeatureManager tree.

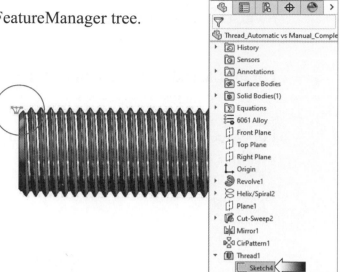

Select **Sketch4** (the thread profile) and press **Control+C** (copy).

Select the **Top** plane from the FeatureManager tree and press **Control+B** (paste).

5. Locating the thread profile:

Click **Sketch6** (the copied sketch) from the FeatureManager tree and select: **Edit Sketch**.

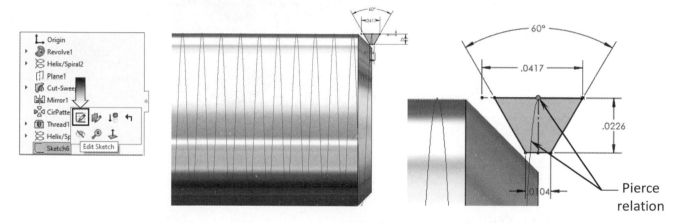

Zoom closer to the upper right corner and drag the thread profile closer to the helix if needed.

Add a **Pierce** relation between the **upper endpoint** of the vertical centerline and the **Helix** as noted.

Inspect your thread profile against the detail view above.

Click **Exit Sketch** or press **Control+Q**.

6. Creating a Swept Cut feature:

Switch to the **Features** tab and click **Swept Cut**.

For Profile and Path, click **Sketch Profile.**

For Sweep Profile, select **Sketch6** either from the FeatureManager tree or from the graphics area.

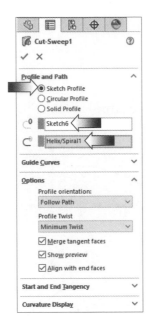
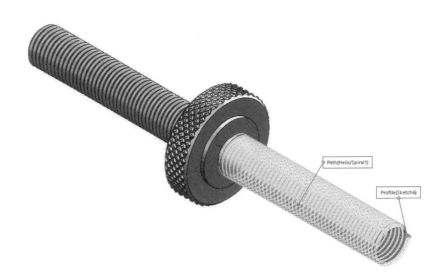

For Sweep Path, select the **Helix**.

Enable the checkboxes
shown in the dialog
box above.

Click **OK**.

Press **Control+7** to change to
the Isometric view.
Inspect your model against the
image shown on the right.

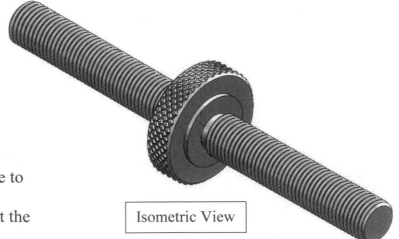

Isometric View

7. Comparing the threads:

Press **Control+1** to change to the **Front** view.

Zoom closer to the center portion of the model so that the threads on both sides can be seen more closely.

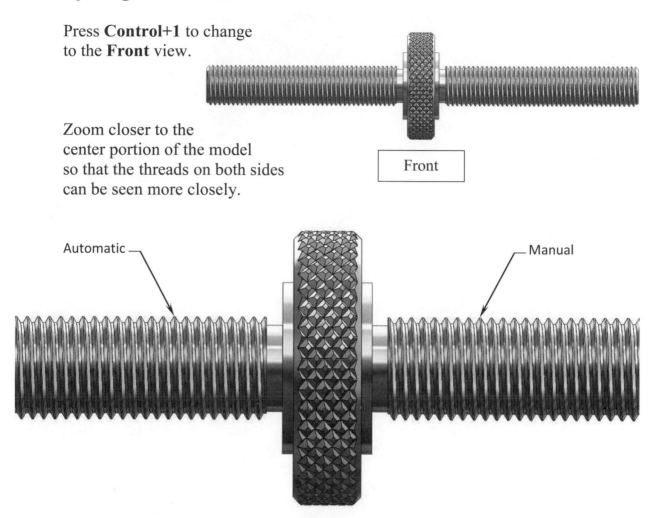

Front

Automatic

Manual

Visually, the threads on both sides are very similar to each other, they matched up very closely.

Both methods can produce very similar results; however, you need to work carefully as the interaction of these options sometimes causes subtle errors.

8. Saving your work:

Select **File, Save as**.

Enter: **Threads_Automatic vs Manual_Completed** for the file name.

Click **Save**.

CHAPTER 10

Bottom Up Assembly

Bottom Up Assembly
Ball Joint Assembly

When your design involves multiple parts, SOLIDWORKS offers four different assembly methods to help you work faster and more efficiently. The first method is called Bottom-Up-Assembly, the second method is Layout-Assembly, the third is Top-Down-Assembly, and the fourth, Master-Modeling.

We will explore the first two methods, the Bottom Up and the Layout Assemblies, in this textbook.

Bottom Up Assembly is when the components are created individually, one by one. They get inserted into an assembly and then mated (constrained) together.

The first component inserted into the assembly will be fixed by the system automatically. If it is placed on the Origin, then the Front, Top, and the Right planes of the first component will automatically be aligned with the assembly's Front, Top, and Right planes.

Only the first component will be fixed by default; all other components are free to move or be reoriented. Each component has a total of 6 degrees of freedom and depending on the mate type, once a mate is assigned to a component, one or more of its degrees of freedom are removed, causing the component to move or rotate only in the desired directions.

All Mates (constraints) are stored in the FeatureManager tree under the Mates group. They can be edited, suppressed, or deleted.

This lesson and its exercises will discuss the use of the Standard Mates, Advanced Mates, and Mechanical Mates.

Bottom Up Assembly
Ball Joint Assembly

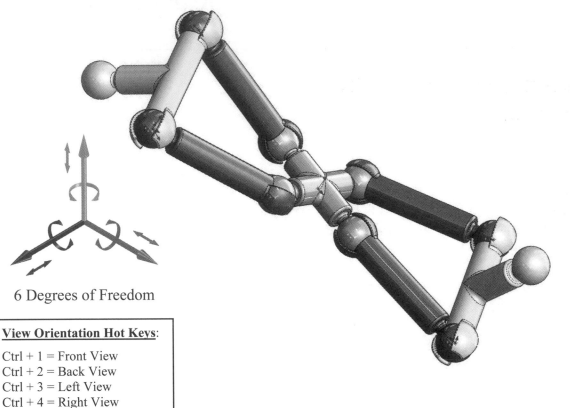

6 Degrees of Freedom

View Orientation Hot Keys:

Ctrl + 1 = Front View
Ctrl + 2 = Back View
Ctrl + 3 = Left View
Ctrl + 4 = Right View
Ctrl + 5 = Top View
Ctrl + 6 = Bottom View
Ctrl + 7 = Isometric View
Ctrl + 8 = Normal To
        Selection

Dimensioning Standards: **ANSI**

Units: **INCHES** – 3 Decimals

Tools Needed:

 Mates

 Move Component

 Rotate Component

 Concentric Mate

 Inference Origins

 Placing New Component

1. Starting a new Assembly template:

Select **File, New, Assembly, OK**.

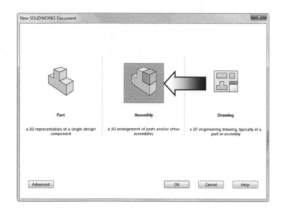

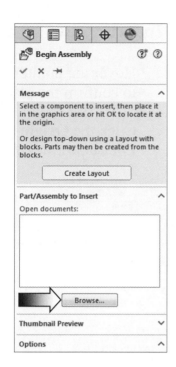

In the Insert Component dialog, click **Browse**.

Select the **Center Ball Joint** document from the training folder and click **Open**.

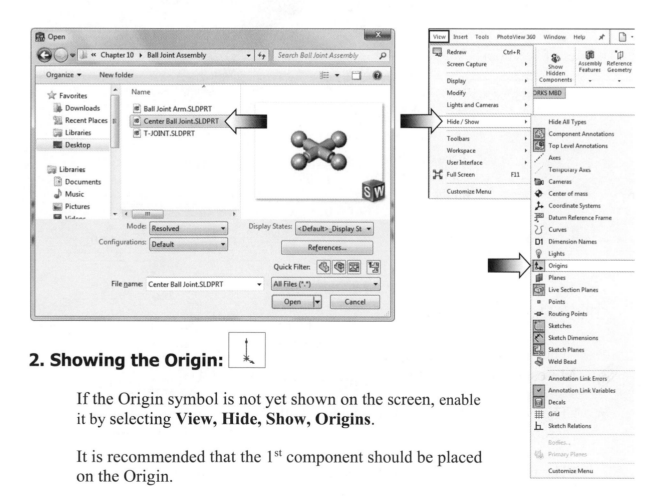

2. Showing the Origin:

If the Origin symbol is not yet shown on the screen, enable it by selecting **View, Hide, Show, Origins**.

It is recommended that the 1st component should be placed on the Origin.

3. Inserting the 1st component (the Parent) on the Origin:

Hover the pointer over the origin; an Origin-Inference symbol 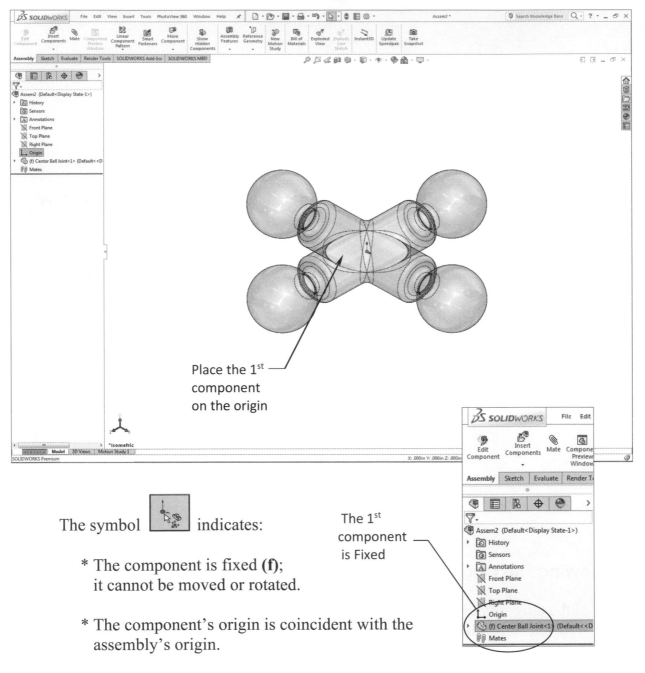 appears to confirm the placement of the 1st component.

Click the **Origin** point to place the component. *(Note: Clicking the Green Checkmark will also place the component on the Origin automatically.)*

Place the 1st
component
on the origin

The symbol   indicates:

The 1st
component
is Fixed

* The component is fixed **(f)**;
  it cannot be moved or rotated.

* The component's origin is coincident with the
  assembly's origin.

* The planes of the component and the planes of the assembly are aligned.

4. Inserting the second component (a child) into the assembly:

Click or select **Insert, Component, Existing Part, Assembly**.

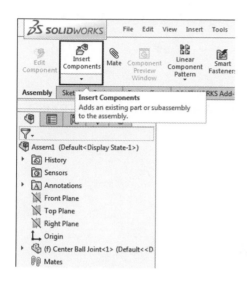
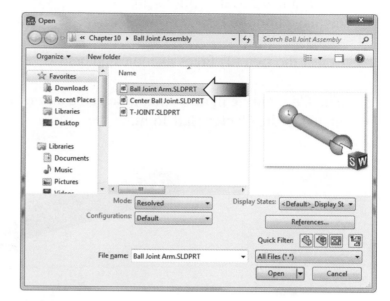

Click **Browse**.

Select **Ball Joint Arm.sldprt** from the Training Files folder and click **Open**.

The preview graphics of the component is attached to the mouse cursor.

Place the Child component above the Parent component as shown below.

5. Mating the components:

Click 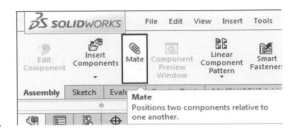 or select **Insert, Mate**.

The Standard Mate options are displayed on the FeatureManager tree.

Select the **two (2) faces** as indicated below.

The **Concentric** mate option is selected automatically and the Ball Joint Arm is mated to the Center Ball Joint.

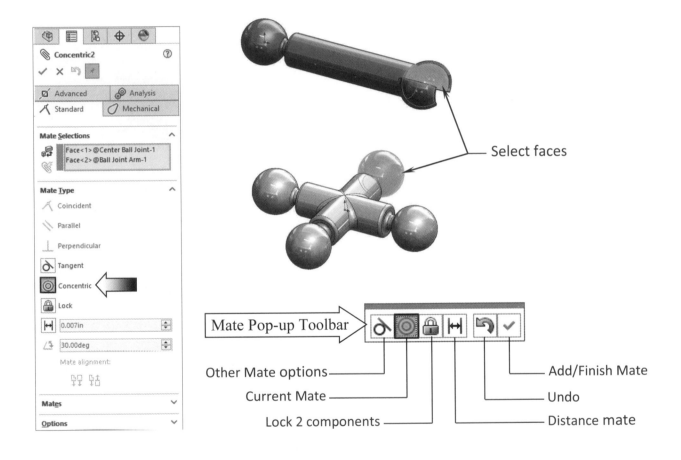

Select faces

Mate Pop-up Toolbar

Other Mate options
Current Mate
Lock 2 components
Add/Finish Mate
Undo
Distance mate

Review the options on the Mate Pop-up toolbar.

Click **OK**.

The **2 faces** are constrained with the concentric mate.

The second component is still free to rotate around the spherical surface.

6. Moving the component:

Click 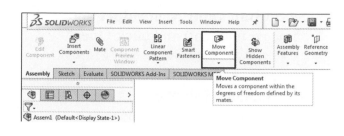 **Move Component**.

Drag the **face** as noted to see its degrees of freedom.

Move the component to the approximate position shown.

Click **OK**.

Drag here

NOTE: Pressing the left mouse button and dragging the component will also move it, but the Move Component command offers more options for moving, pushing, and detecting collisions between the components.

7. Inserting another instance:

Click 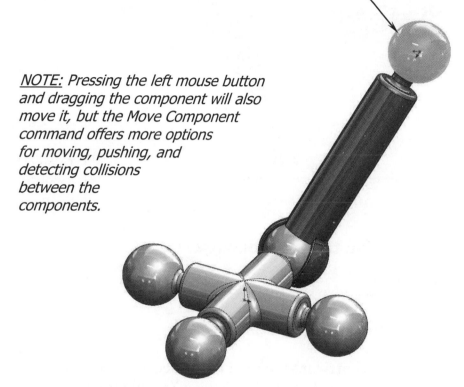 or select **Insert, Component, Existing Part, Assembly**.

Click Browse... .

NOTE: Not all assemblies are fully constrained. In an assembly with moving parts, leave at least one degree of freedom open so that the component can still be moved or rotated.

Select **Ball Joint Arm.sldprt** (the same component) and click **Open**.

Place the new component next to the first one as shown.

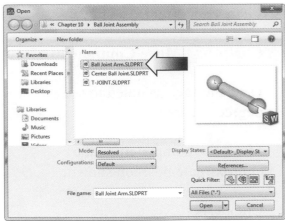

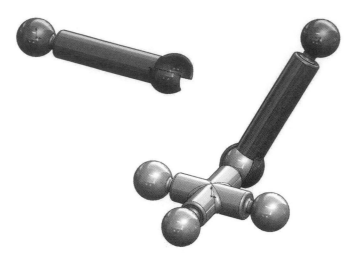

Copy Components

To quickly make a copy of a component: hold down the Control key and drag it.

8. Constraining the components:

Click 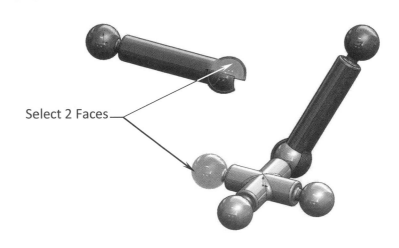 or select **Insert, Mate**.

Select the **two faces** as shown.

The **Concentric** mate option is selected automatically and the system displays the preview graphics of the two mated components.

Click **OK**.

Select 2 Faces

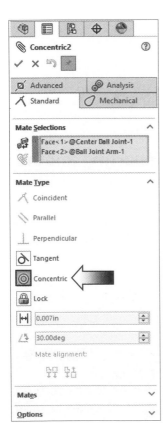

The two selected faces are constrained with a Concentric mate.

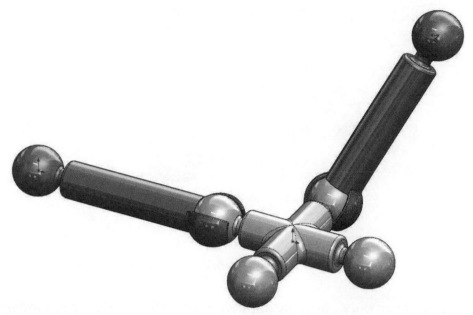

9. Repeating step 5:

Insert (or copy) <u>two more instances</u> of the **Ball Joint Arm** and mate them to the other 2 spheres as shown below.

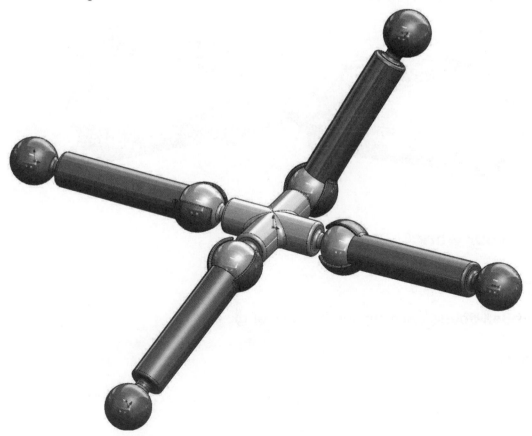

Optional:

Insert (or copy) <u>2 instances</u> of the **T-Joint** component as pictured below and mate it to the assembly.

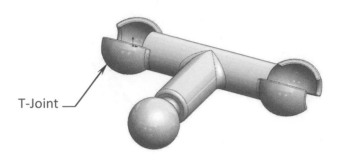

T-Joint

Check the assembly motion by dragging the T-Joint back and forth.

Each component has only one mate applied to it; they are still free to move or Rotate around their constrained geometry.

10. Saving your work:

Select **File / Save As**.

Enter **Ball-Joint-Assembly** for the name of the file.

Click **Save**.

(Open the completed assembly in the training files folder to check your work against it, if needed.)

Questions for Review

1. It is recommended that the 1st component should be fixed on the Origin.
    a. True
    b. False

2. The symbol (f) next to a file name means:
    a. Fully defined
    b. Failed
    c. Fixed

3. After the first component is inserted into an assembly, it is still free to be moved or reoriented.
    a. True
    b. False

4. Besides the first component, if other components are not yet constrained, they cannot be moved or rotated.
    a. True
    b. False

5. You cannot copy multiple instances of a component in an assembly document.
    a. True
    b. False

6. To make a copy of a component, click/drag that component while holding the key:
    a. Shift
    b. Control
    c. Alt
    d. Tab

7. Once a mate is created, its definitions (mate alignment, mate type, etc.) cannot be edited.
    a. True
    b. False

8. Mates can be suppressed, deleted, or changed.
    a. True
    b. False

7. FALSE   8. TRUE
5. FALSE   6. B
3. FALSE   4. FALSE
1. TRUE   2. C

Bottom Up Assembly
Links Assembly

Bottom Up Assembly
Links Assembly

Once the parts are inserted into an assembly document, they are now called components. These components will get repositioned and mated together. This method is called Bottom Up Assembly.

The first component inserted into the assembly will be fixed by the system automatically. If the first component is placed on the Origin then its Front, Top, and Right planes will also be aligned with the planes of the assembly.

Only the first component will be fixed by default; all other components are free to be moved or reoriented. Depending on the mate type, after a mate is assigned to a component, one or more of its degrees of freedom is removed, causing the component to move or rotate only in the desired directions.

Standard mates  are created one by one to constraint components.

Multi-Mates can be used to constrain more than one component, where a common entity is used to mate with several other entities.

This second half of the chapter will teach us the use of the Bottom Up assembly method once again. Some of the components are identical. We will learn how to create several copies of them and then add 2 mates to each one, leaving one degree of freedom open, so that they can be moved or rotated when dragged. There will be an instance number <1> <2> <3>, etc., placed next to the name of each copy to indicate how many times the components are used in an assembly.

Bottom Up Assembly
Links Assembly

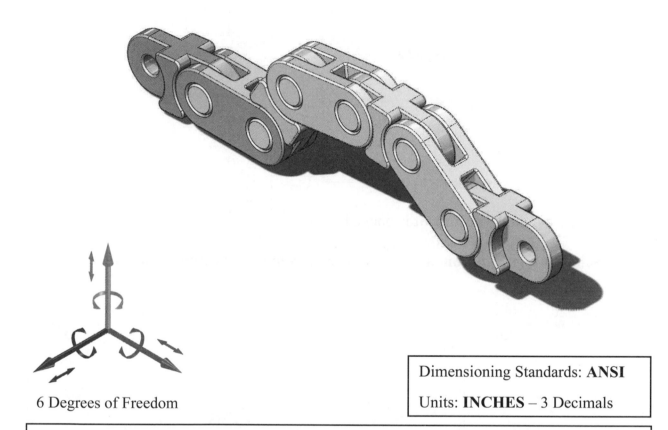

6 Degrees of Freedom

Dimensioning Standards: **ANSI**

Units: **INCHES** – 3 Decimals

Tools Needed:

 Mates

 Insert New Component

 Rotate Component

 Move Component

 Inference Origins

 Place/Position Component

 Align & Anti-Align

 Concentric Mate

 Coincident Mate

1. Starting a new Assembly Template:

Select **File, New, Assembly,** and click **OK**.

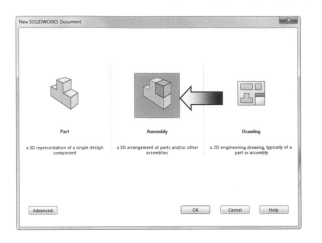

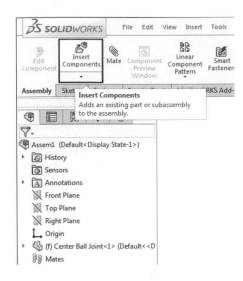

A new assembly document is opened.

Select **Insert Component** command from the Assembly tool tab.

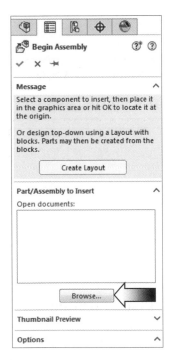

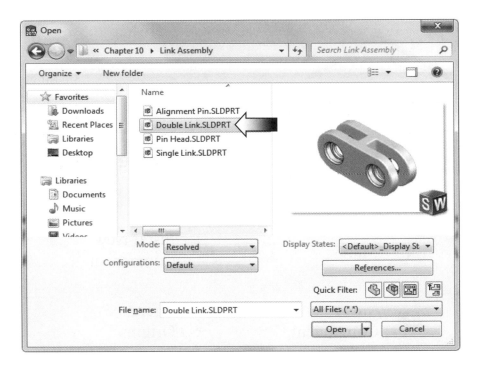

Click the **Browse** button (arrow).

Locate and open a part document named **Double Link.sldprt**.

2. Placing the 1ˢᵗ Component:

Hover the cursor over the assembly's Origin; the double arrow symbol appears indicating the part's origin and the assembly's origin will be mated coincident. Click to place the 1ˢᵗ part on the origin (or simply click the green checkmark).

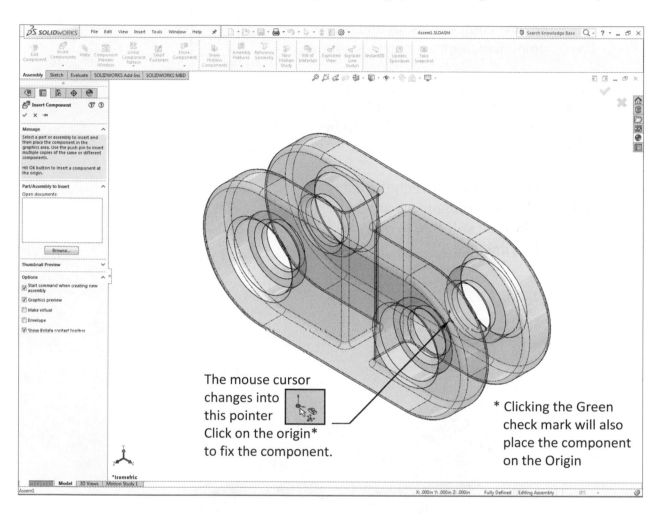

The mouse cursor changes into this pointer Click on the origin* to fix the component.

* Clicking the Green check mark will also place the component on the Origin

---

☼ "Fixing" the 1ˢᵗ component

The first component inserted into the assembly will be fixed automatically by the system. The symbol (f) next to the document's name means the part cannot be moved or rotated.

Other components can be added and mated to the first one using one of three options:

1. Drag & Drop from an opened window.
2. Use the Windows Explorer to drag the component into the assembly.
3. Use the Insert Component command.

3. Adding other components:

Select **Insert, Component, Existing Part/Assembly** (or click 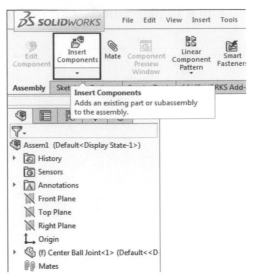) and add five (**5**) more instances of the first component into the assembly document.

Place the new components around the fixed component, approximately as shown.

The Feature Manager tree shows the same name for each component but with an indicator that shows the number of times used **<2>, <3>, <4>**, etc...

The (**-**) signs in front of each file name indicate that the components are under defined; they are still free to move or rotate when dragged.

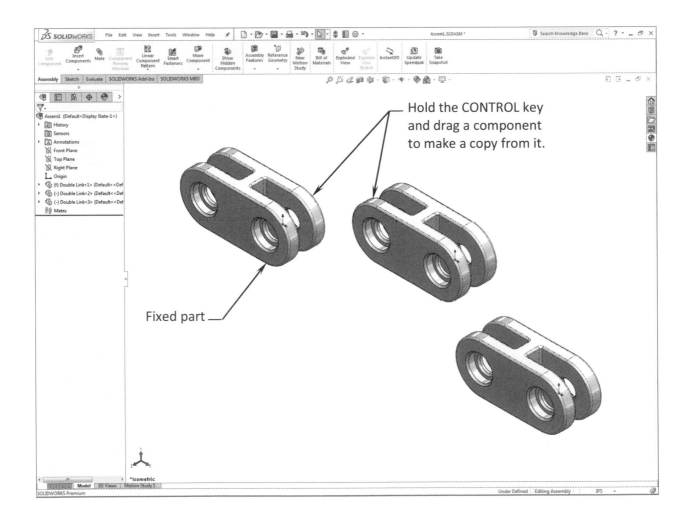

Hold the CONTROL key and drag a component to make a copy from it.

Fixed part

4. Changing colors:

For clarity, change the color of the 1ˢᵗ component to a different color; this way we can differentiate the parent component from the copies.

Click the 1ˢᵗ component and select the **Appearances** button (the beach ball icon) and click the **Edit Part Color** option (arrow).

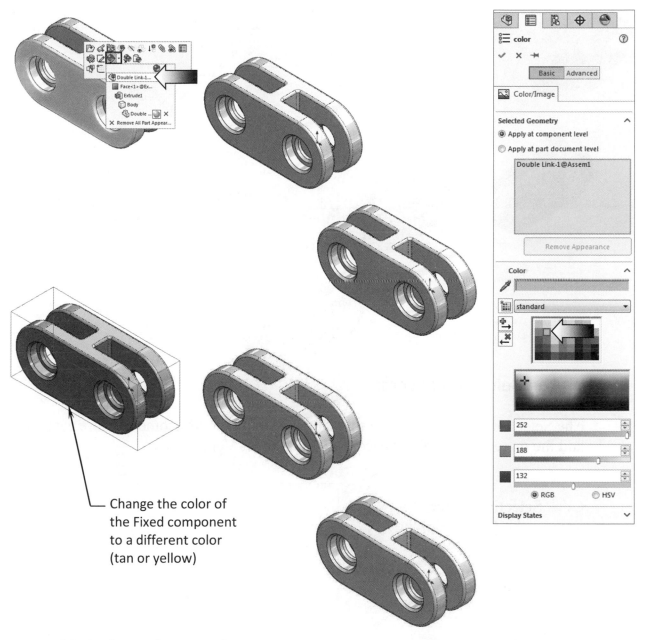

Change the color of
the Fixed component
to a different color
(tan or yellow)

Move the copies away from one
another, approximately as shown.

Click **OK**.

5. Inserting the Single Link into the assembly:

Click **Insert Component** 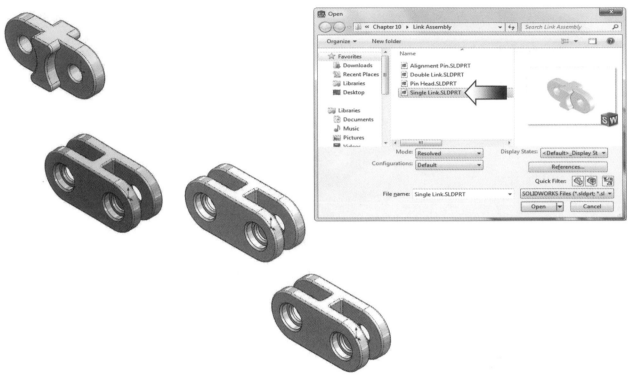 on the Assembly toolbar and select the part **Single Link** from the previous folder.

Place the Single Link approximately as shown.

6. Using the Selection Filters:

The Selection Filters help selecting the right type of geometry more easily to add the mates such as filter Faces, Edges, Axis or Vertices, etc…

Click **Selection Filter** icon, or press the **F5** function key.

Select **Filter Faces** option.

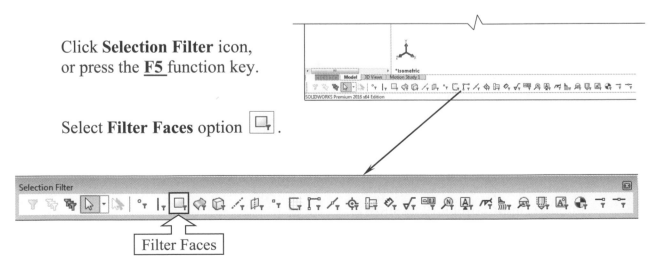

Filter Faces

7. Adding Mates:

Click **Mate** on the Assembly toolbar or select **Insert, Mate**.

Select the faces of the **two holes** as indicated.

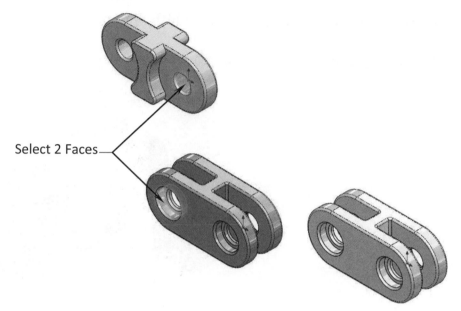

Select 2 Faces

The **Concentric** mate option is selected automatically and the Single Link component moves into its new position.

Click **OK** to accept the concentric mate.

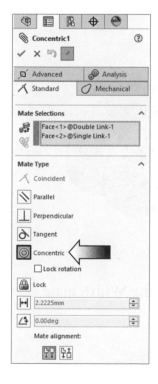

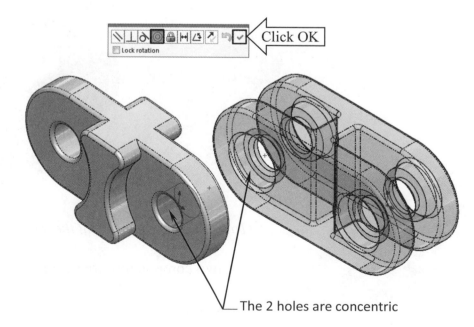

Click OK

The 2 holes are concentric

8. Adding a Width mate:

The Width mate is used to center 2 parts (width of part 1 and groove/tab of part 2).

Click **Mate** on the Assembly toolbar or select **Insert / Mate**.

Expand the **Advanced** section and click the **Width** option.

Select **2 faces** for **each part** as indicated. A total of 4 faces must be selected.

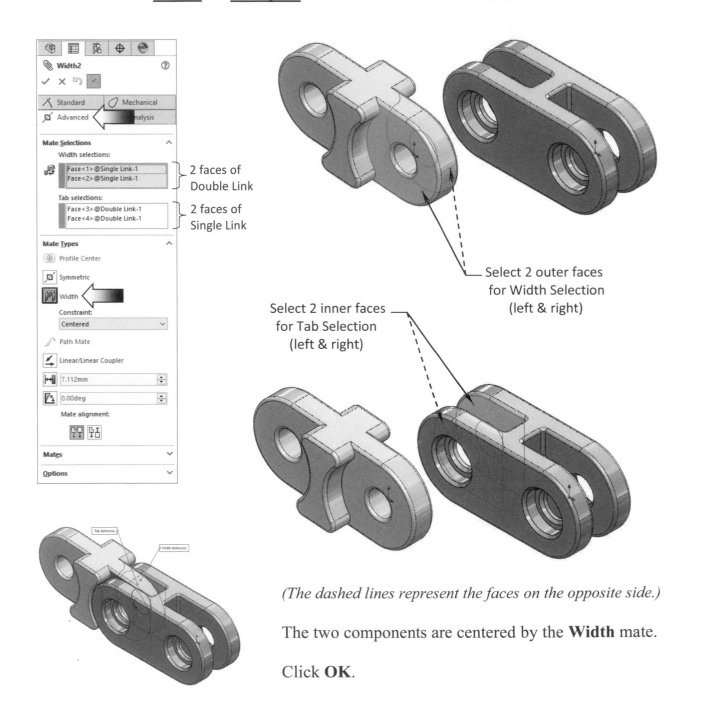

(The dashed lines represent the faces on the opposite side.)

The two components are centered by the **Width** mate.

Click **OK**.

9. Making copies of the component:

Hold down the CONTROL key, click/hold/drag the Single Link to make a copy.

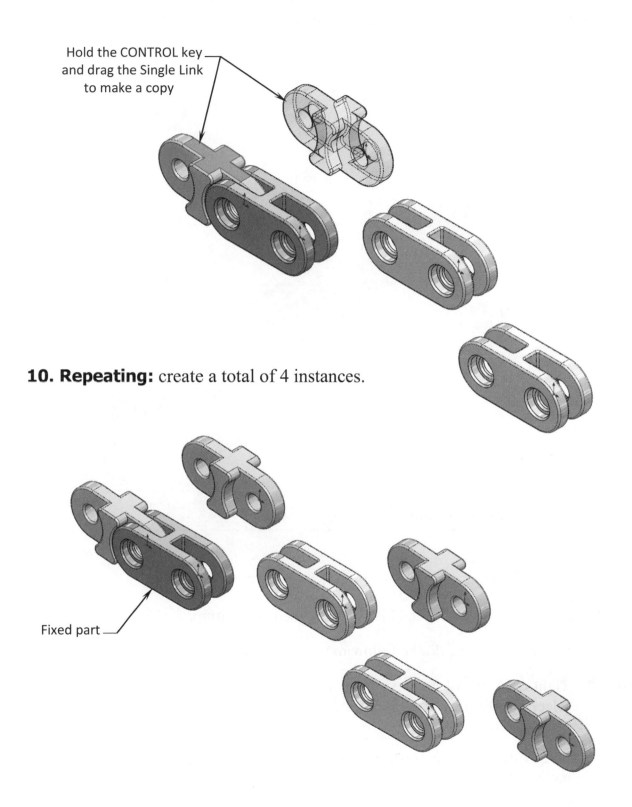

Hold the CONTROL key
and drag the Single Link
to make a copy

10. Repeating: create a total of 4 instances.

Fixed part

Notes:

The <u>correct</u> mates are displayed in Black color at the bottom of the tree.

The <u>incorrect</u> mates are displayed with a red X 🖇️ ⊗ (+) or a yellow exclamation mark 🖇️ ⚠️ (+) next to them.

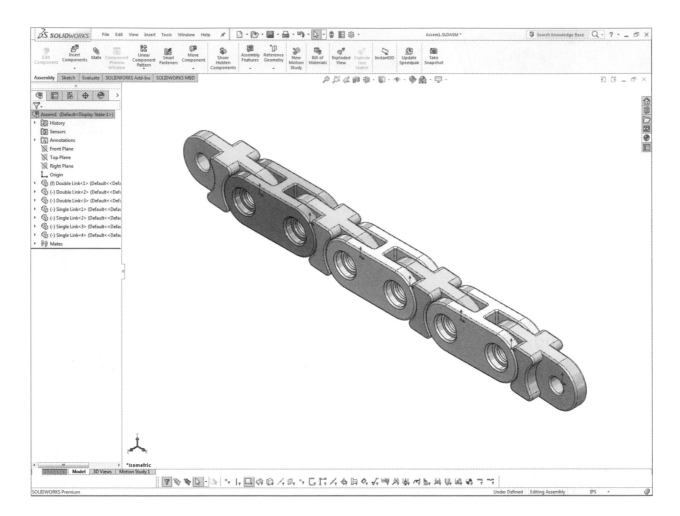

Expand the Mate group ⊟ 🖇️ Mates (click on the **+** sign).

Verify that there are no "Red Flags" under the Mate Group.

If a mate error occurs, do the following:

** Right-click one of the incorrect mates, select Edit-Feature and change the mate selection - **OR** -

** Simply delete the incorrect mates and recreate the new ones.

11. Inserting other components into the Assembly:

Click 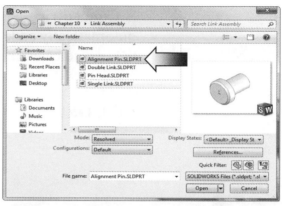 or select **Insert, Component, Existing Part/Assembly.**

Click **Browse**, select **Alignment Pin** and click **Open**.

Place the component on the <u>left side</u> of the assembly as shown below.

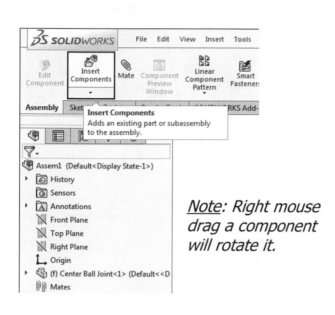

Note: Right mouse drag a component will rotate it.

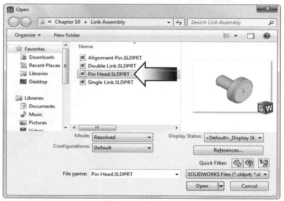

Insert the **Pin Head** and place it on the opposite side.

12. Rotating the Pin Head:

Select the Rotate Component command and rotate the Pin Head to the correct orientation.

Make 5 more copies of both the Pin Head and the Alignment Pin.

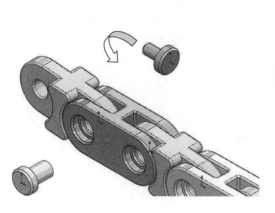

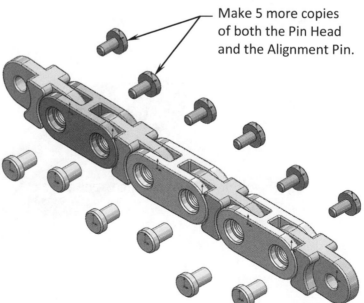

13. Constraining the Alignment Pin:

Click  **Mate** on the Assembly toolbar.

Select the **body** of the Alignment Pin and the **hole** in the Double Link.

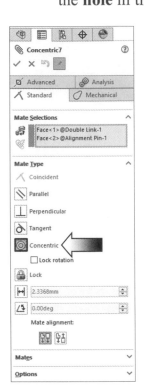

A **Concentric** Mate is automatically added to the two selected faces.

Click **OK**.

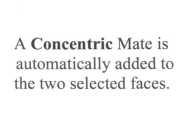

Select 2 Faces to add a **Concentric** mate

Select the planar face of the bore...

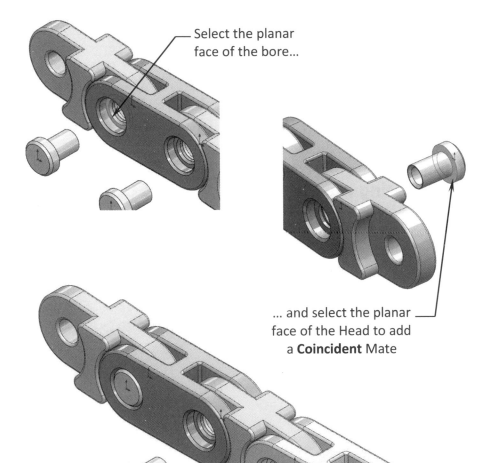

... and select the planar face of the Head to add a **Coincident** Mate

14. Constraining the Pin Head:

Add a **Concentric** mate between the Pin Head and its mating hole.

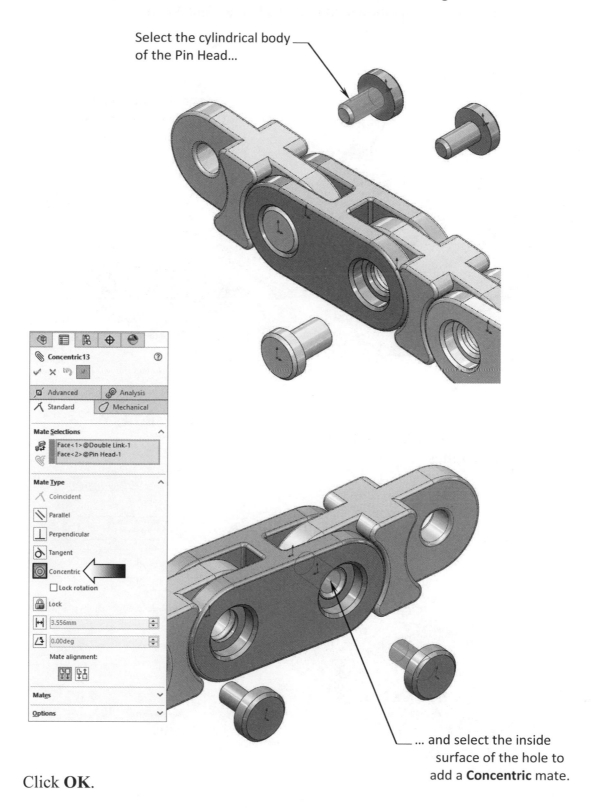

Select the cylindrical body of the Pin Head...

... and select the inside surface of the hole to add a **Concentric** mate.

Click **OK**.

Click 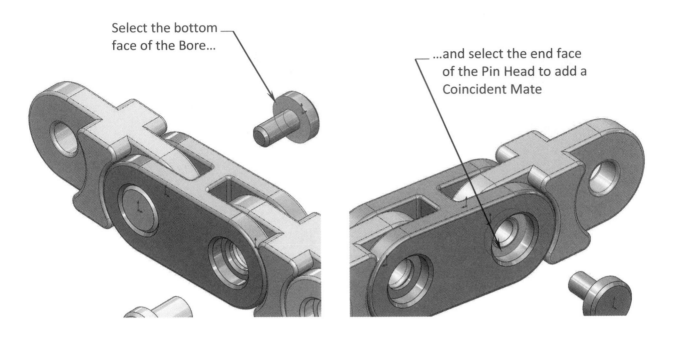 **Mate** again if you have already closed out of it from the last step.

Select the **two faces** as indicated to add a Coincident Mate.

Select the bottom
face of the Bore...

...and select the end face
of the Pin Head to add a
Coincident Mate

A **Coincident** mate is added to the
2 selected entities.

Click **OK**.

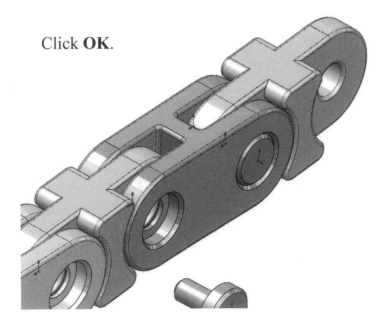

15. Using the Align & Anti-Align Options:

When the Mate command is active, the options **Align** and **Anti-Align** are also available in the Mate dialog box.

Use the alignment options to flip the mating component 180º .

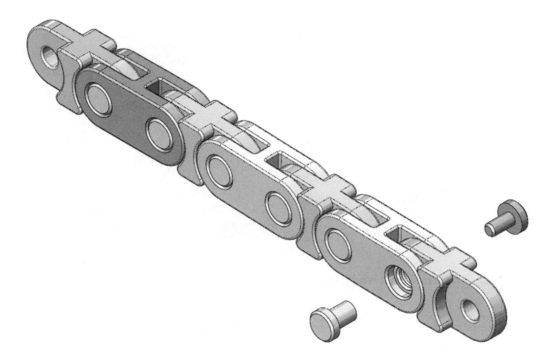

The last two (2) pins will be used to demonstrate the use of Align and Anti-Align options (see step 16).

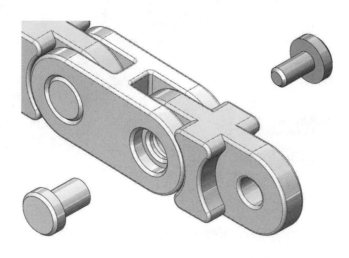

16. Using Align and Anti-Align:

When mating the components using the **Concentric** option, the **TAB** key can be used to flip the component 180° or from **Align** into **Anti-Align.**

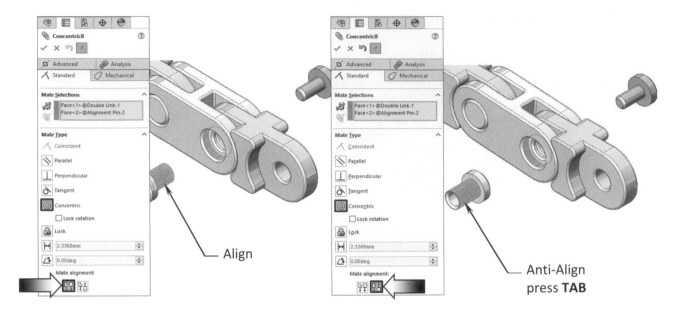

Align

Anti-Align
press **TAB**

17. Viewing the assembly motion:

Drag one of the links to see how the components move relative to each other.

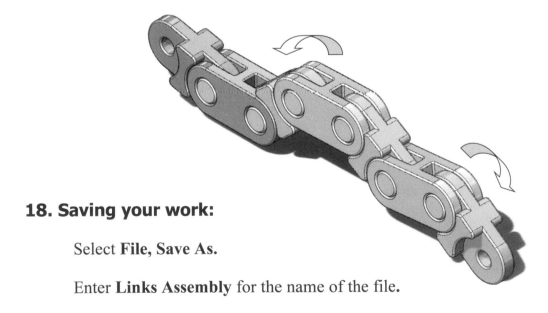

18. Saving your work:

Select **File, Save As.**

Enter **Links Assembly** for the name of the file**.**

Click **Save**.

(Open the pre-built assembly in the Training Files folder to compare your work against it.)

Questions for Review

1. Parts inserted into an assembly document are called components.
   a. True
   b. False

2. Each component in an assembly document has six degrees of freedom.
   a. True
   b. False

3. Once a mate is applied to a component, depending on the mate type, one or more of its degrees of freedom is removed.
   a. True
   b. False

4. Standard mates are created one by one to constrain the components.
   a. True
   b. False

5. A Coincident mate can be used between an edge and a face.
   a. True
   b. False

6. The Align and Anti-Align option can be toggled by pressing:
   a. Control
   b. Back space
   c. Tab
   d. Esc.

7. Mates can be deferred so that several mates can be done and solved at the same time.
   a. True
   b. False

8. Mates can be:
   a. Suppressed
   b. Deleted
   c. Edited
   d. All the above

Exercise: Gate Assembly

Go to The Training Files folder
Gate Assembly Folder.

1. Create an Assembly document
   from the components provided.
2. Create a Mirror plane at **26.125 in**.
   offset from the RIGHT plane.

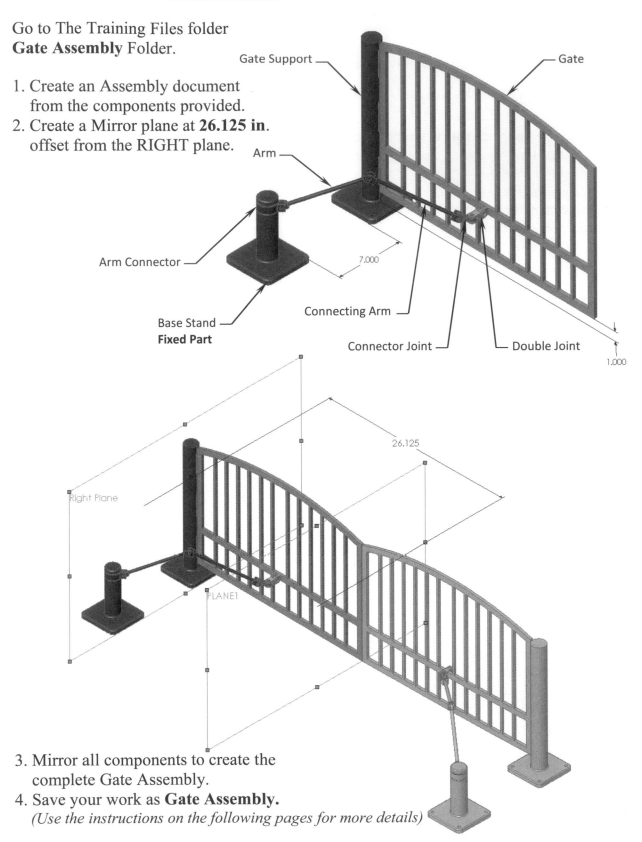

Gate Support

Gate

Arm

Arm Connector

7.000

Connecting Arm

Base Stand
Fixed Part

Connector Joint

Double Joint

1.000

Right Plane

26.125

PLANE1

3. Mirror all components to create the
   complete Gate Assembly.
4. Save your work as **Gate Assembly.**
   (Use the instructions on the following pages for more details)

1. Starting a new assembly:

Click **File / New / Assembly**.

From the Assembly toolbar, select **Insert / Components**.

Select the part **Base Stand** and place it on the assembly's origin.

(Click the green checkmark will also place it on the origin.)

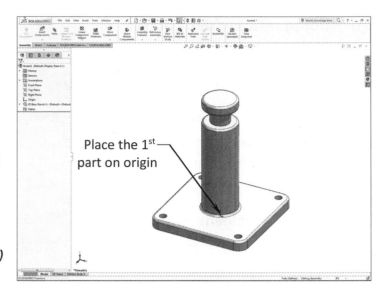

Place the 1ˢᵗ part on origin

2. Inserting other components:

Insert all components (total of 8) into the assembly.

Rotate the **Gate** <u>and</u> the **Gate-Support** approximately **180 degrees**.

Move the Gate and the Gate-Support to the positions shown.

Rotating and positioning the components before mating them helps to see the mating entities a little easier.

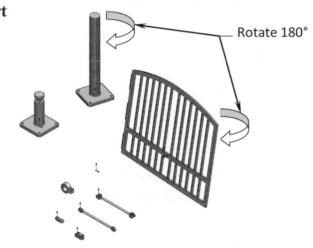

Rotate 180°

3. Adding a Distance mate:

Click **Mate** on the Assembly toolbar.

Select the **2 side faces** of the **Base Stand** and the **Gate-Support** as indicated.

Click the **Distance** button and enter **7.00**in.

Click **OK**.

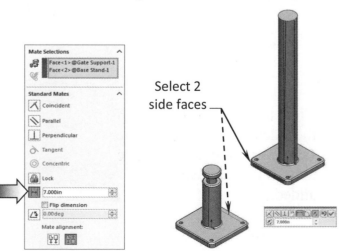

Select 2 side faces

4. Adding a Coincident mate:

Click **Mate** on the Assembly Toolbar, if not yet selected.

Select the **2 front faces** of the **Base Stand** and the **Gate Support** as indicated.

The **Coincident** option is automatically selected.

Click **OK**.

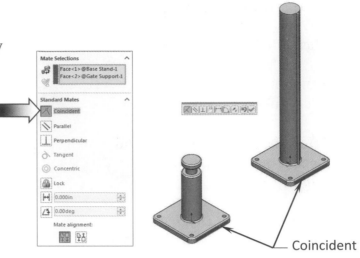

5. Adding other mates:

The **Mate** command should still be active; if not, select it.

Select the **2 bottom faces** of the **Base Stand** and the **Gate Support** as indicated.

The **Coincident** option is automatically selected.

Click **OK**.

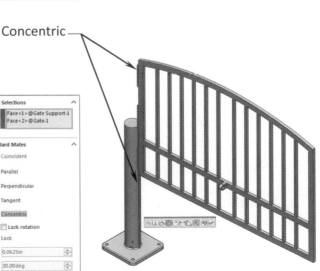

Click the Circular Bosses on the sides of the **Gate** and the **Gate Support**; a **concentric** mate is automatically added.

Click **OK**.

Add a **Distance** mate of
1.000" between the upper
face of the Gate Support and
the bottom face of the Gate,
as noted.

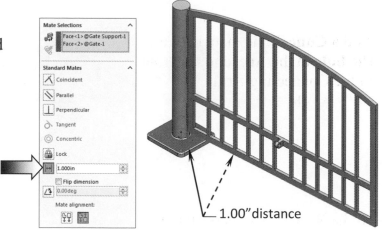

Add a **Concentric** mate
between the **circular face**
of the Base-Stand and
the **center hole** of the
Arm\Connector.

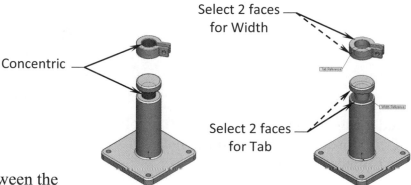

Add a **Width** mate between the
4 faces of the same components.

Add a **Concentric** mate between the
side hole of the Arm Connector and
the **hole** in the Arm.

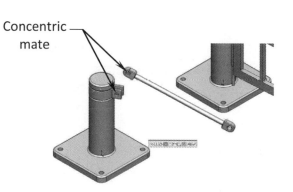

Add another **Width** mate between the
4 faces of these two components.

Click **OK**.

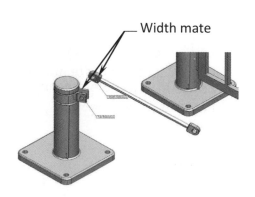

Add a **Concentric** mate between **the hole** in the Arm and **the hole** in the Connecting Arm.

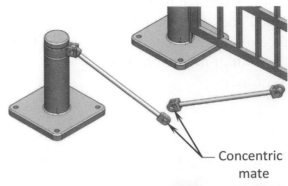

Concentric mate

Add a **Width** mate between the **4 faces** of the same components.

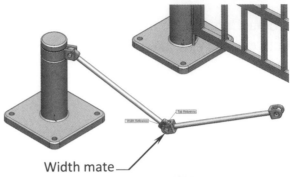

Width mate

Add a **Concentric** mate between **the hole** of the Connecting Arm and **the hole** in the Connector Joint.

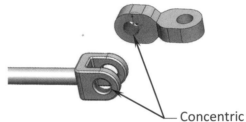

Concentric

Add another **Width** mate between the **4 faces** of these two components.

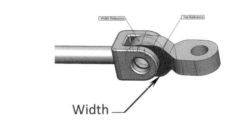

Width

Add a Concentric mate between the **hole** in the Connector Joint and the **hole** in the Double Joint.

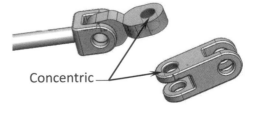

Concentric

Add a **Width** mate between the **4 faces** of these two components.

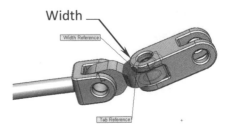

Width

Click **OK**.

Move the components to the position shown.

Since the Double Joint has several components connected to its left side, this time it might be a little easier if we create the Width mate before the Concentric mate.

6. Adding more mates:

Click the **Mate** command again, if not yet selected.

Select the **Advanced** tab.

Click the **Width** option.

Select the **2 faces** of the **Double Joint** for Width selection.

Select the other **2 faces** of the tab on the **Gate** for Tab selection.

Click **OK**.

Change to the **Standard Mates** tab.

Select **the hole** in the Double-Joint and **the hole** on the tab of the Gate.

A **Concentric** mate is added automatically to the selected holes.

Click **OK** twice to close out of the mate mode.

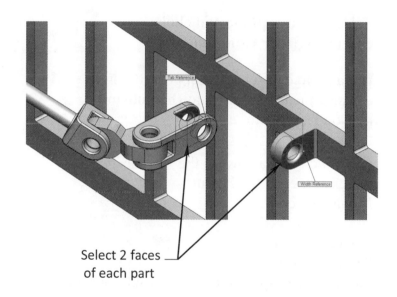

Move the components to approx. here

Select 2 faces of each part

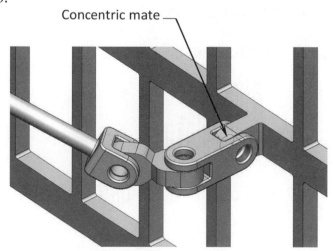

Concentric mate

7. Creating an Offset-Distance plane:

Select the **Right** plane of the assembly and click the **Plane** command, or **select Insert / Reference Geometry / Plane**.

The **Offset Distance** button is selected by default.

Enter **26.125**in. for distance.

Click **OK**.

8. Mirroring the components:

Select the **new plane** and click **Insert / Mirror Components**.

Expand the Feature tree and **select all components** from there.

Click the **Next** arrow on the upper corner of the Feature-Manager tree.

Select the part **Gate** from the list and click the **Create Opposite Hand Version** button.

Click **OK** and drag one of the gates to test out your assembly.

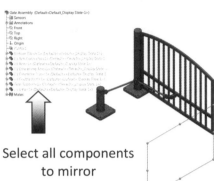

Select all components to mirror

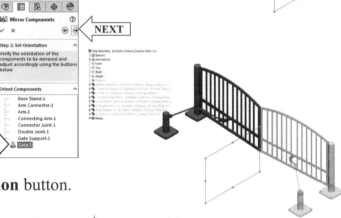

9. Save and Close all documents.

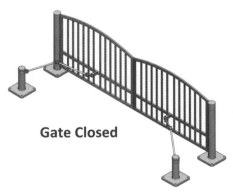

Gate Closed

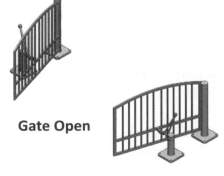

Gate Open

CHAPTER 11

Using Advanced Mates

Using Advanced Mates
Rack and Pinion

Besides the standard mates such as Concentric, Coincident, Tangent, Angle, etc., which must be done one by one, SOLIDWORKS offers an alternative option that is much more robust called Smart-Mates.

There are many advantages for using Smart-mates. It can create several mates at the same time, depending on the type of entity selected. For example, if you drag a circular edge of a hole and drop it on a circular edge of another hole, SOLIDWORKS will automatically create 2 mates: a concentric mate between the two cylindrical faces of the holes and a coincident mate between the two planar faces.

Using Smart-mates, you do not have to select the Mate command every time. Simply hold the ALT key and drag an entity of a component to its destination; a smart-mate symbol will appear to confirm the types of mates it is going to add.
At the same time, a mate context-toolbar will also pop up offering some additional options such as Flip Mate Alignment, change to a different mate type, or simply undo the selection.

When using the Alt+Drag option, the Tab key is used to flip the mate alignments; this is done by releasing the Alt key and pressing the Tab key while the mouse button is still pressed. This option works well even if your assembly is set to lightweight.

In addition to the Alt+Drag, if you hold the Control key and Drag a component, SOLIDWORKS will create an instance of the selected component and apply the smart-mates to it at the same time. Using this option, you will have to click the Flip Mate Alignment button on the pop-up toolbar to reverse the direction of the Mate; the Tab key does not work.

This lesson will teach us the use of the Smart-mate options as well as the advanced and standards mate options.

Using Advanced Mates
Rack & Pinion

Rack & Pinion Mates

The Rack and Pinion mate option allows linear translation of the Rack to cause circular rotation in the Pinion, and vice versa.

For each full rotation of the Pinion, the Rack translates a distance equal to π multiplied by the Pinion diameter. You can specify either the diameter or the distance by selecting either the Pinion Pitch Diameter or the Rack Travel / Revolution options.

1. Open an assembly document:

From the Training Files folder, browse to the Rack & Pinion folder and open the assembly document named **Rack & Pinion Mates.sldasm**

The assembly has two components, the **Rack** and **Gear1**. Some of the mates have been created to define the distances between the centers of the two components.

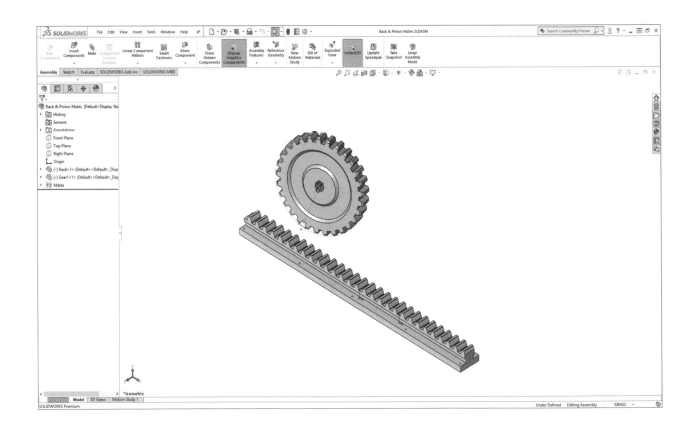

2. Adding standard mates:

Click the **Mate** command [Mate] from the assembly toolbar or select **Insert, Mate**.

Expand the FeatureManager tree and select the **2 Front planes** for both components. A **Coincident** mate is automatically created for the two selected planes.

Click **OK**.

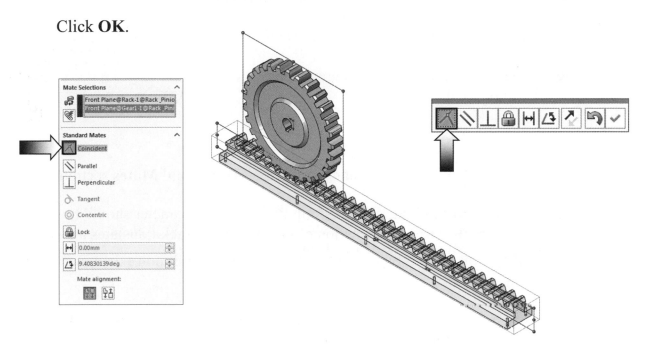

Click the **Mate** command again if it is not already selected.

Select the **2 Right planes** for both components and click the **Parallel** mate option.

Click **OK**.

This mate is used to locate the starting position of the Gear1, but it needs to be suppressed prior to adding the Rack & Pinion mate.

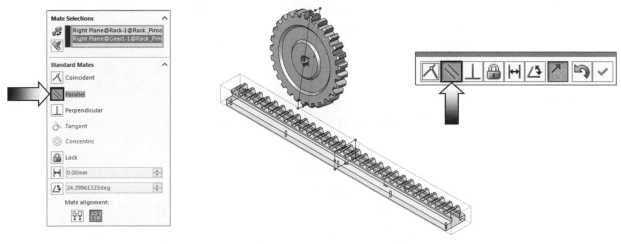

3. Suppressing a mate:

Expand the **Mate Group** (click the **+** sign) at the bottom of the FeatureManager tree.

Right-click the Parallel mate and select the **Suppress** button from the pop-up window (arrow).
(Move the rack so that it does not interfere with the gear.)

The Parallel mate icon should turn grey to indicate that it is suppressed.

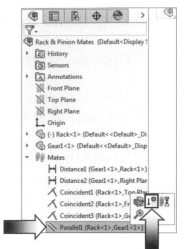

4. Adding a Mechanical mate:

Move down the Mate properties and expand the **Mechanical Mates** section (arrow).

Click the **Rack Pinion** button. The option Pinion Pitch Diameter should be selected already. For each full rotation of the pinion, the rack translates a distance equal to π (pi) multiplied by the pinion diameter, and the Pinion's diameter appears in the window.

In the Mate Selections, highlight the Rack section and select the **Bottom-Edge** of one of the teeth on the Rack as indicated.

Click in the Pinion section and select the **Construction-Circle** on the gear as noted.

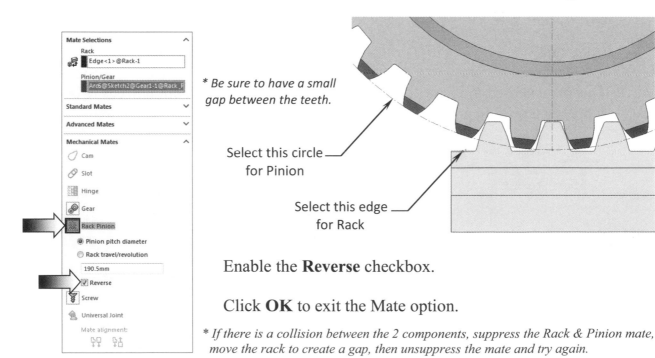

* Be sure to have a small gap between the teeth.

Select this circle for Pinion

Select this edge for Rack

Enable the **Reverse** checkbox.

Click **OK** to exit the Mate option.

* If there is a collision between the 2 components, suppress the Rack & Pinion mate, move the rack to create a gap, then unsuppress the mate and try again.

5. Testing the mates:

Change to the Isometric orientation
(Control + 7).

Drag either the Rack or the Gear1
back and forth and see how the
two components will move
relative to each other.

Hide the construction circle.
Next, we will create an animation
using a linear motor to drive the motions.

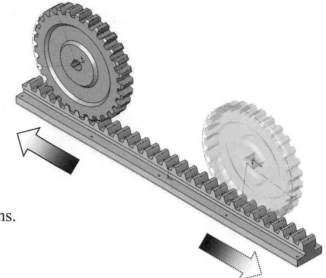

6. Creating a Linear Motion:

Remain in the Isometric view; click the **Motion Study1** tab at the
lower left corner of the screen.

The screen is split into 2 viewports showing the Animation program on the bottom.

Click the **Motor** button 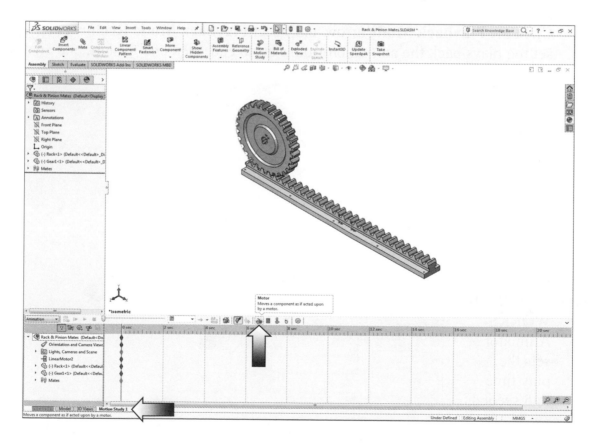 (arrow) on the Motion Study toolbar.

Under the **Motor Type**, select the **Linear Motor (Actuator)** option.
Under the **Motion** section, use the default **Constant Speed** and set the speed to **50mm/s** (50 millimeters per second).

Click **OK**.

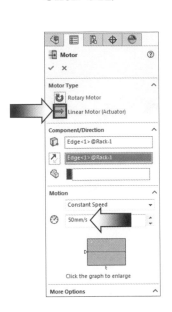

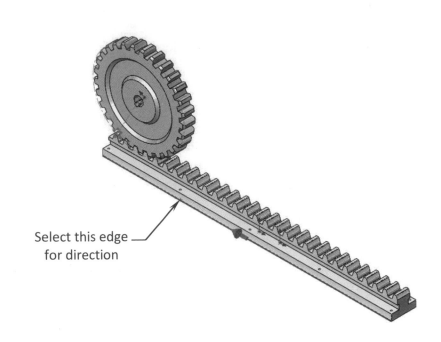

Select this edge for direction

7. Creating a Linear Motion:

Click the **Playback Mode** arrow and select **Playback Mode Reciprocate**. This setting plays back the animation from start to end, then end to start, and continues to repeat.

Click the **Play** button to view the animation.

By default the SOLID-WORKS-Animator sets the play time to 5 seconds.

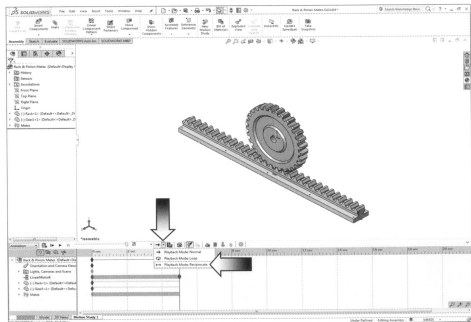

To change the **Playback Time**, drag the diamond key to **10 second** spot (arrow).
To change the **Playback Speed**, click the drop-down arrow and select **5X** (arrow).

Click the **Play** 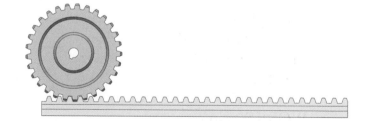 button again to re-run the animation.

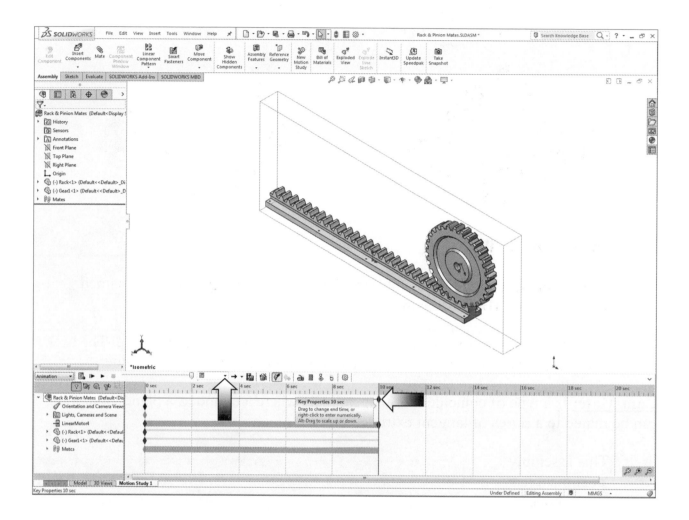

8. Saving your work:

Click **File / Save As**.

Enter **Rack & Pinion Mates**
for the name of the file.

Overwrite the old file when prompted.

Click **Save**.

Limit & Cam Mates

1. Opening a part file:

Browse to the Training Files folder and open the assembly document named:
Limit & Cam Mates.sldasm

Limit mates: Allow components to move within a specified distance or angle. The user specifies a starting distance or angle as well as a maximum and minimum value.

Cam Mate: Is a type of coincident or tangent mate where a cylinder, a plane, or a point can be mated to a series of tangent extruded faces.

This assembly document has 2 components. The **Part1** has been fixed by the Inplace1 mate and the **Part2** still has 6 degrees of freedom.

We will explore the use of Limit and Cam mates when positioning the Part2.

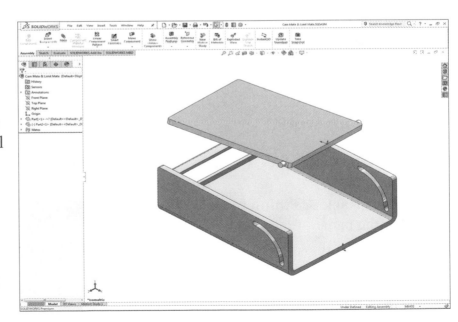

2. Adding a Width mate:

Click the **Mate** command from the Assembly tab.

Click the **Advanced Mates** tab and select the **Width** option.

The Width mate aligns the two components so that the Part2 is centered between the faces of the Part1 (Housing). The Part2 can translate along the center plane of the Housing and rotate about the axis that is normal to the center plane. The width mate prevents the Part2 from translating or rotating side to side.

For the **Width Selection**, select the **2 side faces** of the **Part2** (arrow).

For the **Tab Selection**, select the **2 side faces** of the **Part1** (Housing).

It does not matter which component you select, first or second,

just be sure to select both faces of the same part before selecting the next set.

Click **OK**.

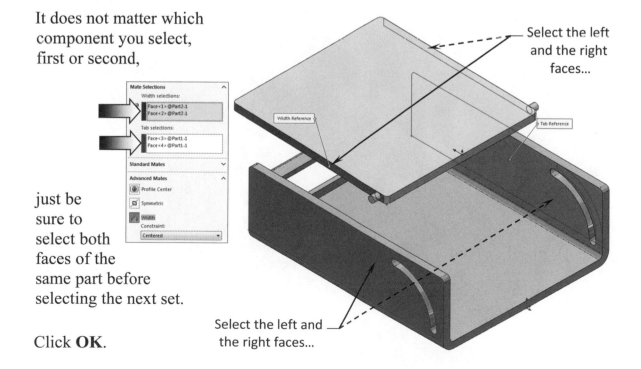

Select the left and the right faces...

Select the left and the right faces...

3. Adding a Cam mate:

Click the **Mechanical Mates** tab next to the Standard tabs.

Select the **Cam** mate button from the list.

A Cam mate forces a cylinder, a plane, or a point to be coincident or tangent to a series of tangent extruded faces.

Note: *The Slot-Mate can also be used to achieve the same result.*

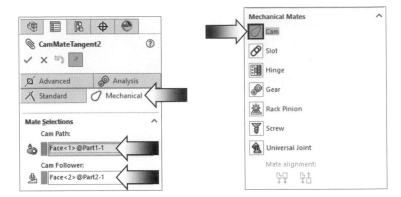

For the **Cam Path**, select one of the faces of the slot. All connecting faces are selected automatically.

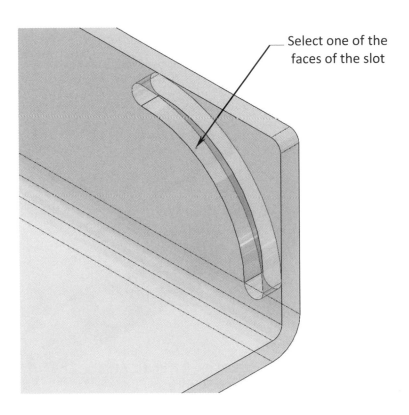

Select one of the faces of the slot

For the **Cam Follower**, select the <u>cylindrical face</u> of one of the pins.

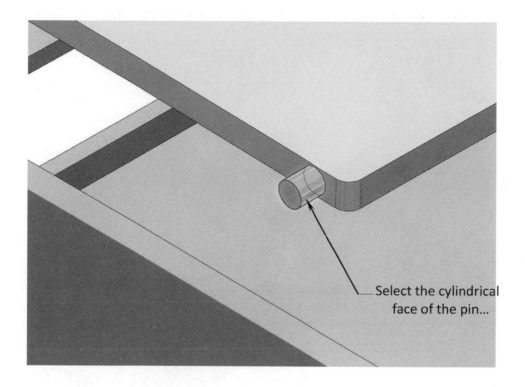

Select the cylindrical
face of the pin...

Zoom in and check the alignment between the pin and the slot.

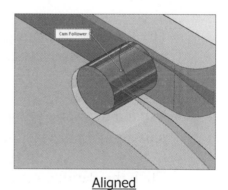

Toggle
between the
2 mate
alignment
buttons
(arrow) to
make sure the two
components are
properly oriented.

Anti-Aligned

Aligned

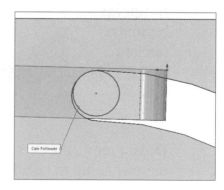

Click **OK**.

4. Adding a Parallel mate:

Click the **Standard Mates** tab (arrow) and select the **Parallel** option from the list.

Select the **2 faces** as indicated to make parallel.

Click **OK**.

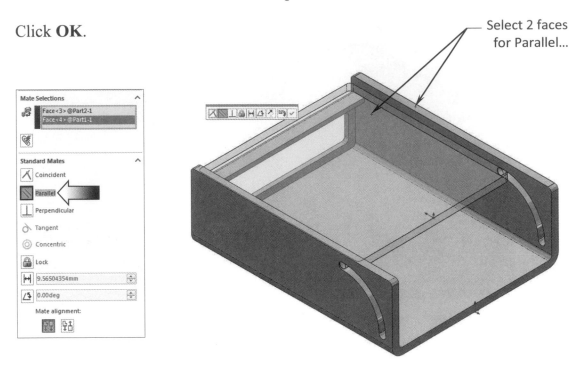

Select 2 faces
for Parallel...

The Parallel mate is used to precisely rotate the Part2 to its vertical position, but we will need to suppress it so that other mates can be added without over defining the assembly.

Expand the Mate group by clicking the plus sign (+) next to the Mates folder.

Click the Parallel mate and select the **Suppress** button (arrow).

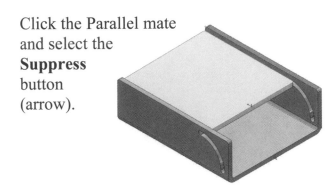

Do not move the component; a Limit mate is going to be added next.

5. Adding a Limit mate:

Click the **Advanced Mates** tab (arrow) and select the **Angle** option from the list.

Select the **2 faces** of the 2 components as indicated.

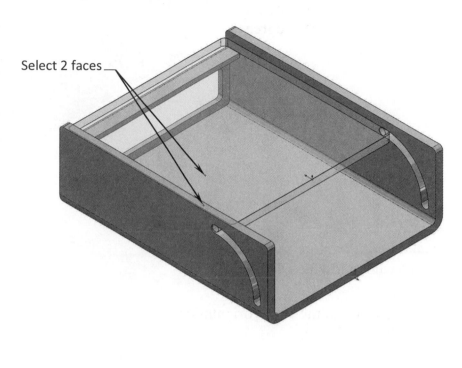

Select 2 faces

Enter the following:

> **90.00 deg** for "starting" Angle.
> **0.00 deg** for **Maximum** value.
> **0.00 deg** for **Minimum** value.

Click **OK**.

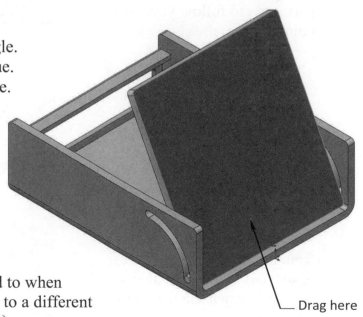

Test the mates by dragging the Part2 up and down.

It may take a little getting used to when moving a cam part. (Changing to a different orientation may make it easier.)

Drag here

Drag the right end of the Part2 downward. It should stop when it reaches the 90 degree angle.

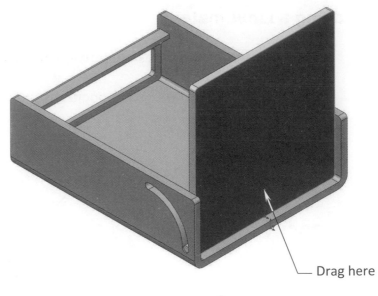

Drag here

Let us explore some other options when moving a cam part.

Change to the Front orientation (Control + 1).

Drag the **circular face** of the Pin upward. Start out slowly at first, then move a bit faster when the part starts to follow your mouse pointer.

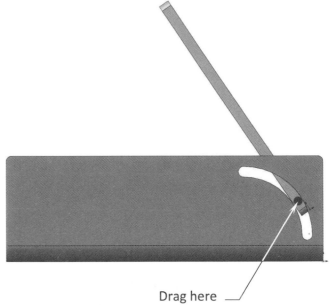

Drag here

Drag here

Now try dragging the same part from the side as shown here. Also, start out slowly then speed up a little when the part starts to catch on.

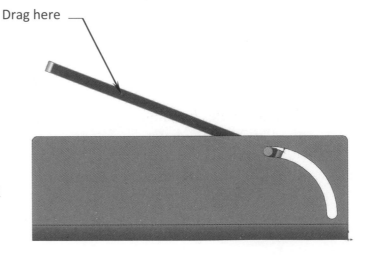

6. Saving your work:

Click **File, Save As**.

Enter **Limit & Cam Mates** for file name.

Click **Save**.

Overwrite the document if prompted.

(In the Training Files folder, locate the Built-Parts folder; open the pre-built assembly and check your work against it, if needed)

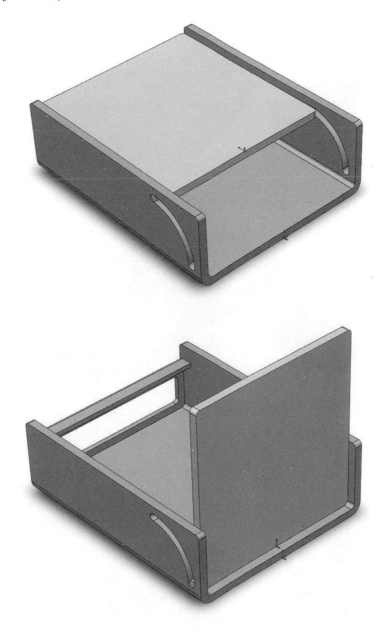

Exercise: Cam Followers*

1. Copying the Cam Followers Assembly folder:

Go to The Training Files folder.
Copy the entire folder named **Cam Followers** to your desktop.

* A Cam-Follower mate is a type of tangent or coincident mate. It allows you to mate a cylinder, a plane, or a point to a series of tangent extruded faces, such as you would find on a cam. You can make the profile of the cam from lines, arcs, and splines if they are tangent and form a closed loop.

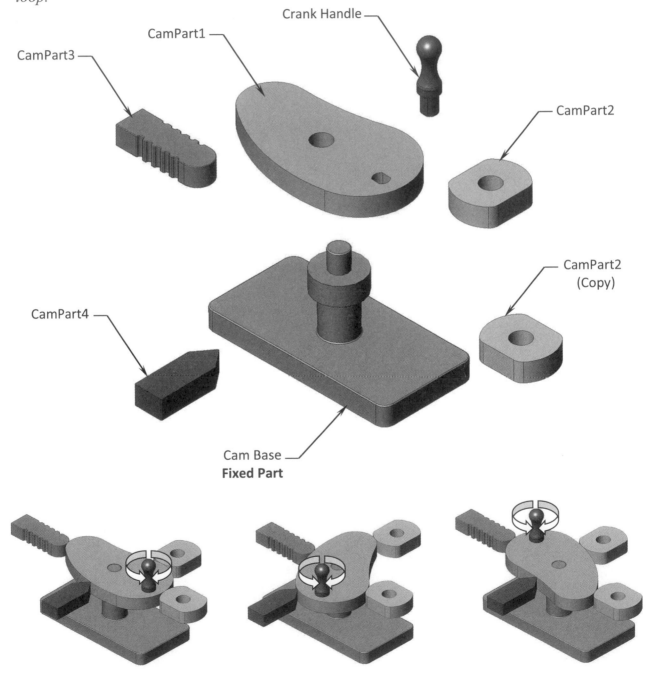

2. Assembling the components using the Standard Mates:

Create a **Concentric** mate between the **shaft** and the **hole** as shown below.

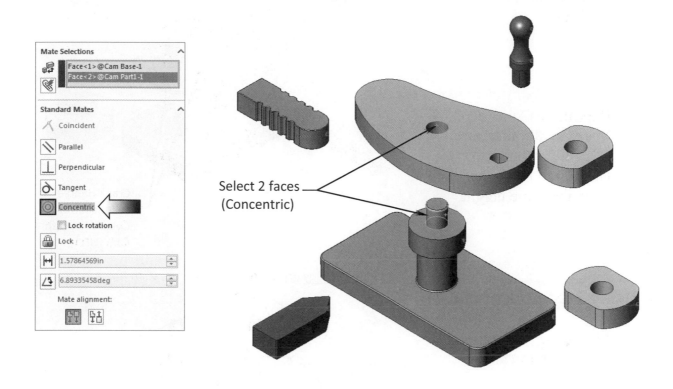

Create a **Coincident** mate between the **two faces** as shown below.

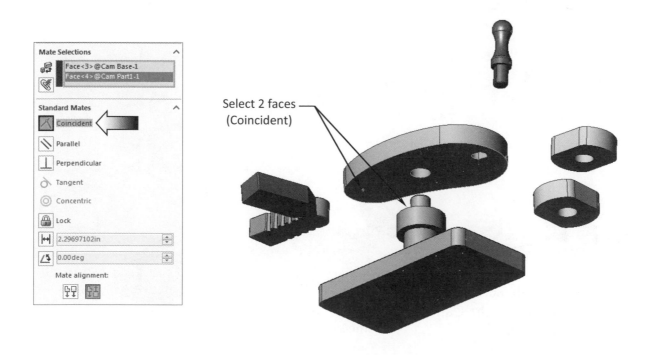

Create a **Coincident** Mate between the **2 upper surfaces** of the 2 components as indicated.

Repeat the Coincident Mates to all other parts (except for the Crank-Handle) to bring them up to the same height.

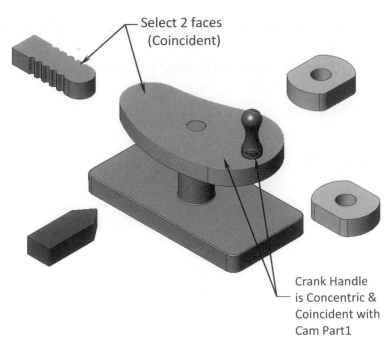

Select 2 faces (Coincident)

Crank Handle is Concentric & Coincident with Cam Part1

3. Using Mechanical Mates:

Locate the **Mechanical Mates** section and expand it.

Select the **Cam** option (arrow).

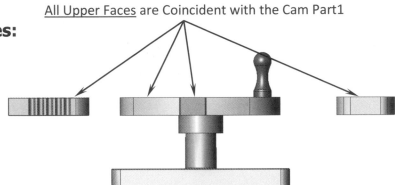

All Upper Faces are Coincident with the Cam Part1

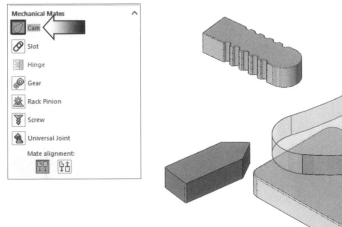

For Cam Path, select one of the side-faces of the Cam Part as indicated.

Select one of the side faces

For **Cam Follower**, select the curved face on the left side of the Cam Part2 as noted.

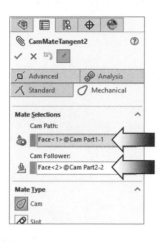

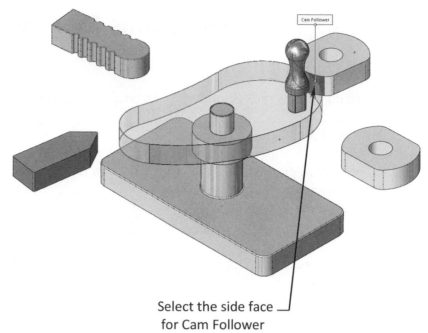

Cam Follower

Select the side face for Cam Follower

NOTES:

In order for the components to "behave" properly when they are moved or rotated, their center-planes should also be constrained.

Create a Coincident Mate between the FRONT plane of the CamPart1 and the FRONT plane of the CamPart3, and then repeat the same step for the others.

4. Viewing the Cam Motions:

Drag the Crank Handle clockwise or counter-clockwise to view the Cam-Motion of your assembly.

Drag the Crank Handle to test the Cam Motions

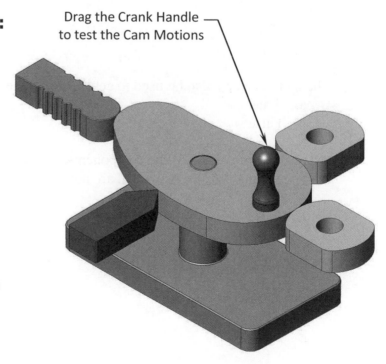

5. Saving your Work:

Click **File, Save As**.

Enter **Cam Follower** for the name of the file.

Click **Save**.

Questions for Review

1. When adding a rack & pinion mate, only a linear edge is needed for directions.
   a. True
   b. False

2. The rack & pinion mate will still work even if there is an interference between the components.
   a. True
   b. False

3. In a cam mate, more than one face can be selected to use as the cam follower.
   a. True
   b. False

4. In a cam mate, the alignment between the follower and the path can be toggled in or out.
   a. True
   b. False

5. In SOLIDWORKS Animator, a motor can be used to help control the movements of an assembly.
   a. True
   b. False

6. Using the Angle-Limit mate, the maximum & minimum extents can only be set from zero to 90°.
   a. True
   b. False

7. The slot mate can also be used to achieve the same results as the cam mate.
   a. True
   b. False

8. After mates are added to the components, their motions are restricted. They can only move or rotate along the directions that are not yet constrained.
   a. True
   b. False

1. FALSE   2. TRUE
3. FALSE   4. TRUE
5. TRUE   6. FALSE
7. TRUE   8. TRUE

Using Gear Mates

Gear mates force two components to rotate relative to one another about selected axes.

Valid selections for the axis of rotation for gear mates include cylindrical and conical faces, axes, and linear edges.

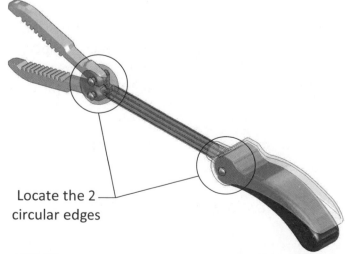

1. Opening an assembly document:

Browse to the Training folder and open the assembly document named: **Gear Mates.sldasm**

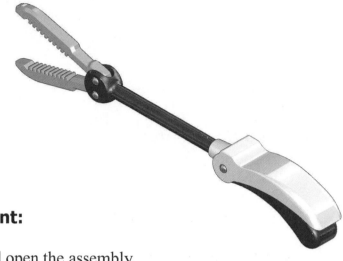

2. Changing the transparency:

Click the **Stem** and select: **Change Transparency** (arrow).

Also change the **Handle** to Transparent as noted.

Change to Transparent

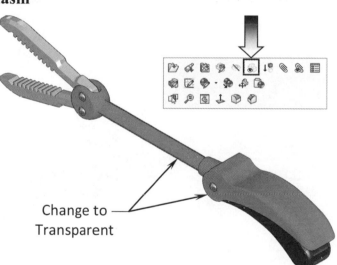

3. Locating the edges:

Locate the circular edges of the **Jaw-1** and the **Handle**.

Those edges will be used to create the Gear mate.

Locate the 2 circular edges

4. Suppressing the Angle mate:

The Angle mate that was used to locate the Jaws at 30° will need to be suppressed prior to adding the gear mate.

Locate the **Angle1** mate under the Mates Group and Suppress it.

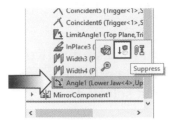

5. Adding a Gear mate:

Click **Mate** 🖇.

Expand the **Mechanical Mates** section and click the **Gear** button.

Select the circular edge of the **Jaw-1** and the circular edge of the **Handle**. (Hide the Pin if that would make it easier to select the circular edges.)

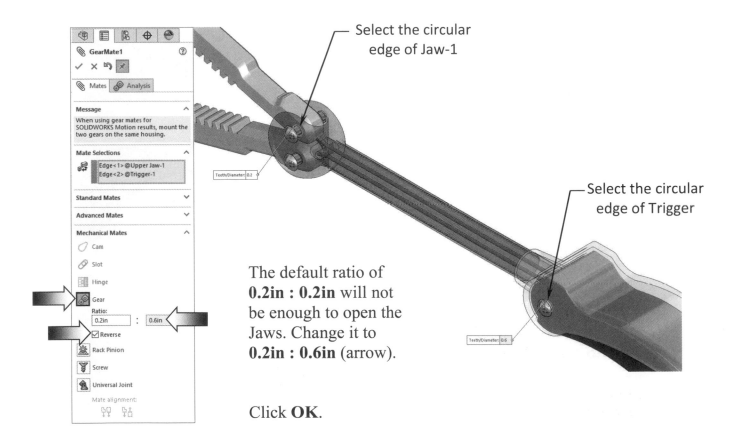

Select the circular edge of Jaw-1

Select the circular edge of Trigger

The default ratio of **0.2in : 0.2in** will not be enough to open the Jaws. Change it to **0.2in : 0.6in** (arrow).

Click **OK**.

(*Note: To open the jaws even wider change the ratio to **0.2in : 1.0in**.*)

6. Testing the Gear mate:

Change the Stem and the Handle back to their default shading (no transparency).

Drag the **Trigger** up and down to test out the Gear Mate.

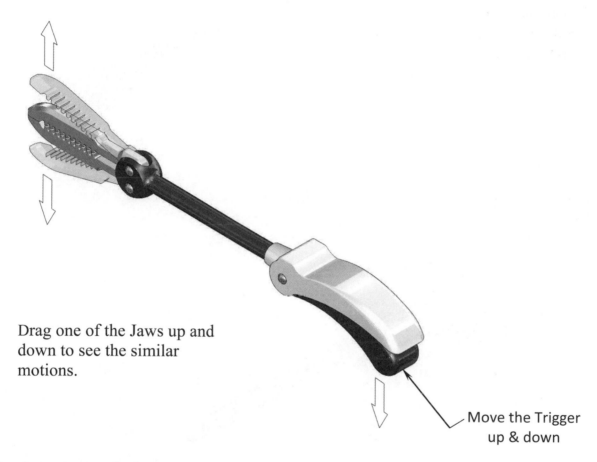

Drag one of the Jaws up and
down to see the similar
motions.

Move the Trigger
up & down

7. Saving your work:

Select: **File, Save As**.

Enter **Gear Mates Assembly_Completed** for the file name.

Press **Save**.

Close all documents.

OPTIONAL:

Create an exploded view similar to the one shown below.

Change the Stem, Handle, and Trigger to Transparent.

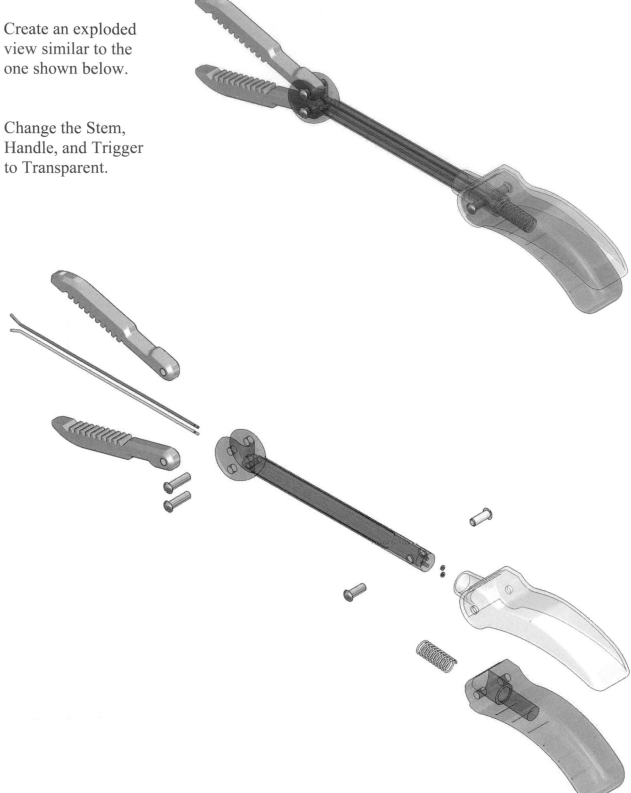

Using Limit Mates

Limit Mates allow components to move within a range of values for distance and angle mates. You specify a starting distance or angle, and a maximum and minimum value.

1. Opening an assembly document:

Browse to the Training Folder and open the assembly document named: **Limit Mates.sldasm**

For the purpose of this exercise, the component Arm Extension is under defined, it still can move up and down. We will use a mate called Limit Distance to constrain the linear movement of the Arm Extension.

2. Adding a Limit Distance mate:

Click **Mate** .

Expand the **Advanced Mates** section and select: **Limit Distance**.

For Mate Selections, select the bottom face of the Arm Extension and the inside face of the Base as indicated.

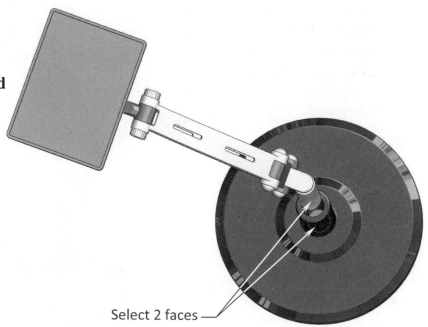

Select 2 faces —

...next

The Arm Extension moves to touch the inside face of the Base (do not click OK). For Travel Distance, enter **4.00in**.

For Maximum Value, enter **4.00in**.

For Minimum value, enter **0.00in**.

Click **OK**.

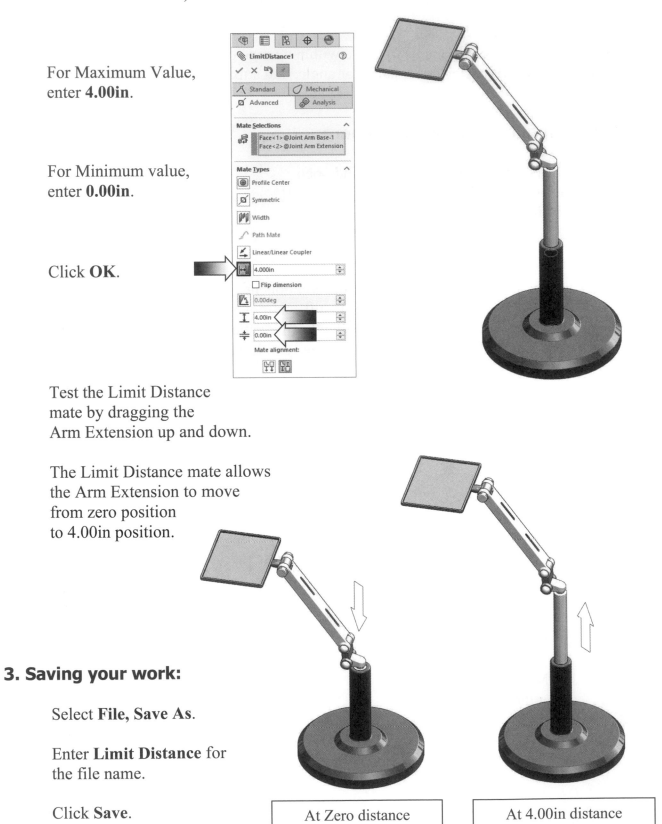

Test the Limit Distance mate by dragging the Arm Extension up and down.

The Limit Distance mate allows the Arm Extension to move from zero position to 4.00in position.

3. Saving your work:

Select **File, Save As**.

Enter **Limit Distance** for the file name.

Click **Save**.

| At Zero distance | At 4.00in distance |

Exercise: Bottom Up Assembly

<u>Files location:</u> Training Files folder.
Bottom Up Assembly Folder.
<u>Copy</u> this folder to your Desktop.

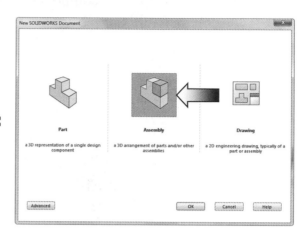

1. Starting a new Assembly document:

Select **File, New, Assembly**

2. Inserting the Fixed component:

Click the **Insert Component** command
from the Assembly toolbar.

Locate the component **Molded Housing**, open and place it on the assembly's origin.

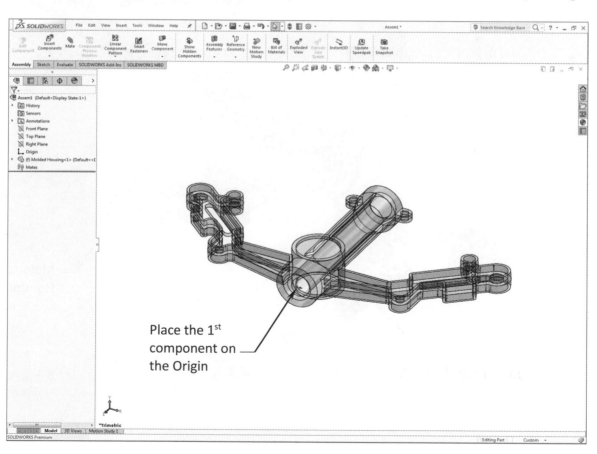

Place the 1st
component on
the Origin

Note: *To show the Origin symbol, click* ***View / Origins.***

3. Inserting the other components:

Insert the rest of the components as labeled from the Bottom Up Assembly folder into the assembly.

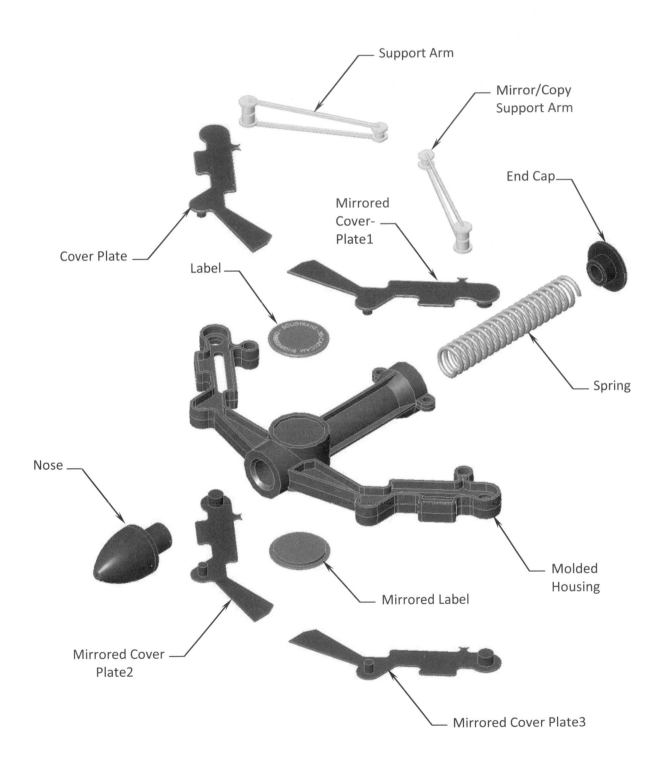

4. Mating the components:

Assign the Mate conditions such as Concentric, Coincident, etc., to assemble the components.

The finished assembly should look like the one pictured below.

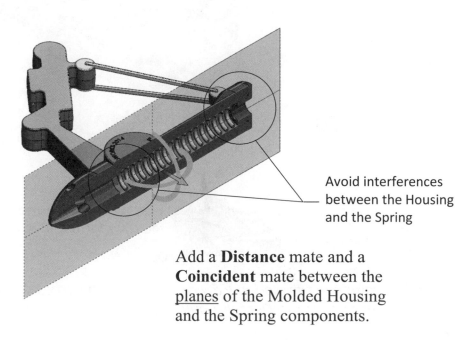

5. Verifying the position of the Spring:

Select the <u>Right</u> Plane and click **Section View**.

Position the Spring approximately as shown to avoid assembly's interferences.

Click-off the section view option when finished.

Avoid interferences between the Housing and the Spring

Add a **Distance** mate and a **Coincident** mate between the <u>planes</u> of the Molded Housing and the Spring components.

6. Creating the Assembly Exploded View:

Click [Exploded View] or Select **Insert / Exploded View**.

Select a component either from the Feature tree or directly from the Graphics area.

Drag one of the three Drag-Handles to move the component along the direction you wish to move and click **Done**.

Repeat the same step to explode all components.

For editing, right-click on any of the steps on one of the Explode-Steps and select Edit-Step.

To simulate the move/rotate at the same time, first drag the direction arrow to move, then drag the ring to rotate.

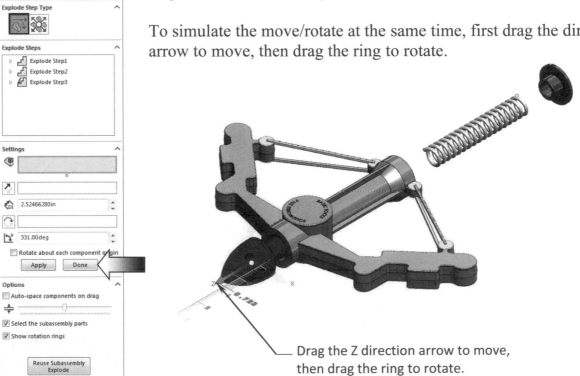

Drag the Z direction arrow to move, then drag the ring to rotate.

<u>Note:</u> *The completed exploded view can be edited by accessing the Configuration-Manager under the Default Configuration.*

Configuration Manager tree

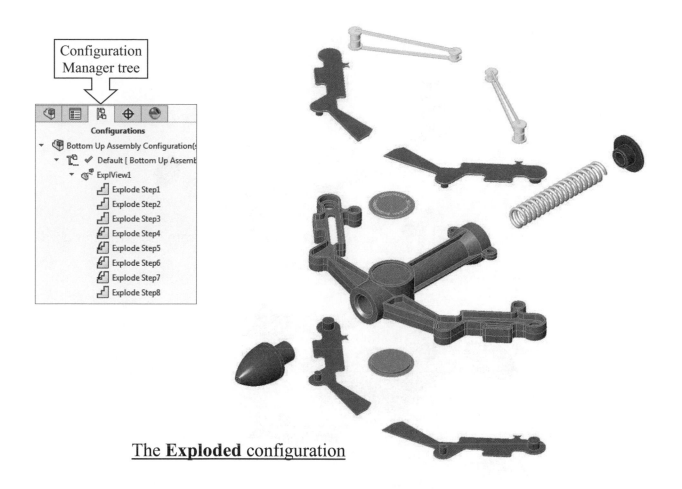

The **Exploded** configuration

The **Default** configuration

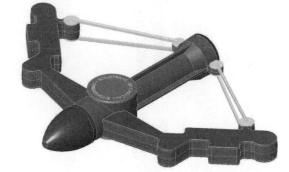

7. Saving your work:

Select **File, Save As**.

Enter **Bottom Up Assembly Exe** for the name of the file.

Click **Save**.

(From the Training Files folder, locate the Built-Parts folder; open the pre-built assembly to check your work against it.)

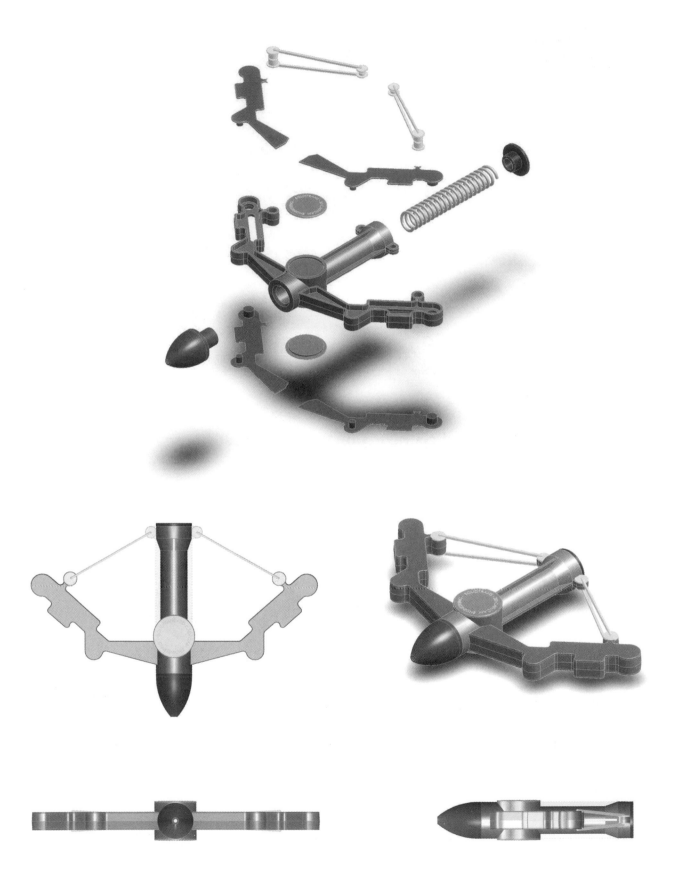

Level 1 Final Exam: Assembly Motions

1. Open the assembly document named: **Assembly Motions Level 1 Final.sldasm**
2. Assemble the components using the reference images below.
3. Create an Assembly Feature and modify the Feature Scope so that only 3 components will be affected by the cut (shown).
4. Drag the Snap-On Cap to verify the assembly motions. All components should move or rotate except for the Main Housing.

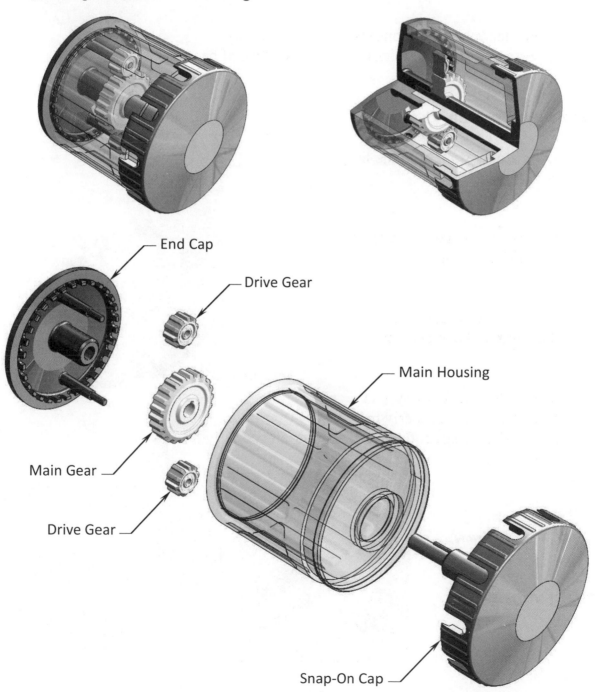

End Cap

Drive Gear

Main Housing

Main Gear

Drive Gear

Snap-On Cap

Testing the Assembly Motions:

Drag the Snap-On Cap in either direction to verify the motions of the assembly.

The Main Gears, the two Drive Gears, and the Snap-On Cap should rotate at the same time.

If the mentioned components are not rotating as expected, check your mates, and recreate them if needed.

Creating a Section View:

Open a **new sketch** on the <u>face</u> as indicated.

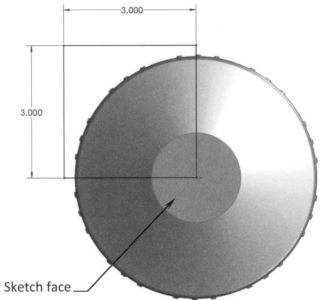

Sketch a rectangle and add the dimensions shown.

Create an Extruded Cut **Through All** components.

Sketch face

Verify that the Keyway on the Snap-On Cap is properly mated to the Main Gear, and there is no interference between them.

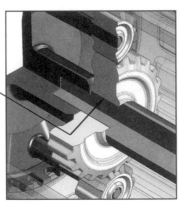

Keyway

Saving your work:

Save your work. as **Level 1 Final**.

CHAPTER 12

Kinematics – Assembly Motions

Kinematics
Assembly Motions

To describe motion, kinematics studies the paths of points, lines and other geometric objects in space as well as some of their properties such as velocity and acceleration.

Mechanical, robotics, and biomechanic engineering use it to describe the motion of systems composed of joined parts such as an engine, a robotic arm or the skeleton of the human body.

Kinematic analysis is the process of measuring the kinematic quantities used to describe motion.
In engineering, kinematic analysis may be used to find the range of movement for a given mechanism and working in reverse. Kinematic synthesis designs a mechanism for a desired range of motion.

A four-bar linkage and a pumpjack system are examples of kinematic linkages where components in an assembly are connected to manage forces and movement.

The example in this lesson uses 3 (simplified) linear actuators and other mechanisms to demonstrate the kinematics motions which can be used in many different applications.

After all components are correctly mated, the components can only be moved or rotated along the directions that were defined to control their movements.

Kinematics
Assembly Motions

View Orientation Hot Keys:

Ctrl + 1 = Front View
Ctrl + 2 = Back View
Ctrl + 3 = Left View
Ctrl + 4 = Right View
Ctrl + 5 = Top View
Ctrl + 6 = Bottom View
Ctrl + 7 = Isometric View
Ctrl + 8 = Normal To
       Selection

Dimensioning Standards: **ANSI**

Units: **INCHES** – 3 Decimals

Tools Needed:

 Mate

 Concentric Mate

 Coincident Mate

 Animation Wizard

 Motors

 Add New Key

1. Opening an assembly document:

Browse to the Training Files folder and open an assembly document named: **Kinematics Assembly.sldasm**.

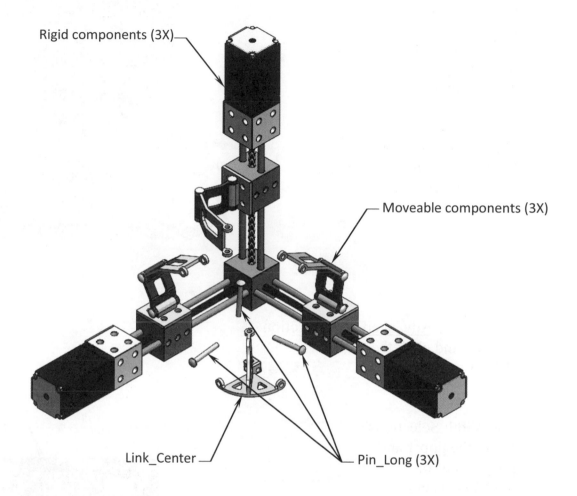

The rigid components have already been fixed ahead of time. Only the Link_Center and the 3 Pin_Long will need to be constrained.

The Link_Center component needs to be mated to the other links using the Concentric and the Width mates. After all the link components are mated, then 3 instances of the Pin_Longs will be assembled to keep the links connected.

Kinetics Motions can be viewed by dragging the Link_Center along any direction. The motions can also be recorded using either the Record Video option or with SOLIDWORKS Animator.

2. Constraining the Link_Center:

Click **Mate**.

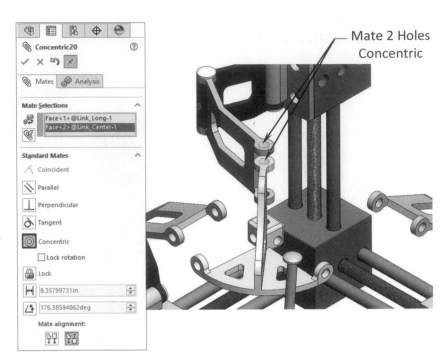

Mate 2 Holes Concentric

Select the <u>hole</u> in the **Link_Center** and the <u>hole</u> in the **Link_Long** as indicated.

Click **OK** to accept the **Concentric** mate but do not exit out of the Mate mode.

Expand the <u>Advanced Mates</u> section (arrow) and click the **Width** mate button.

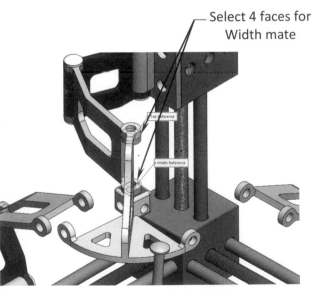

Select 4 faces for Width mate

For Width Selections, select the <u>upper</u> and <u>lower faces</u> of the **Link_Long** component.

For Tab-Selections, select the <u>2 inner</u> <u>faces</u> of the **Link_Center** component as noted.

Click **OK** to accept the **Width** mate but do not exit the mate.

3. Mating the 1st Pin_Long:

First, we will align
the Pin with the hole
using a Concentric
mate.

Select the body of
Pin and the upper
hole of the **Center_
Link** as noted.

Click **OK** to
accept the
Concentric mate.

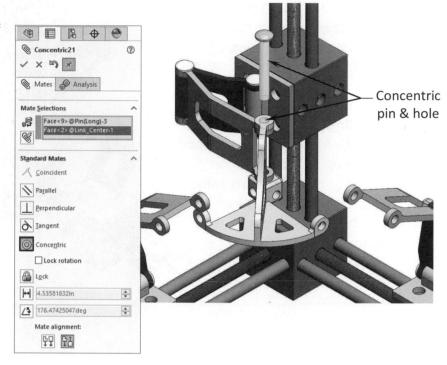

Concentric
pin & hole

Select 2 faces for
Coincident mate

And next, we will
make the Pin head
flush with the
Link_Center using
a Coincident mate.

Select the lower
face of the **Pin head**
and the upper face
of the **Center_
Link**.

Click **OK** to accept
the **Coincident** mate
but do not exit out
of the mate mode.

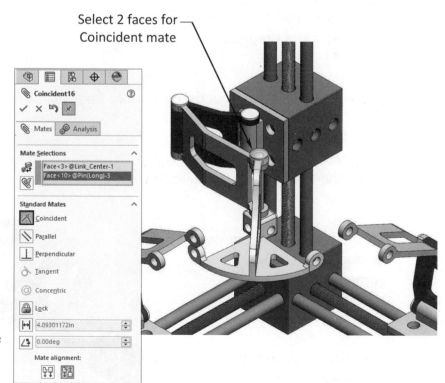

4. Mating the last 2 pins:

Repeat the last step and mate the last 2 pins to connect the Link_Center to the other Links.

Note: The Pin_Long is supposed to be threaded into the square block at the corner of the Center_Link, but due to the larger file size, the threads feature has been removed to improve the computer performance.

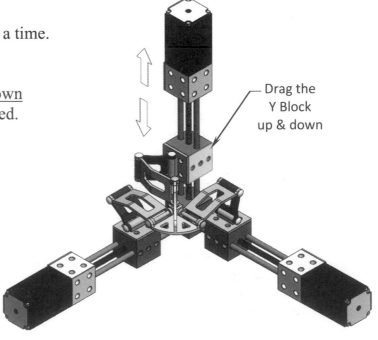

Repeat step 3 and mate the last 2 pins

5. Testing the kinematics motions:

Each component still has at least one degree of freedom which allows them to translate or rotate when dragged.

We will test out one direction at a time.

Drag the square block up and down along the **Y direction** as indicated.

As the square block moves up and down, other components also move with it.

We will not detect the collisions for now.

Drag the Y Block up & down

<u>Drag</u> the next square block <u>in and out</u> along the **Z direction** as indicated.

The other components are also moved along their X,Y,Z directions, when the square block is moved.

Move the components to a position where they are not colliding with each other.

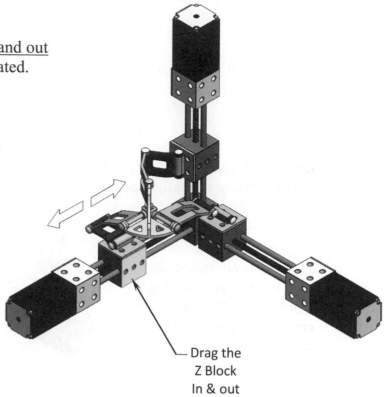

Drag the
Z Block
In & out

<u>Drag</u> the next square block <u>left and right</u> along the **X direction** as indicated.

Other components are also moved along with the square block.

The last 3 steps show how components are moved by dragging but without reference to the causes of motion.

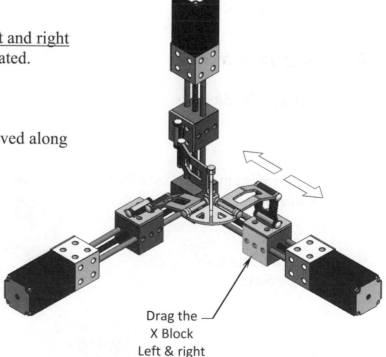

Drag the
X Block
Left & right

Lastly, we will test out all 3 directions at the same time.

<u>Drag</u> the **Center_Link** in a circular fashion, move it in all directions.

This time, all moveable components are moving along the X,Y,Z directions, simulating circular motions of all connected components.

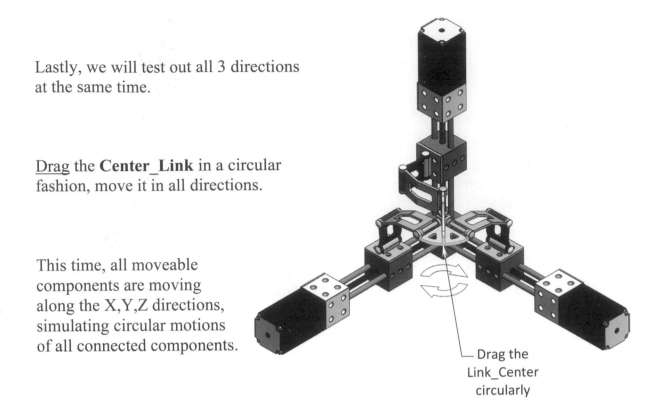

Drag the Link_Center circularly

<u>Note:</u> If you drag a component too far it may collide with others. When this happens, simply drag the square block for that particular direction away from the others; start with the Y direction block.

6. Saving your work:

Select **File, Save As**.

Enter: **Kinematics Assembly_ Completed** for the file name.

Click **Save**.

Close all documents.

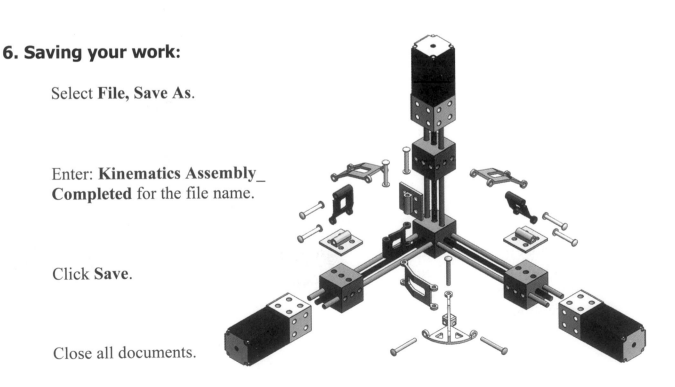

Exercise: Oscillating Mechanism
Bottom Up Assembly

Mates create geometric relationships between assembly components.
As you add mates, you define the allowable directions of linear or rotational motion of the components.

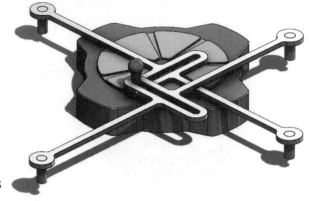

A coincident mate forces two planar faces to become coplanar. The faces can move along one another, but cannot be pulled apart.
A concentric mate forces two cylindrical faces to become concentric. The faces can move along the common axis, but cannot be moved away from this axis.

Mates are solved together as a system. The order in which you add mates does not matter; all mates are solved at the same time. You can suppress mates just as you can suppress features.

1. Starting with an assembly template:

Click **File, New**.

Select an **Assembly** template from the dialog box and click **OK.**

Switch to the **Assembly** tab and click: **Insert Component**.

Browse to the Training Folder and select the part document named: **Round Base.sldprt**.

Click the **Green checkmark** to place this component on the **Origin**.

This is the Parent component.
Other components will be inserted and mated to the parent component.

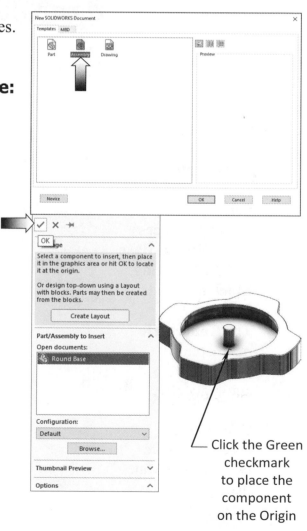

Click the Green checkmark to place the component on the Origin

2. Inserting other components:

Click **Insert Component**.

Insert the **Rotating Block** and 2 instances of the **Link** components.

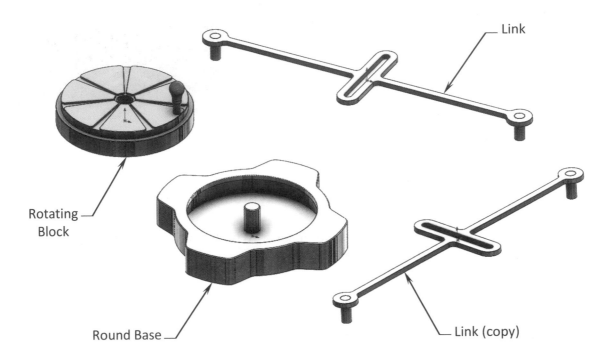

Link

Rotating Block

Round Base

Link (copy)

3. Adding mates:

Click **Mate**.

Select the two cylindrical faces of the **Round Base** and the **Rotating Block** as noted.

The **Concentric** mate option is selected automatically.

Click **OK** to accept the mate but keep the Mate dialog box open.

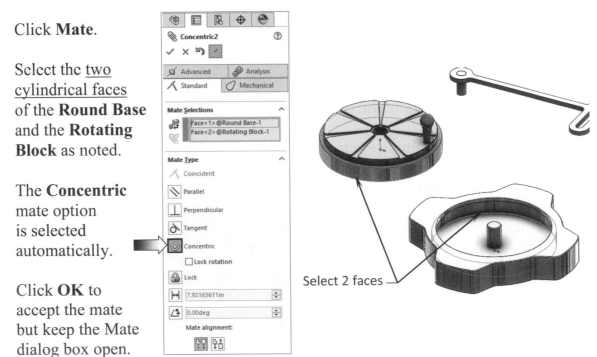

Select 2 faces

The Rotating Block should fit inside the Round Base.

Next, select the <u>inside face</u> of the **Round Base** and the <u>bottom face</u> of the **Rotating Block**.

The **Coincident** mate option is selected.

Click **OK**.

Note: For Slot and Cam mates, only one face of the Slot or the Cam feature needs to be selected.

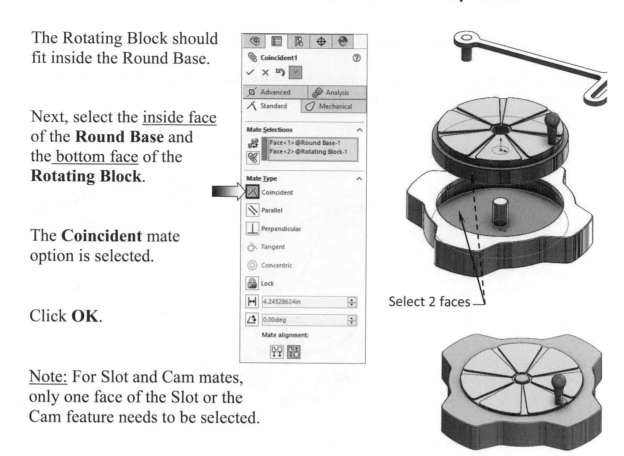

Select 2 faces

Switch to the **Mechanical** tab (arrow) and click the **Slot** mate button.

Select the <u>cylindrical face</u> of the **red knob** and <u>1 face</u> of the **slot** as noted below.

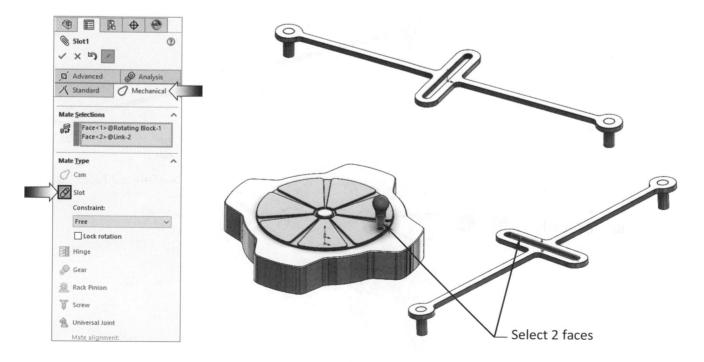

Select 2 faces

Switch back to the **Standard** mates tab (arrow).

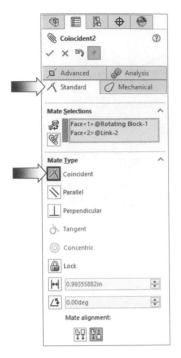

Select the <u>upper face</u> of the **Rotating Block** and the <u>bottom face</u> of the **Link**.

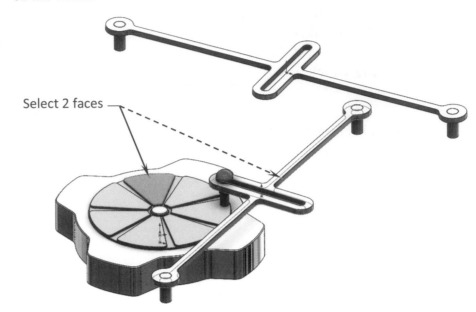

Select 2 faces

A **Coincident** mate is added to the two planar faces.

Switch to the **Mechanical** mates tab (arrow).

Click the **Slot** mate button and select the <u>cylindrical face</u> of the **red knob** and the <u>inside face</u> of the slot in the 2nd **Link** as indicated below.

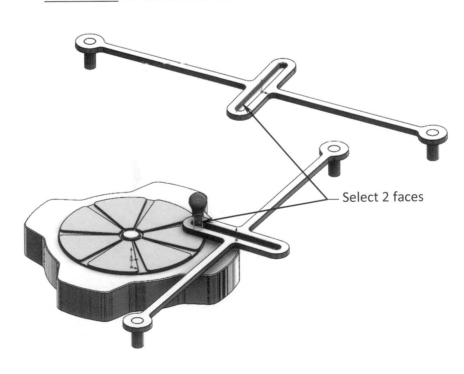

Select 2 faces

Switch back to the **Standard** mates tab.

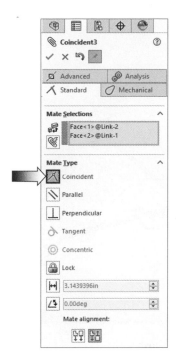

Select the <u>upper face</u> of the **1ˢᵗ Link** and the <u>bottom face</u> of the **2ⁿᵈ Link**.

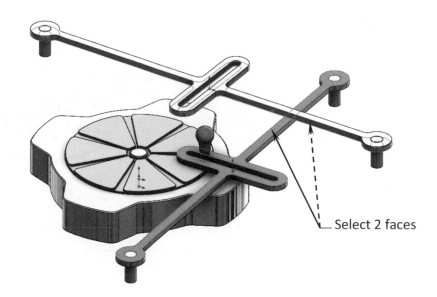

Select 2 faces

A **Coincident** mate is added to the two selected faces.

For the Parallel mate below, expand the FeatureManager tree and select the **Front** plane and the <u>front face</u> of the **2ⁿᵈ Link**.

Click the **Parallel** mate button.
Your assembly should look similar to the image shown below.

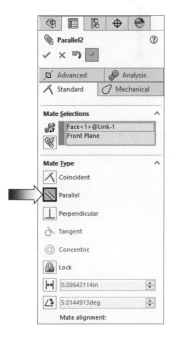

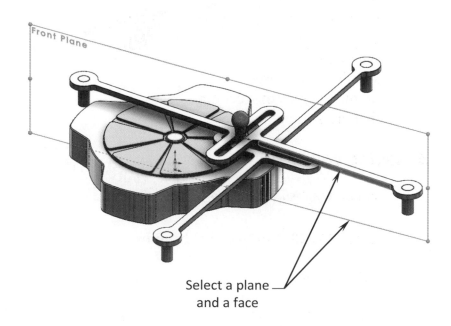

Select a plane
and a face

For the 2nd Parallel mate, select the **Right** plane from the FeatureManager tree and the side face of the **1st Link**.

Click the **Parallel** mate button.

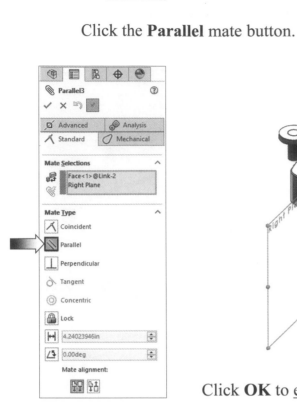

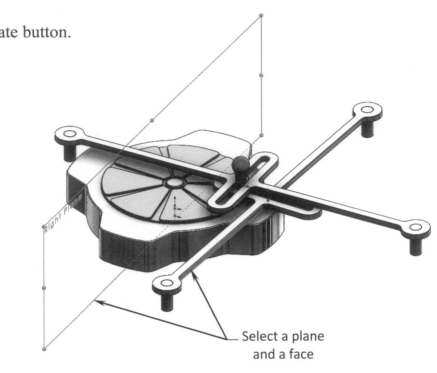

Select a plane and a face

Click **OK** to exit the Mate mode.

4. Testing the motions:

Drag the knob

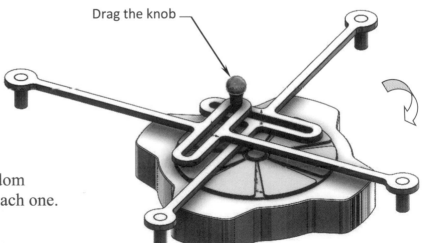

Drag the red knob in a circular fashion to test out the motions.

All components should move based on the degrees of freedom that are still open for each one.

5. Saving your work:

Select **File, Save As**.

Enter: **Oscillating Assembly.sldasm** for the file name.

Click **Save**.

SOLIDWORKS Animator – The Basics

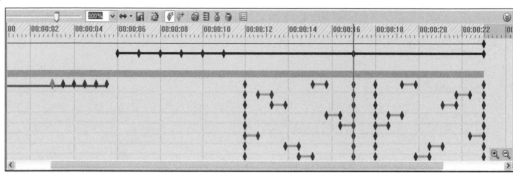

You can use animation motion studies to simulate the motion of assemblies.

The following techniques can be used to create animated motion studies:

* Create a basic animation by dragging the timeline and moving components.
* Use the Animation Wizard to create animations or to add rotation, explodes, or collapses to existing motion studies.
* Create camera-based animations and use motors or other simulation elements to drive the motion.

This exercise discusses the use of all techniques mentioned above.

1. Opening an existing Assembly document:

Open a file named: **Egg Beater.sldasm** from the Training Files folder.

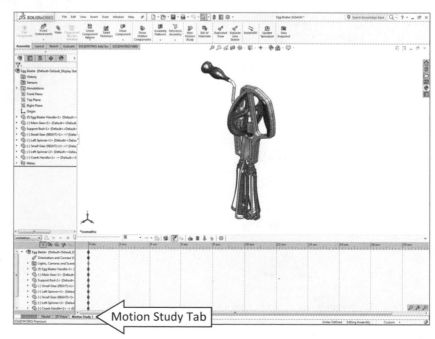

Click the **Motion Study** tab to switch to the SOLIDWORKS-Animation program (Motion Study1).

2. Adding a Rotary Motor:

Click the **Motor** icon from the **Motion Manager** toolbar.

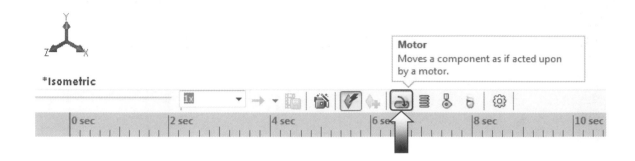

Under **Motion Type**, click the **Rotary Motor** option (arrow).

Select the **Circular edge** of the Main Gear as indicated for **Direction**.

Click **Reverse direction** (arrow).

Under **Motion**, select **Constant Speed**.

Set the speed to **30 RPM** (arrow).

Click **OK**.

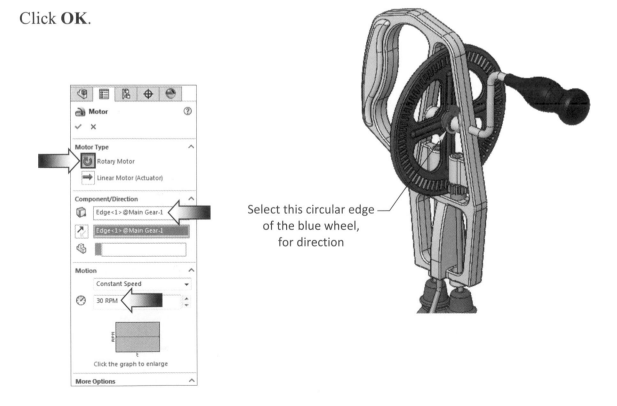

Select this circular edge
of the blue wheel,
for direction

3. Viewing the rotary motions:

Click the **Calculate** button on the **Motion Manager** (arrow).

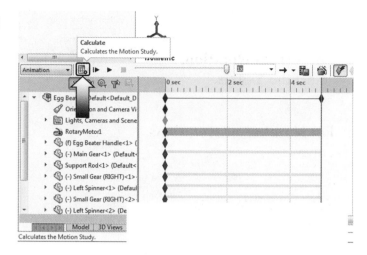

The motor plays back the animation and by default stops at the 5-second timeline.

4. Using the Animation Wizard:

Click the **Animation Wizard** button on the **Motion Manager** toolbar.

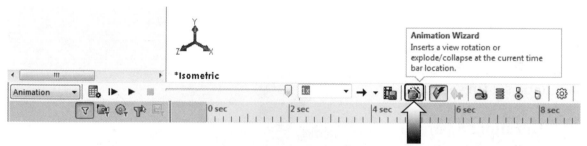

Select the **Rotate Model** Option from the **Animate Type** dialog.

As noted in this dialog box, in order to animate the Explode or Collapse of an assembly, an exploded view must be created prior to making the animation.

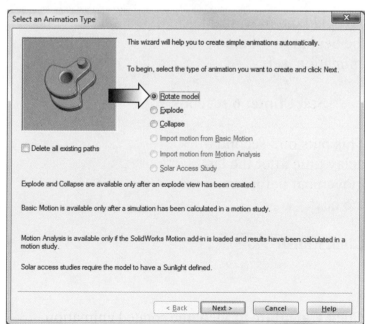

Click **Next** [Next >].

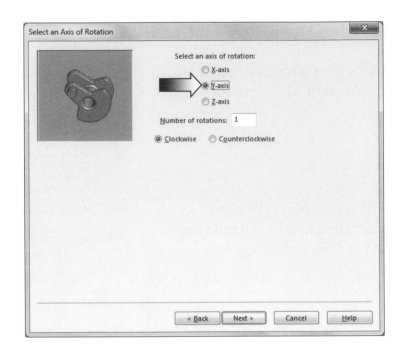

For **Axis of Rotation**, select the **Y-axis**.

For **Number of Rotations**, enter **1**.

Select the **Clockwise** option.

Click **Next** .

To control the speed of the animation, the duration should be set (in seconds).

  * **Duration: 5** (arrow)

To delay the movement at the beginning of the animation, set:

  * **Start time: 6** seconds

This puts one-second of delay time after the last movement before animating the next.

Click **Finish** .

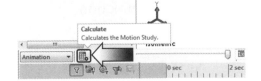

Click **Calculate** to view the rotated animation.

5. Animating the Explode of an Assembly:

Click the **Animation-Wizard** button once again.

Select the **Explode** option (arrow).

Click **Next** Next > .

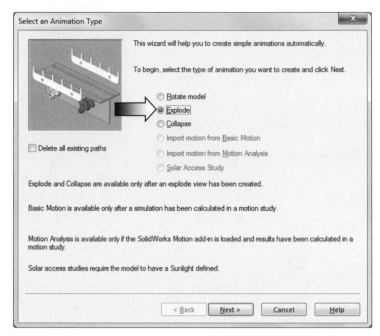

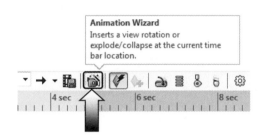

Use the same speed as the last time.

    * Duration: 5 seconds (arrow).

Also add one-second of delay time at the end of the last move.

    * Start Time: 12 seconds (arrow).

Click **Finish** Finish .

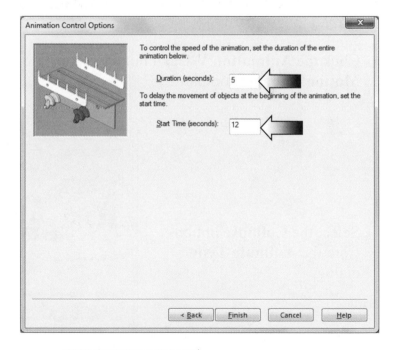

Click **Calculate** to view the new animated motions.

To view the entire animation, click **Play from Start**.

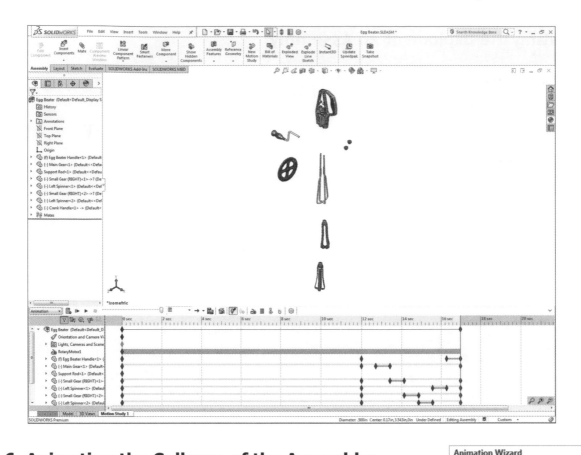

6. Animating the Collapse of the Assembly:

Click the **Animation Wizard** button on the **MotionManager** toolbar.

> **Animation Wizard**
> Inserts a view rotation or explode/collapse at the current time bar location.

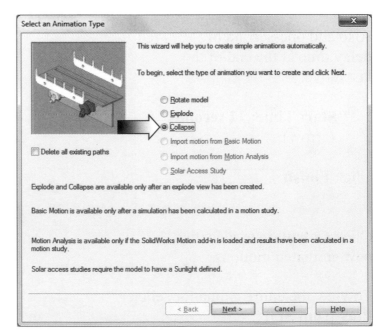

Select the **Collapse** option from the **Animate Type** dialog.

Click **Next** [Next >] .

Set the **Duration** to
5 seconds.

Set the **Start Time** to
18 seconds.

Click **Finish**.

Click **Calculate** to view the
new animated movements.

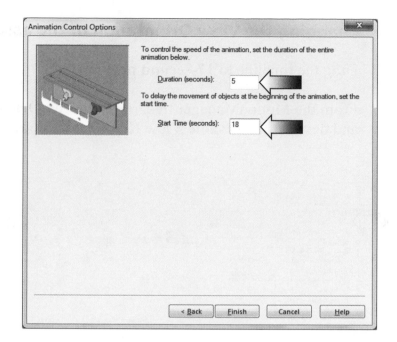

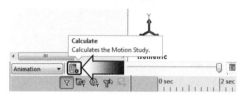

Click the **Play from Start** button to view the
entire animation. Notice the change in the view?
We now need to change the view orientation.

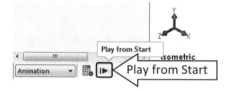

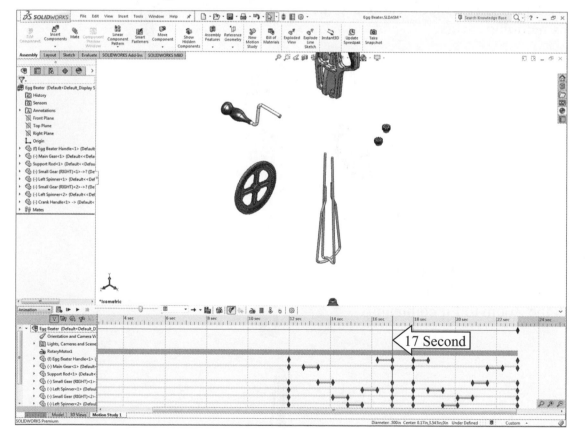

7. Changing the View Orientation of the assembly at 17-second:

Drag the timeline to **17-second position**.

From the MotionManager tree, right-click on **Orientation and Camera Views** and deselect the **Disable View Key Creation** (arrow).

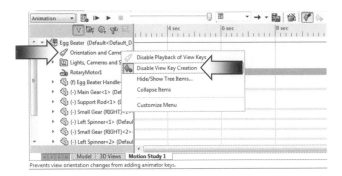
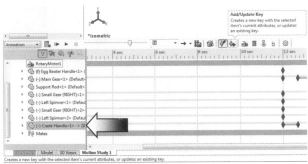

Select the Crank Handle from the Animation tree and click the **Add/Update Key** (arrow).

This key will allow modifications to the steps recorded earlier.

Press the **F** key on the keyboard to change to the full screen – OR – click the Zoom to Fit command.

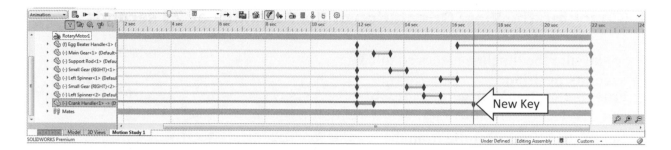

The change in the view orientation from a zoomed-in position to a full-screen has just been recorded. To update the animation, press **Calculate.**

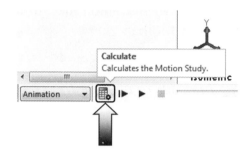

The animation plays back showing the new Zoomed position.

<u>Note:</u> *This last step demonstrated that a key point can be inserted at any time during the creation of the animation to modify when the view change should occur. Other attributes like colors, lights, cameras, shading, etc., can also be defined the same way.*

When the animation reaches the 23 second timeline, the assembly is collapsed. We'll need to add another key and change the view orientation to full screen once again.

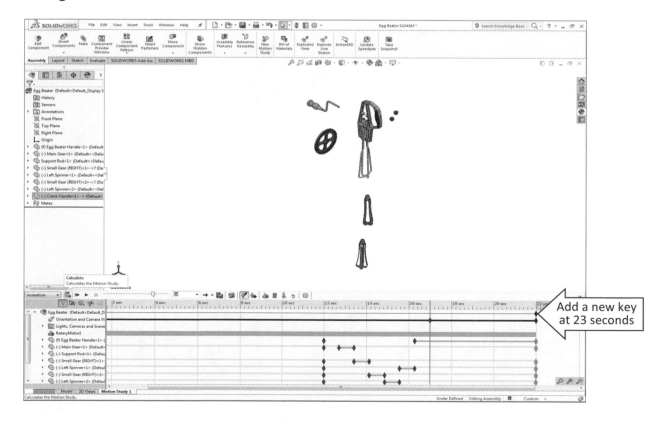

8. Changing the View Orientation of the Assembly at 23 seconds:

Make sure the timeline is moved to **23 seconds position**.

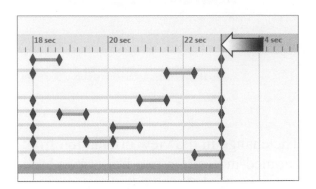

Select the **Crank Handle** from the Animation tree and click the **Add / Update Key** button on the Motion Manager toolbar.

Note: _In order to capture the changes in different positions, a new key should be added each time. We will need to go back to the full screen so that more details can be viewed._

Press the **F** key on the keyboard to switch to the full screen or click Zoom to Fit.

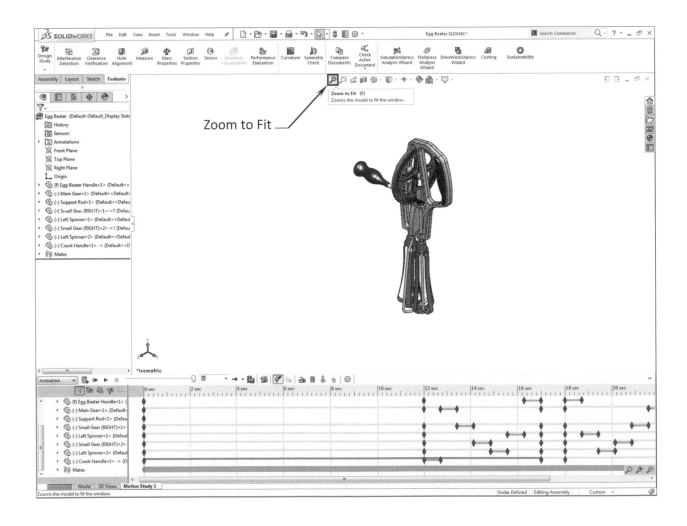

Similar to the previous step, the change in the view orientation has just been captured.

To update the animation, press **Calculate.**

The system plays back the animation showing the new zoom to fit position.

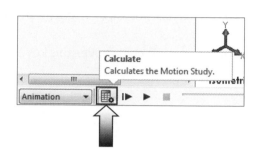

Save your work before going forward to the next steps.

9. Creating the Flashing effects:

Move the timeline to **3-second position**.

Select the **Crank Handle** from the Animation tree and click the **Add/Update Key** button.

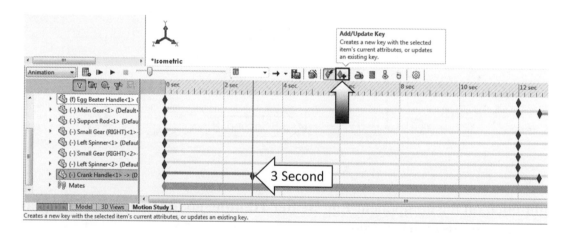

We are going to make the Handle flash 3 times.

Right-click the **Egg-Beater-Handle** from the Animation tree, go to **Component Display** and select **Wireframe**.

Click **Calculate** .

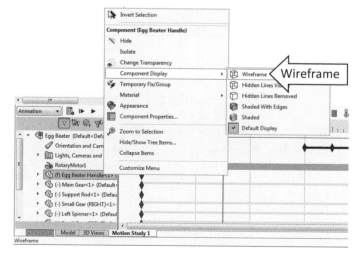

The Handle changes to Wireframe when the animation reaches the 3 second timeline and stays that way until the end.

Hover the cursor over the key point to display its key properties.

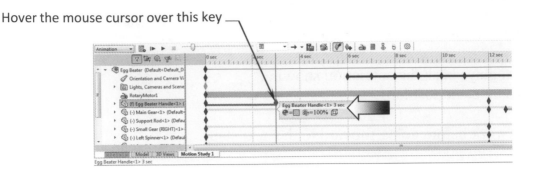

Next, move the timeline to the **3.5 second position**. Select the **Crank Handle** and click the **Add / Update Key** button.

This time we are going to change the Handle back to the shaded mode.

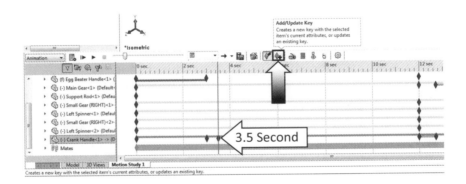

Right-click on the component **Egg Beater Handle**, go to **Component Display**, and select **Shaded with Edges**.

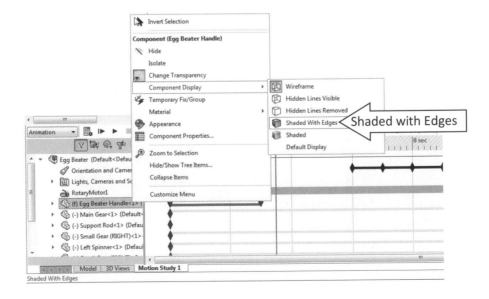

Click **Calculate** 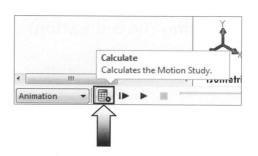 .

Since the change only happens within ½ of a second, the handle looks like it was flashing.

Repeat the same step a few times, to make the flashing effect look more realistic.

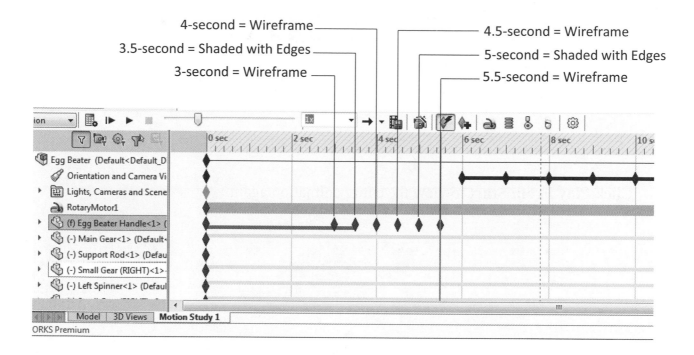

4-second = Wireframe

3.5-second = Shaded with Edges

3-second = Wireframe

4.5-second = Wireframe

5-second = Shaded with Edges

5.5-second = Wireframe

Use the time chart listed above and repeat the step number 9 at least 3 more times.

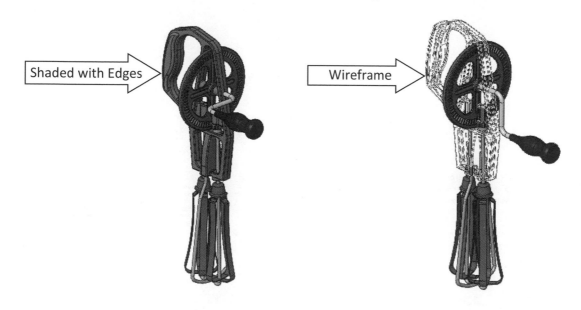

Shaded with Edges

Wireframe

10. Looping the animation:

Click the Playback Mode arrow and select **Playback Mode: Reciprocate** (arrow).

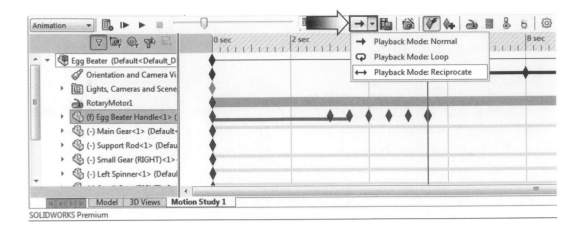

Click **Play From Start** to view the entire animation again.

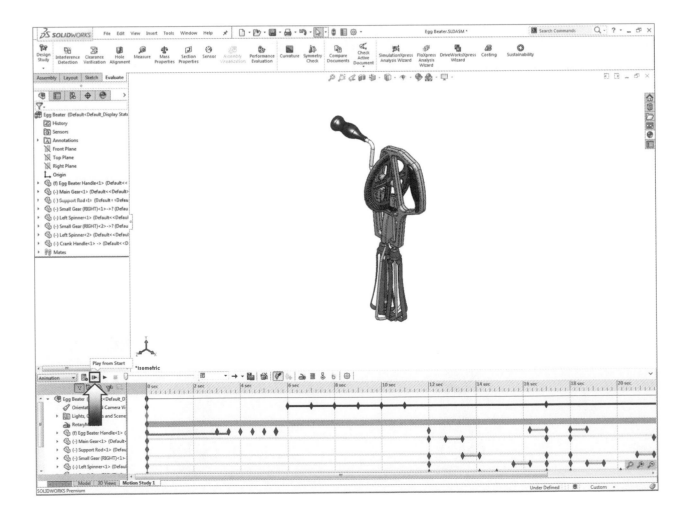

11. Saving the animation as AVI: (Audio Visual Interleaving)

Click the **Save Animation** icon on the MotionManager toolbar.

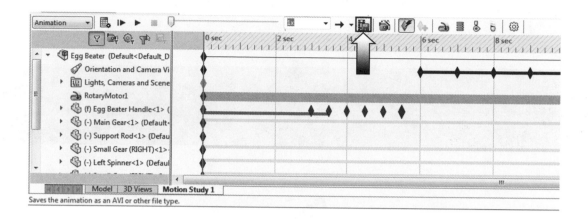

Saves the animation as an AVI or other file type.

Use the default name
(Egg Beater) and other
default settings.

Click **Save**.

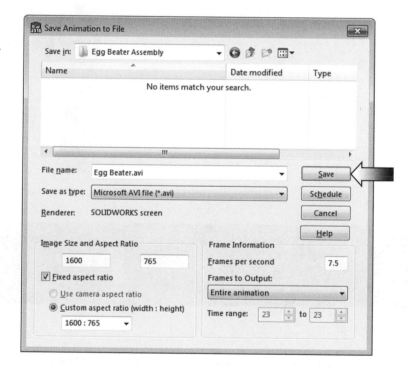

The **Image Size** and
Aspect Ratio (grayed-
out) adjusts size and
shape of the display.
It becomes available
when the renderer is
PhotoWorks buffer.

Compression ratios
impact image quality.
Use lower compression ratios
to produce smaller file sizes of lesser image quality. Use the **Full Frames** compressor.

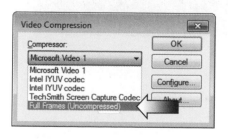

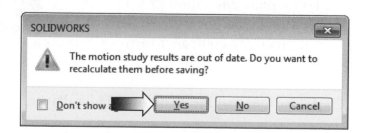

12. Viewing the Egg Beater AVI with Windows Media Player:

Exit the SOLIDWORKS program, locate and launch the **Windows Media Player**.

Open the **Egg Beater.AVI** from Windows Media Player.

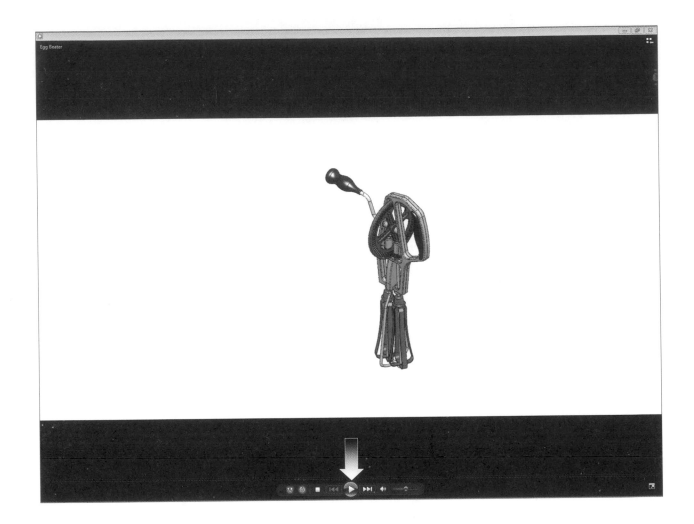

Click the **Play** button to view the animated AVI file.

To loop the animation in Windows Media Player, click the **Repeat** button on the controller (arrow).

Close and **exit** the Window Media Player.

CHAPTER 13

PhotoView 360 Basics

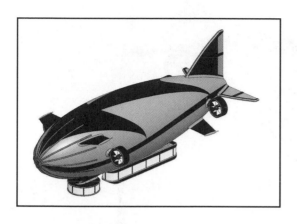

PhotoView 360 allows the user to create photo-realistic renderings of the SOLIDWORKS models. The rendered image incorporates the appearances, lighting, scene, and decals included with the model. PhotoView 360 is available with SOLIDWORKS Professional or SOLIDWORKS Premium only.

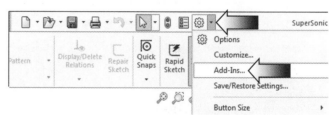

1. Activating PhotoView 360:

Open the document named **Blimp for Rendering.sldprt** from the Training Files folder.

Click **Tools / Add-Ins**…

Select the **PhotoView 360** checkbox.

Click **OK**.

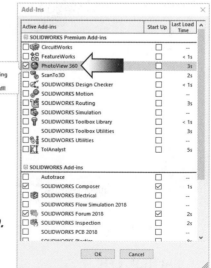

NOTE: _Enable the checkbox on the right of the PhotoView 360 add-ins if you wish to have it available at startup. Otherwise, only activate it for each use._

2. Setting the Appearance: (right-click the Sketch tab and select **Tabs, Render Tools**).

Click the **Edit Appearance** button from the **Render** tool tab (arrow).

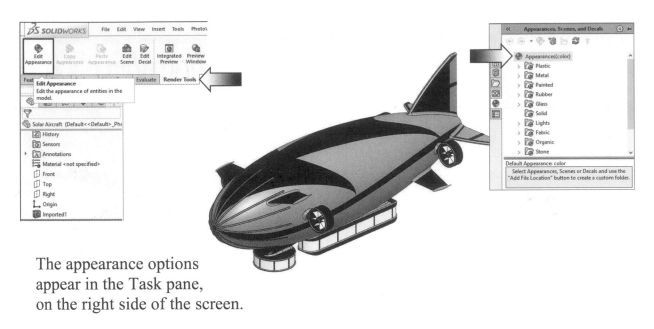

The appearance options
appear in the Task pane,
on the right side of the screen.

Expand the **Appearances** folder, the **Painted** folder, the **Car** folder and double-click
the **Metallic Cool Grey** appearance to apply it to the model (arrows).

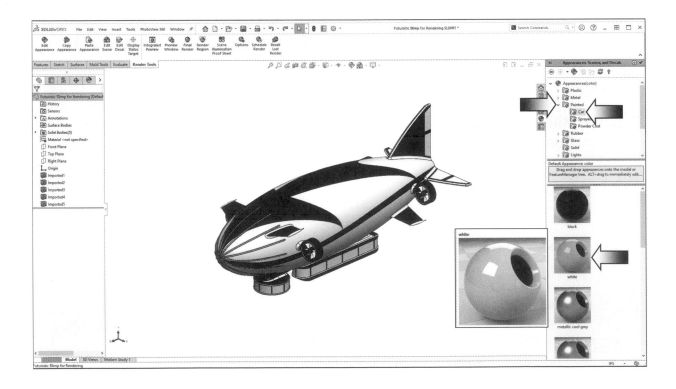

Click **OK.**

3. Setting the Scene:

Click the **Edit Scene** button from the Render tool tab (arrow).

A scene is a
combination
of lighting,
background,
& foreground.
It is a 2D
image
between the
model and
the scene
environment.

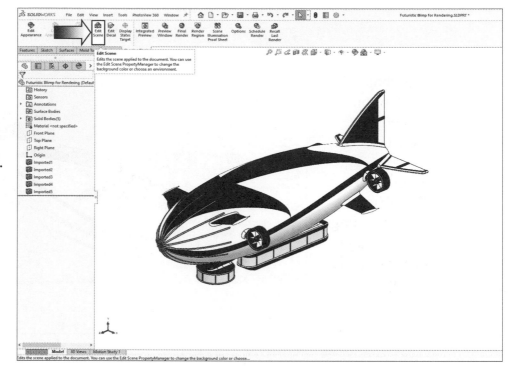

Expand the **Scenes** folder, the **Basic Scenes** folder (arrows) and <u>double-click</u> the
Backdrop - Studio Room scene to apply it. Also set the <u>other settings</u> as indicated.

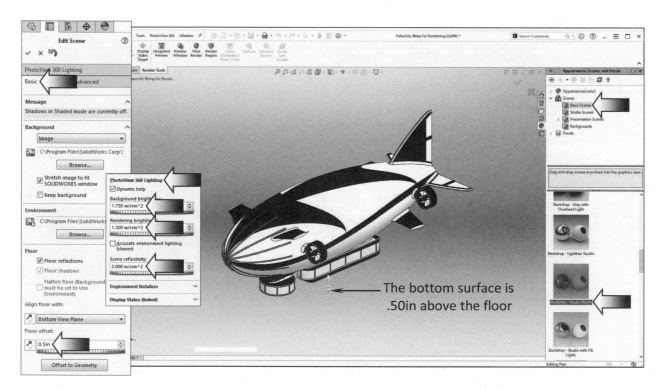

The bottom surface is
.50in above the floor

4. Setting the Image Quality options:

Click the **Options** button from the Render tool tab (arrow).

The PhotoView 360 Options Property-Manager controls settings for PhotoView 360, including output image size and render quality.

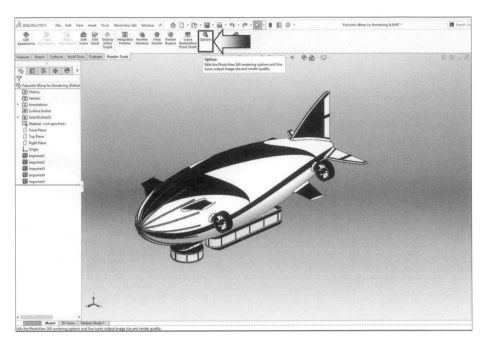

Set the **Output Image** to **1280x1024** from the drop-down list and set the other options as indicated with the arrows. Click **OK** when done.

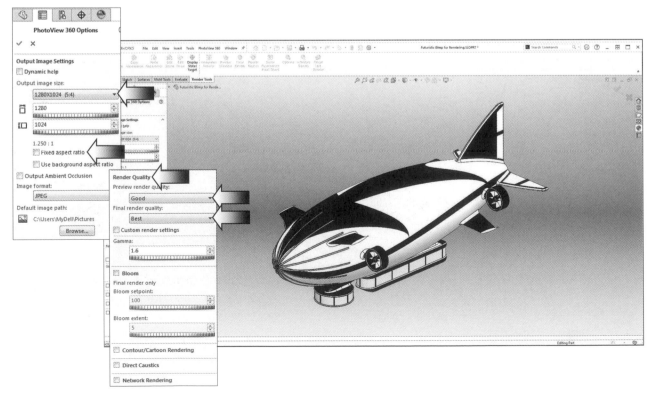

5. Rendering the image:

Click the **Final Render** button from the Render tool tab (arrow).

Select the
**"Turn On
Perspective
View"**
option in the
dialog box.
A more
realistic
rendering
is produced
with
perspective
view
enabled.

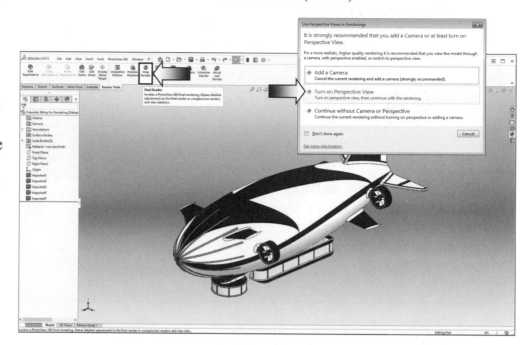

6. Saving the image:

After the
rendering is
completed,
select the
option zoom
50% (or 100%)
from the
drop-down
list (arrow).

Click **Save-
Image**
(arrows).

Select the **JPEG** format from the Save-as-Type drop-down list.

Enter a file name and click **OK**.

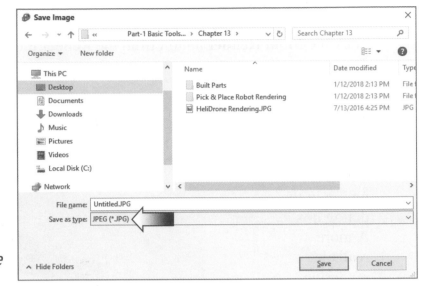

**NOTE:** Different file formats may reduce the quality of the image and at the same time may increase or decrease the size of the file.

Close all documents when finished.

Exercise: HeliDrone Assembly

PhotoView 360 is a SOLIDWORKS add-in that produces photo-realistic renderings of SOLIDWORKS models.

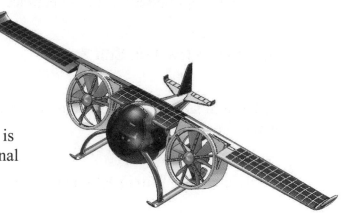

The rendered image incorporates the appearances, lighting, scene, and decals included with the model. PhotoView 360 is available with SOLIDWORKS Professional or SOLIDWORKS Premium.

1. Opening an assembly document:

Open an assembly document named **Helidrone.sldasm**.

Enable the **PhotoView 360** (Tools, Add-Ins) and click the **Appearances** tab.

Expand the **Metal** folder. Click the **Platinum** folder and drag/drop the **Polished Platinum** appearance to the drone's body (arrows).

Select the **Apply to Imported5** option (arrow).

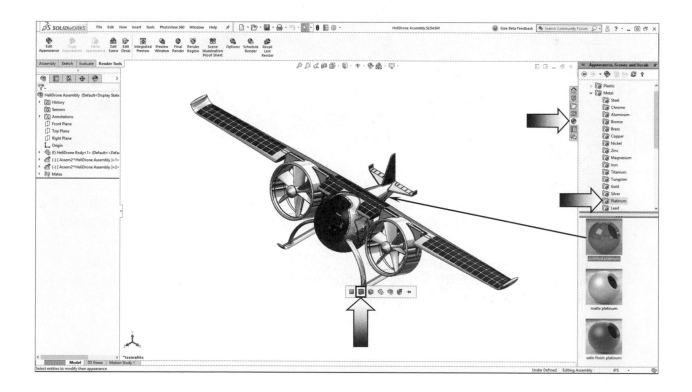

2. Applying the Scene:

The scene is applied after the appearance is set.

Click **PhotoView 360, Edit Scene**.

On the right side of the screen, in the Task Pane, expand the **Scene** and the **Basic Scene** folders.

Double-click the **Warm Kitchen** scene to apply its settings to the assembly.

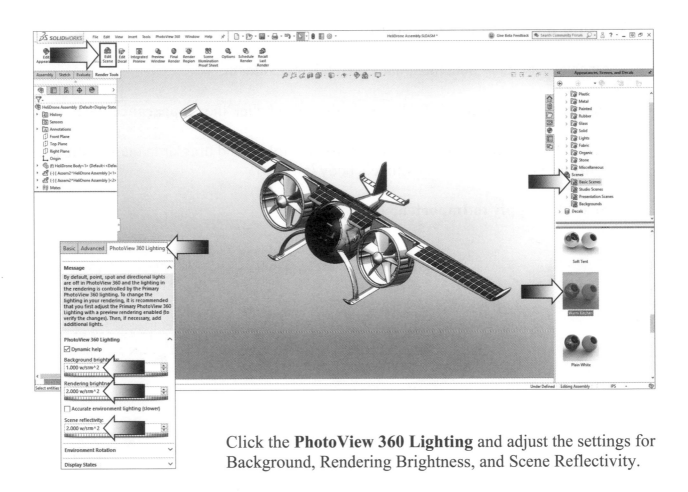

Click the **PhotoView 360 Lighting** and adjust the settings for Background, Rendering Brightness, and Scene Reflectivity.

Use the options in the **Edit Scene** sections, on the left side of the screen, to modify the lightings, floor reflections, and environment rotation.

Click **OK** to exit the Scenes settings.

3. Setting the Render Region:

The render region provides an accelerator that lets you render a subsection of the current scene without having to zoom in or out or change the window size.

Select the **Render Region** option from the PhotoView 360 pull down menu.

Drag the handle point at one of the corners of the rectangle to define the region to render.

4. Setting the Image Size:

Select **Options** under the PhotoView 360 pull down.

The height and width of the render image can be selected from some of the pre-set image sizes.

Select the **1920x1080** (16:9) from the Image Size pull down options. (The larger the image size the more time required.)

Select **Final Render** from the PhotoView 360 and select the **Turn on Perspective View** option, when prompted.

Click **Save Image** and select the **JPEG** format to save the rendering.

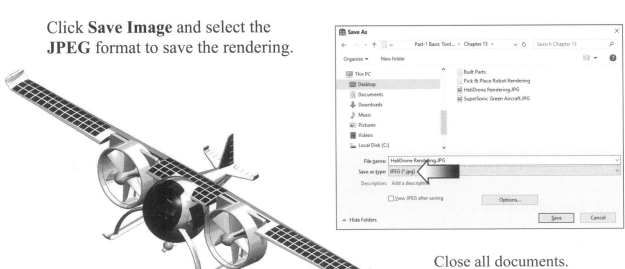

Close all documents.

Rendering the Screen with Ambient Occlusion

Ambient occlusion is a global lighting method that adds realism to models by controlling the attenuation of ambient light due to occluded areas.

When Ambient Occlusion is enabled, objects appear as they would on a hazy day. Ambient occlusion is available in all scenes when you use RealView graphics.

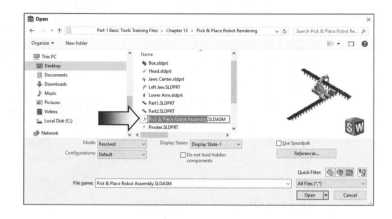

This exercise discusses the use of the Ambient Occlusion and other options such as lighting and background settings that can help enhance the quality of the rendering.

1. Opening an assembly document:

Select **File, Open**.

Browse to the Training Files folder and open an assembly document named:
Pick & Place Robot Assembly.sldasm

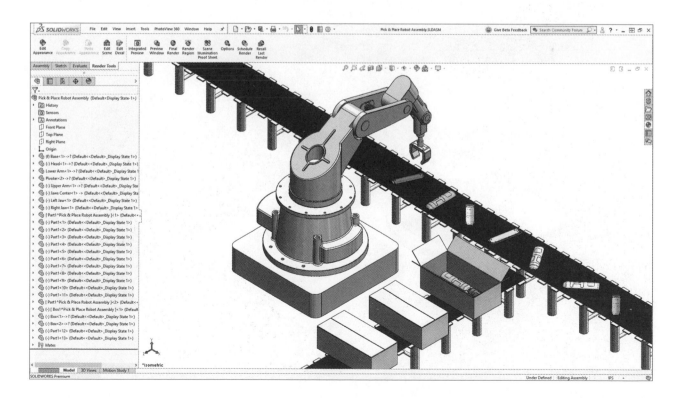

2. Changing the scene:

Scenes provide a visual backdrop behind a model. In SOLIDWORKS, they provide reflections on the model.

With PhotoView360 added in, scenes provide a realistic light source, including illumination and reflections, requiring less manipulation of lighting. The objects and lights in a scene can form reflections on the model and can cast shadows on the floor.

Expand the **Apply Scene** section and double click the **Factory** scene (arrow).

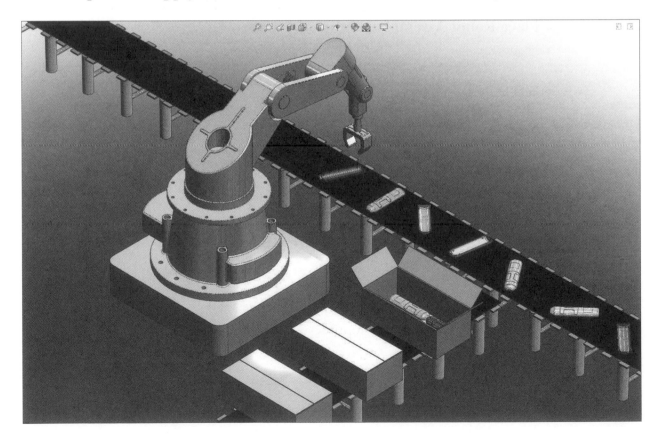

3. Retrieving a Named View:

Custom view orientations can be
created as a Named View and
saved within a SOLIDWORKS
document.

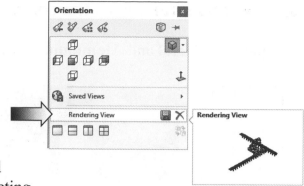

The Named Views can be retrieved
by pressing the Space Bar and selecting
them under the Saved Views list.

Press the **Space Bar** and select the view named **Rendering View** (arrow).

Expand the **View Settings** section and enable the
RealView Graphics option
(if applicable).

RealView Graphics is only available with
graphics cards that support RealView Graphics
display.

Also enable the **Perspective** option (arrow).

When you create a PhotoView 360 rendering, for the most realistic results, you
should add a camera with perspective or at least turn on
perspective view. Except when you explicitly want
to render using an orthographic view, viewing a
model from a perspective view is more realistic.

A perspective view allows PhotoView 360 to
more accurately calculate reflections, shadows,
and lighting. In addition, a view through a
camera with perspective gives you the highest
possible level of control for your viewing
position, field of view, and focal length.

Click the **Appearances, Scenes, and Decals** tab and
expand the **Appearances** folder (arrow).

4. Applying appearances to the components:

Expand the **Metal** and **Aluminum** folders.

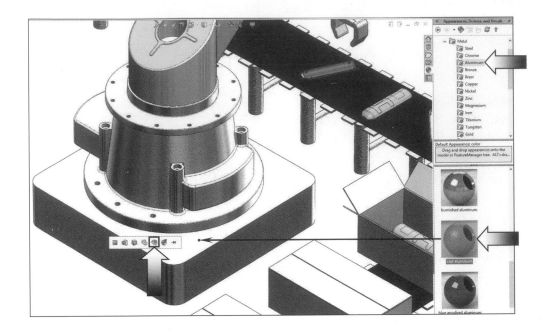

Drag and drop the **Cast Aluminum** appearance onto the **Base** and select the option: **Component Level** (arrow).

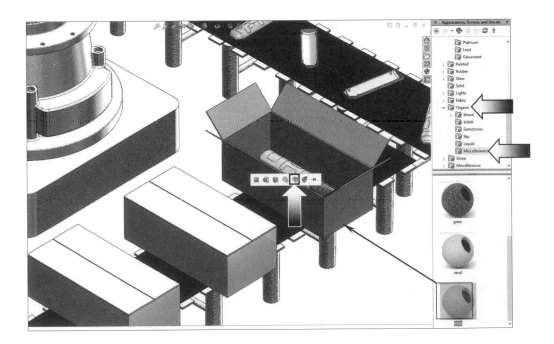

Expand the **Organic, Miscellaneous** folders. **Drag and drop** the **Skin** appearance to the **Box** and select the option: **Component Level** (arrow).

Continue with applying the appearances to the other components as indicated below.

(The Factory scene is temporarily removed for clarity only.)

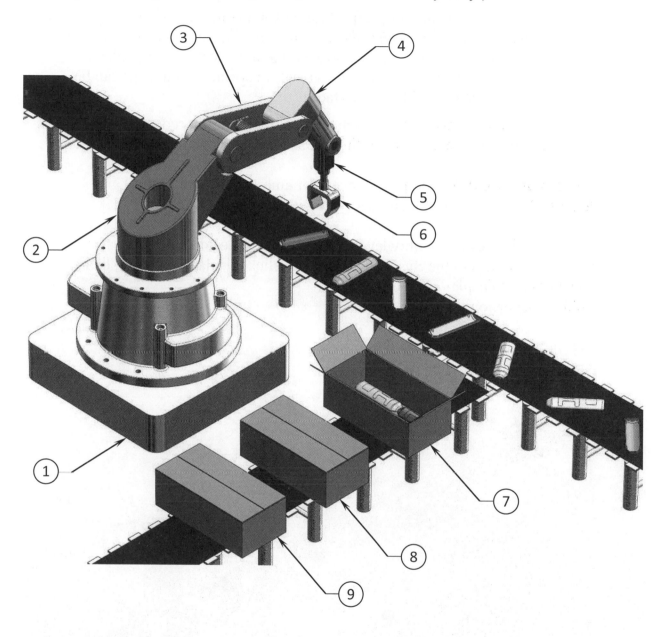

1. Metal / Aluminum / Cast Aluminum

2. Painted / Car / Sienna

3. Metal / Steel / Cast Stainless Steel

4. Metal / Titanium / Brushed Titanium

5. Metal Nickel / Polished Nickel

6. Metal / Chrome / Chromium Plate

7-8-9: Organic / Miscellaneous / Skin

The effects of ambient occlusion depend on other factors such as the nature of the model itself, the appearances you apply to the model, the scene, and the lighting. You might need to adjust some of these other variables to get the desired result.

    * Ambient occlusion often looks best on solid color or non-reflective surfaces. Highly reflective surfaces, in contrast, can diminish the shadows in occluded areas. However, the methods used to render ambient occlusion can sometimes result in artifacts, and these can be more visible on solid color surfaces.

    * Too many lights, or lighting that is too bright, might reduce the impact of ambient occlusion by hiding the shadows it creates.

You can use two different quality levels for ambient occlusion: draft and default. Draft displays models faster but with less visual fidelity.

To change the display quality level for Ambient Occlusion, click: Tools > Options. On the System Options tab, click Display/Selection, then select or clear Display draft quality ambient occlusion.

Expand the View Settings and select: **Ambient Occlusion** (arrow).

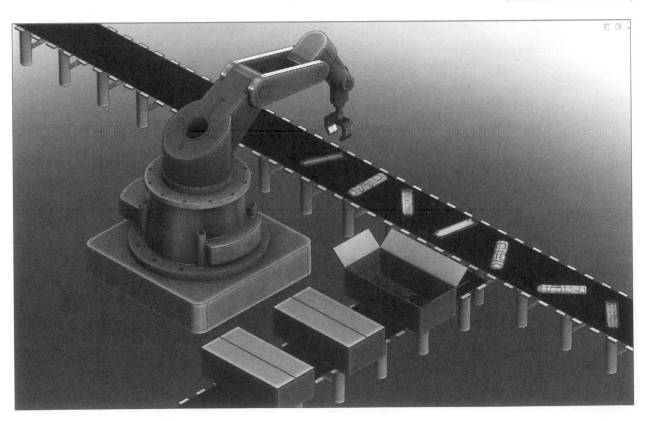

5. Applying appearances to the components:

You can adjust the direction, intensity, and color of light in the shaded view of a model. You can add light sources of various types, and modify their characteristics to illuminate the model as needed.

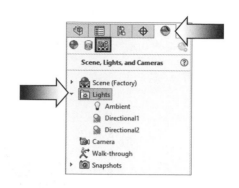

Switch to the **DisplayManager** tab and expand the **Lights** folder (arrow).

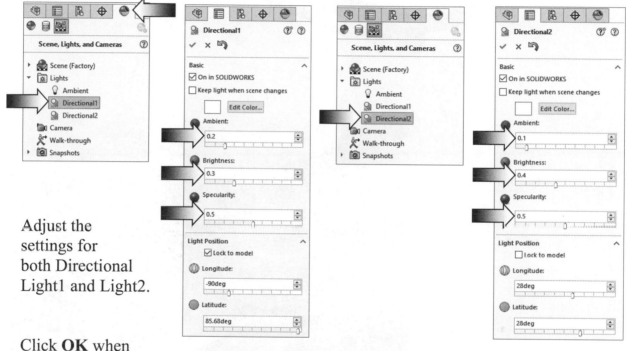

Adjust the settings for both Directional Light1 and Light2.

Click **OK** when finished.

6. Saving your work:

Select **File, Save As**

For the file name, enter **Pick & Place Robot Assembly (Completed)**.

Press **Save**.

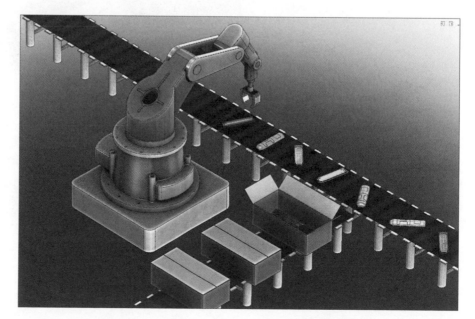

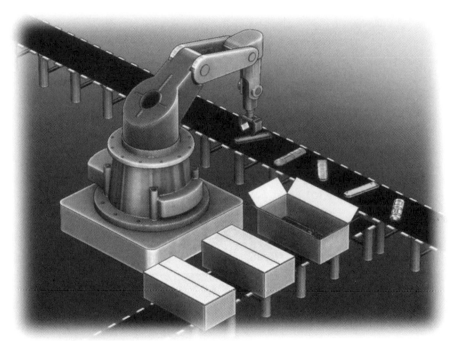

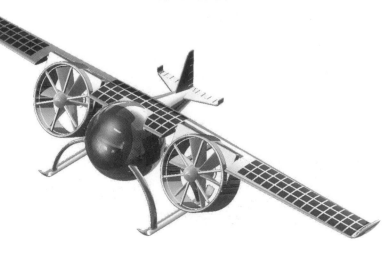

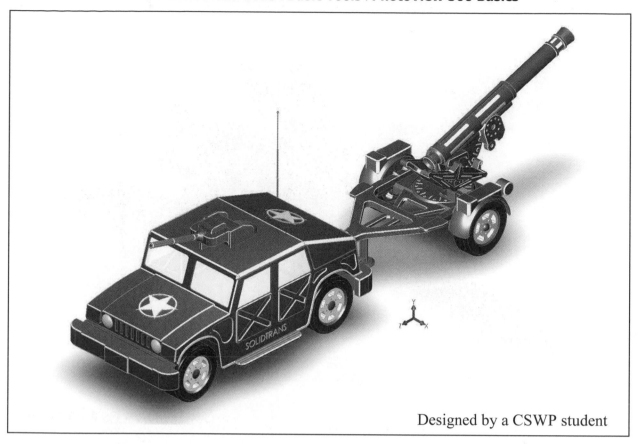

Designed by a CSWP student

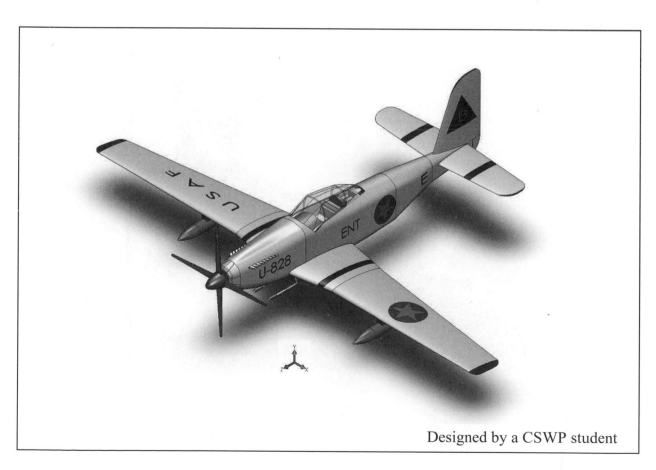

Designed by a CSWP student

Designed by a CSWP student

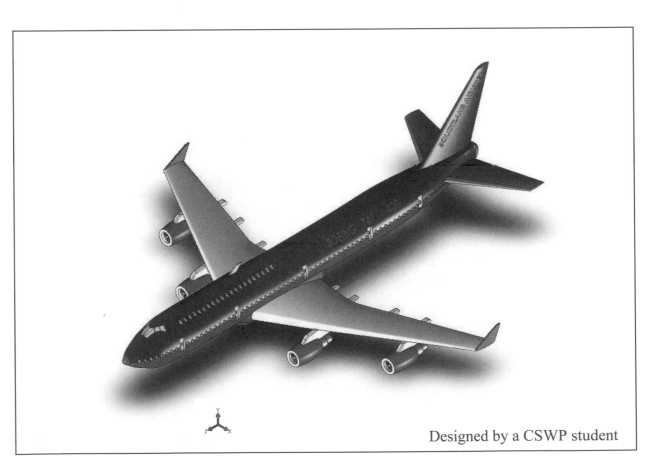

Designed by a CSWP student

CHAPTER 14

Drawing Preparations

Drawing Preparations
Customizing the Document Template

Custom settings and parameters such as ANSI standards, units, number of decimal places, dimensions and note fonts, arrow styles and sizes, line styles and line weights, image quality, etc., can be set and saved in the document template for use with the current drawing or at any time in the future.

All Document Templates are usually stored either in the templates folder or in the Tutorial folder:

* (C:\Program Data\SolidWorks\SolidWorks2023\Templates)
* (C:\Program Files\SolidWorks Corp\SolidWorks\Lang\English\Tutorial)

By default, there are 2 "layers" in every new drawing. The "top layer" is called the **Sheet** layer, and the "bottom layer" is the **Sheet Format**.

The **Sheet** layer is used to create the drawing views and annotations. The **Sheet-Format** layer contains the title block information, revision changes, BOM-anchor, etc.

The 2 layers can be toggled back and forth by using the FeatureManager tree, or by right-clicking anywhere in the drawing and selecting Edit Sheet Format or Edit Sheet.

When the settings and parameters are completed they will get saved in the Document Template with the extension **.drwdot** (Drawing Document Template).

This chapter will guide us through the settings and the preparations needed prior to creating a drawing.

Drawing Preparations
Customizing the Document Template

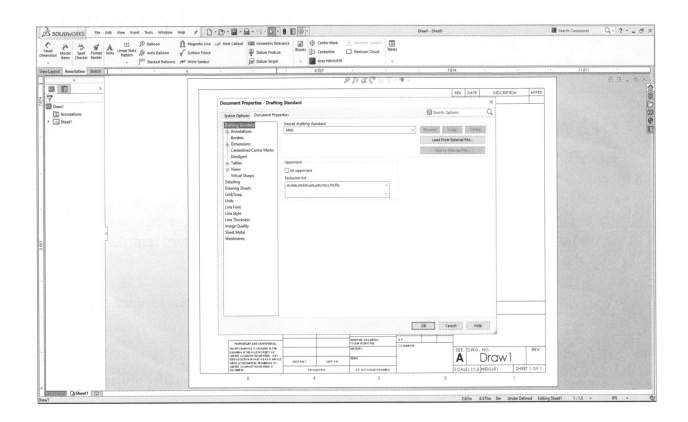

Dimensioning Standards: **ANSI**	Third Angle Projection
Units: **INCHES** – 3 Decimals	

Tools Needed:

New Drawing

Options...

Edit Sheet Format

Edit Sheet

Drawing Templates (*.drwdot)

1. Setting up a new drawing:

Select **File / New / Draw** (or Drawing) / **OK**.

Change the option **Novice** to **Advanced** as indicated below.

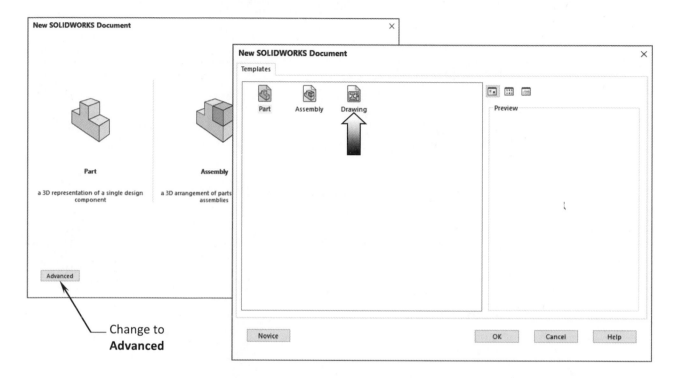

Change to
Advanced

By selecting either **Advanced** or **Novice** (Circled), you can switch to the appropriate dialog box to select the templates.

Set the following options:

    * **Scale: 1:2**

    * **Third Angle Projection**

    * For View Label and Datum
      Label: Change both to **A**.

For Standard Sheet Size, choose:
C-Landscape.

Deselect the **Only Show Standard
Format** checkbox (arrow).

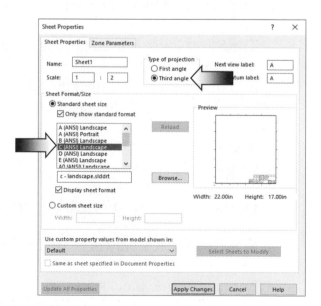

Click **Apply Change**. *(Right-click and select Properties to edit the sheet settings).*

The SOLIDWORKS Drawing User Interface:

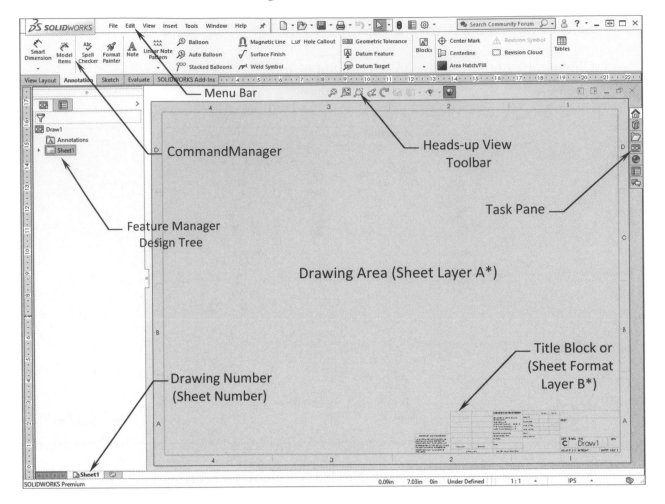

*A. By default, the **Sheet** layer is active and placed on top of the **Sheet Format** layer.

> The **Sheet** layer is used to create drawing views, dimensions, and annotations.

* B. The "bottom layer" is called the **Sheet Format** layer, which is where the revision block, the title block and its information are stored.

> The Sheet Format layer includes some links to the system properties and the custom properties.

> OLE objects (company's logo) such as .Bmp or .Tif can be embedded here.

> SOLIDWORKS drawings can have different Sheet Formats or none.

> Formats or title blocks created from other CAD programs can be opened in SOLIDWORKS using either DXF or DWG file types, and saved as SOLIDWORKS Sheet Format.

2. Switching to the Sheet Format layer:

Right-click inside the drawing and select **Edit Sheet Format**.

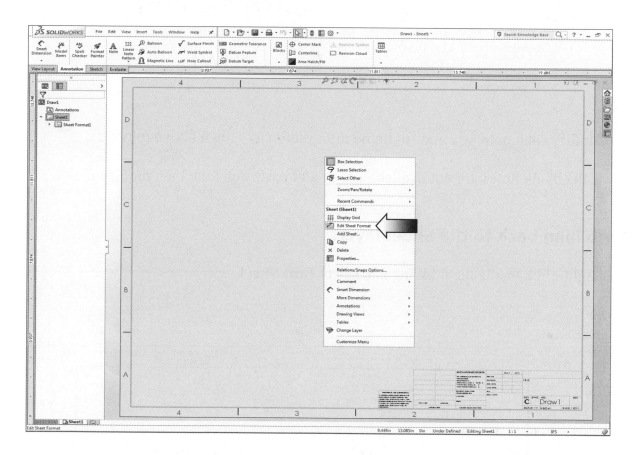

Using the SOLIDWORKS drawing templates, there are "blank notes" already created for each field within the title block.

Double-click the blank note in the Company Name field and enter: **SOLIDWORKS**.

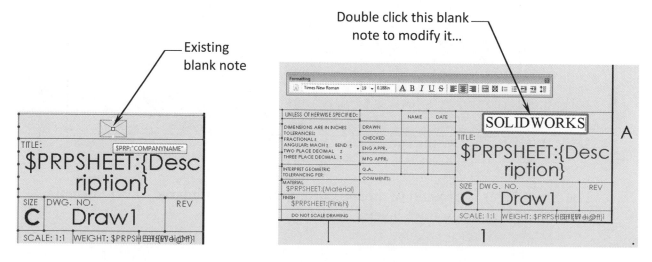

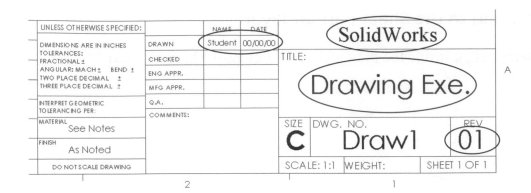

Modify each note box and fill in the information as shown above (circled).

If the blank notes are not available, copy and paste any note then modify it.

3. Switching back to the Sheet layer:

Right-click inside the drawing and select **Edit Sheet**.

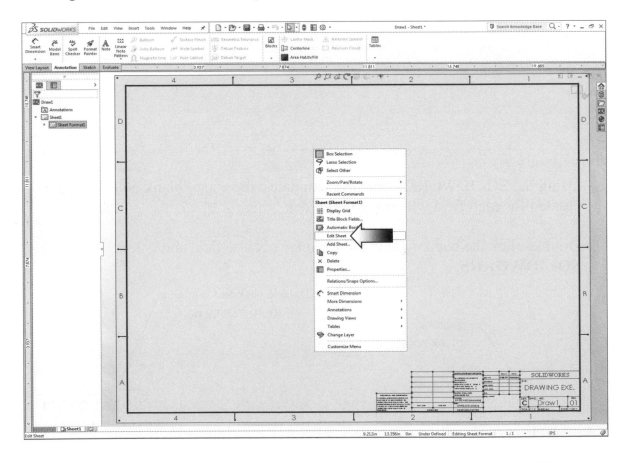

The Sheet layer is brought to the top; all information within the Sheet Format layer is kept on the bottom layer.

4. Setting up the Drawing Options:

Go to **Tools / Options**, or click the Options icon.

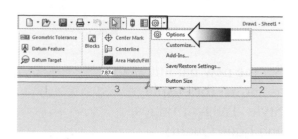

Select the **Drawings** option from the list.

Enable and/or disable the drawing options by clicking on the checkboxes as shown below.

<u>*NOTE:*</u>
For more information about these options, refer to Chapter 2 (Document Templates).

These parameters are examples for use in this textbook only. You may have to modify them to work with your application or your company's standards.

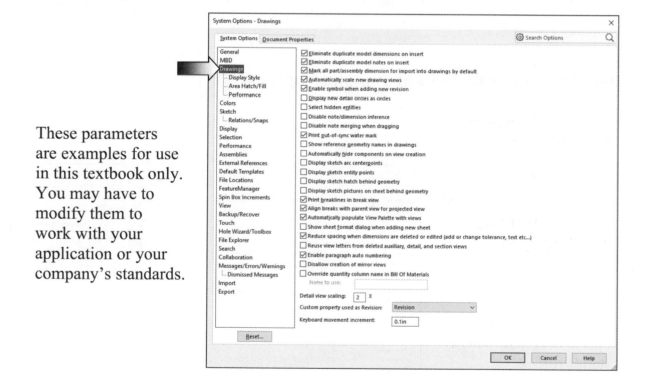

The parameters that you are changing here will be saved as the default system options, and they will affect the current as well as all future documents.

Once again, enable only the checkboxes as shown in the dialog above.

Select the **Display Style** option (arrow).

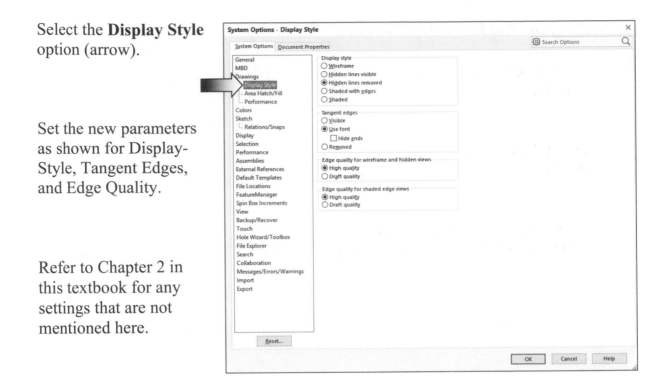

Set the new parameters as shown for Display-Style, Tangent Edges, and Edge Quality.

Refer to Chapter 2 in this textbook for any settings that are not mentioned here.

5. Setting up the Document Template options:

Click the **Document Properties** tab.

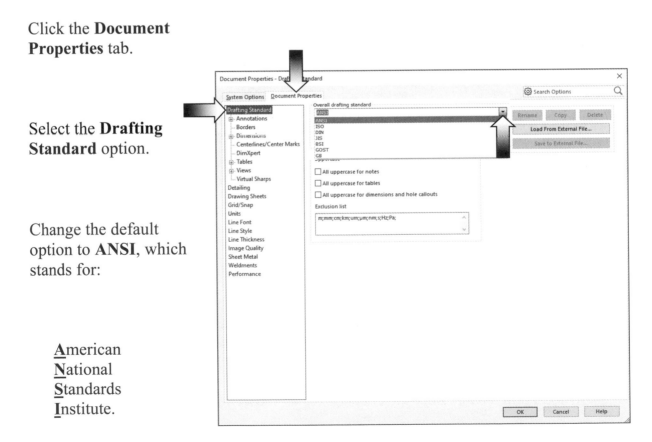

Select the **Drafting Standard** option.

Change the default option to **ANSI**, which stands for:

American
National
Standards
Institute.

Select the **Annotations** option (arrow).

Set the new parameters as shown.

Click the **Font** button and select the following:

> **Century Gothic**
> **Regular**
> **12 points**

Set the Note, Dimension, Surface Finish, Weld-Symbol, Tables, and Balloon to match the settings for Annotations.

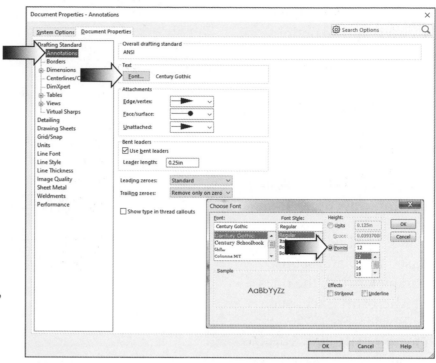

Select the **Dimensions** option (arrow).

Set the parameter in this section to match the ones shown in this dialog box.

If the Dual Dimension Display is selected, be sure to set the number of decimal places for both of your Primary and Secondary dimensions.

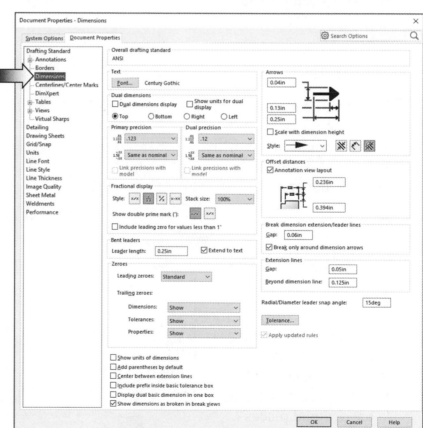

NOTE: *Skip to the Units option below if you need to change your Primary unit from Millimeter to Inches, or vice versa.*

Select the **Centerlines/ Center Marks** option.

Set the options as indicated for Center-line Extension, Center Marks Size and Slot Center Marks.

Select the **Tables / Bill-of Materials** option.

The B.O.M. is an Excel based template; set its parameters like the ones in Microsoft Excel such as Border, Font, Zero Quantity Display, Missing Component, Leading and Trailing Zeros, etc.

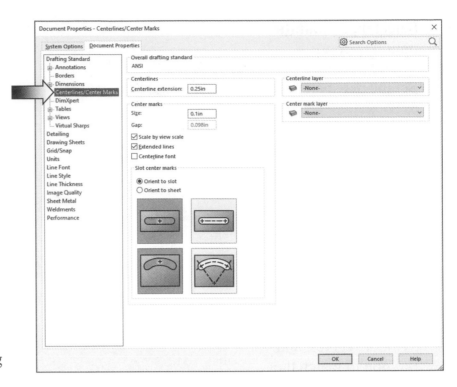

Leading Zeros:

* Standard: Leading zeros appear according to the overall drafting standard.
* <u>Show</u>: Shows zeros before decimal points are shown.
* <u>Remove</u>: Leading zeros do not appear.

Trailing Zeros:

* <u>Smart</u>: Trailing zeros are timed for whole metric values.
* <u>Standard</u>: Trailing zeros appear according to ASME standard.
* <u>Show</u>: Trailing zeros are displayed according to the decimal places specified in Units.
* <u>Remove</u>: Trailing zeros do not appear.

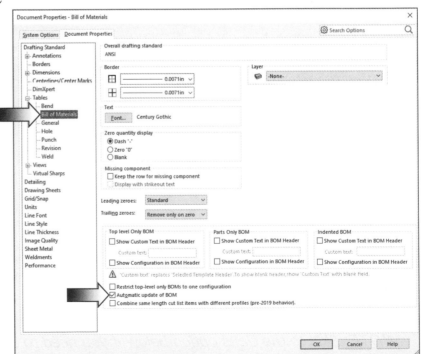

Select the **Tables / Revision** option.

Set the document-level drafting settings for revision table such as Border, Font, Alphabet Numerical Control, Multiple Sheet Style and Layer.

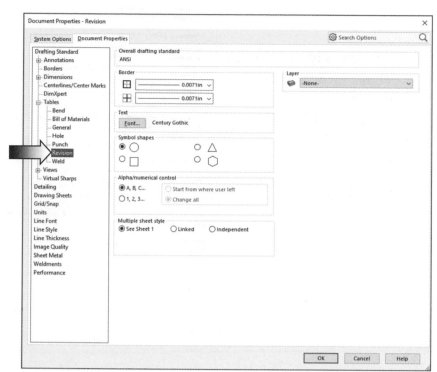

Select the **Views / Detail** option.

Set the options for ANSI View Standard, Circle, Fonts, Label and Border.

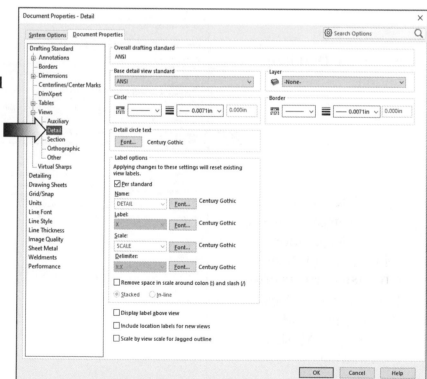

Select the **Views / Section** option.

Set the Line Style, Line Thickness, Fonts, Label, Scale, Layer, and Section Arrow Size.

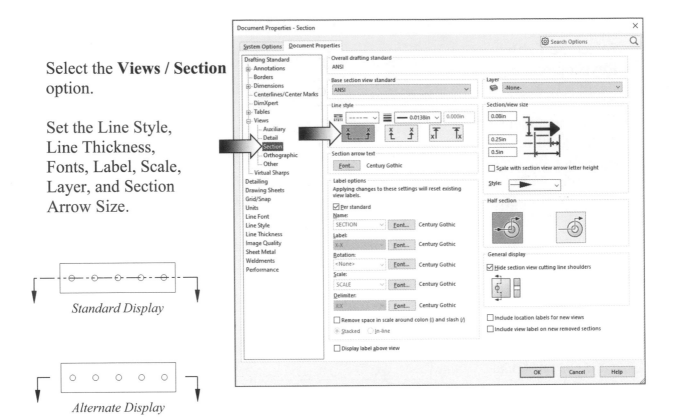

Standard Display

Alternate Display

Select the **Detailing** option.

Select the checkboxes for the Display Filters, Import Annotations, Auto Insert on View Creation, Area Hatch Display, View Break Break Lines, and Dimensions Marked For Drawing.

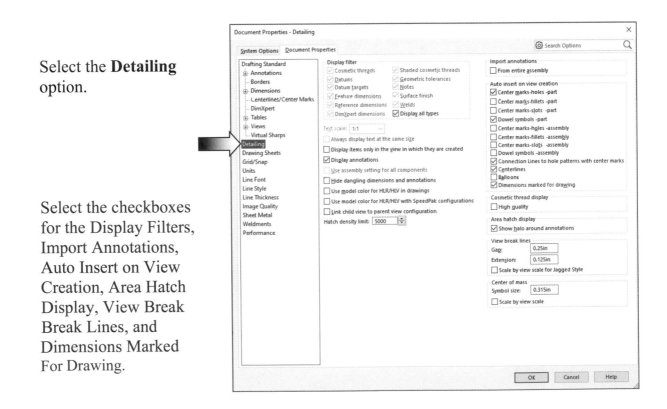

Select the **Drawing-Sheet** option.

Enable the check-box **Use Different Sheet Format** if sheet 2 uses a different (or partial) title block.

This property lets you automatically have one sheet format for the first sheet and a separate sheet format for all additional sheets.

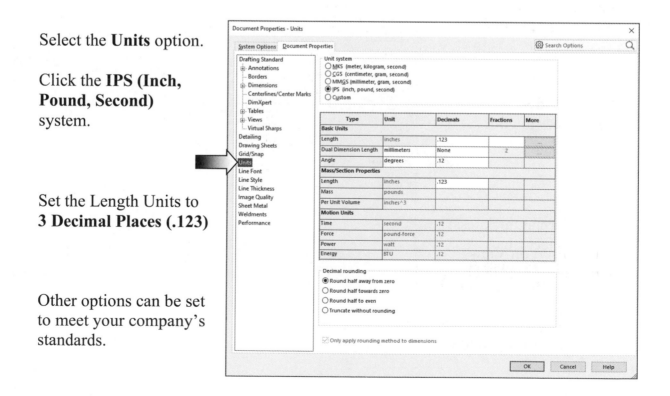

Select the **Units** option.

Click the **IPS (Inch, Pound, Second)** system.

Set the Length Units to **3 Decimal Places (.123)**

Other options can be set to meet your company's standards.

Select the **Line Font** option.

Set the Style and Weight of lines for various kinds of edges in drawing documents only.

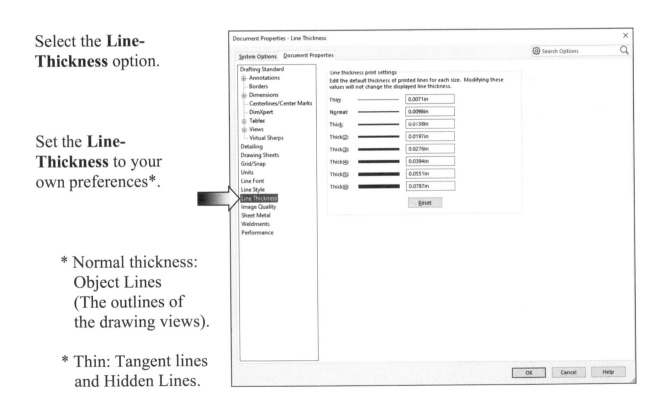

Select the **Line-Thickness** option.

Set the **Line-Thickness** to your own preferences*.

* Normal thickness:
  Object Lines
  (The outlines of
  the drawing views).

* Thin: Tangent lines
  and Hidden Lines.

Select the **Image-Quality** option.

Drag the Deviations Slider to the right to improve the shaded quality (0.00800in Deviation).

For Wireframe, drag the slider up to about 70%-75%.

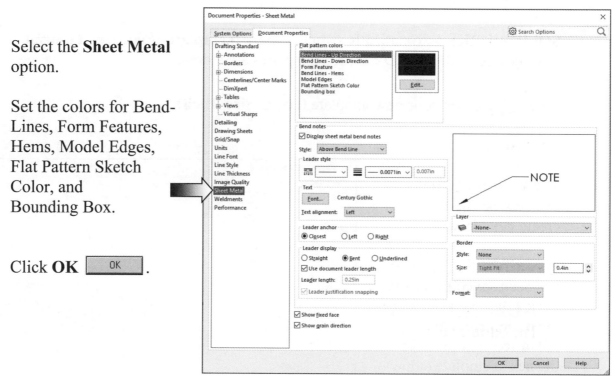

Select the **Sheet Metal** option.

Set the colors for Bend-Lines, Form Features, Hems, Model Edges, Flat Pattern Sketch Color, and Bounding Box.

Click **OK** OK .

These settings control the tessellation of curved surfaces for shaded rendering output. A higher resolution setting results in slower model rebuilding but more accurate curves.

6. Saving the Document Template:

Go to **File / Save As** and change the Save-As-Type to **Drawing Templates** (arrow).

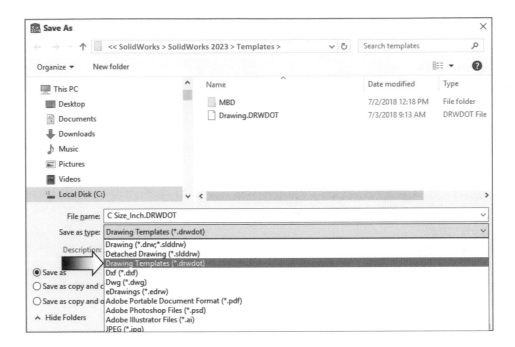

SOLIDWORKS automatically switches to the **Templates** folder (or **Tutorial** folder), where all SOLIDWORKS templates are stored.

Enter a name for your new template (i.e., **C Size_Inch**).

Click **Save**.

The Drawing Template can now be used to create new drawings.

To verify if the Template has been saved properly, click **File / New**. Change the option Novice to **Advance**, and look for the new Template (arrow).

Close the document template.

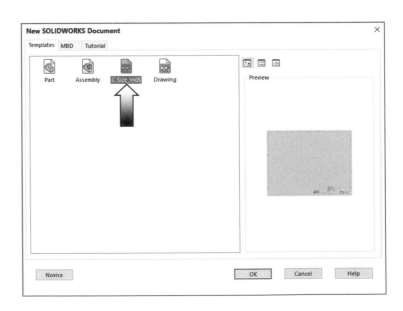

Questions for Review

1. Custom settings and parameters can be set and saved in the Document Template.
    a. True
    b. False

2. Document Templates are stored either in (C:\Program Data\SolidWorks\Data\ Templates) OR (C:\Program Files\SolidWorks Corp\Lang\English\Tutorial).
    a. True
    b. False

3. To access the sheet format and edit the information in the title block:
    a. Right click in the drawing and select Edit Sheet Format.
    b. Right click the Sheet Format icon from the Feature Tree and select Edit Sheet-Format.
    c. All the above.

4. The **Sheet** layer is where the Title Block and Revisions information are stored.
    a. True
    b. False

5. The **Sheet Format** layer is used to create the drawing views, dimensions, and annotations.
    a. True
    b. False

6. Information in the Title Block and Revision Block are Fixed; they cannot be modified.
    a. True
    b. False

7. The Document Template is saved with the file Extension:
    a. DWG
    b. DXF
    c. SLDPRT
    d. DRWDOT
    e. SLDDRW

7. D
5. FALSE   6. FALSE
3. C   4. FALSE
1. TRUE   2. TRUE

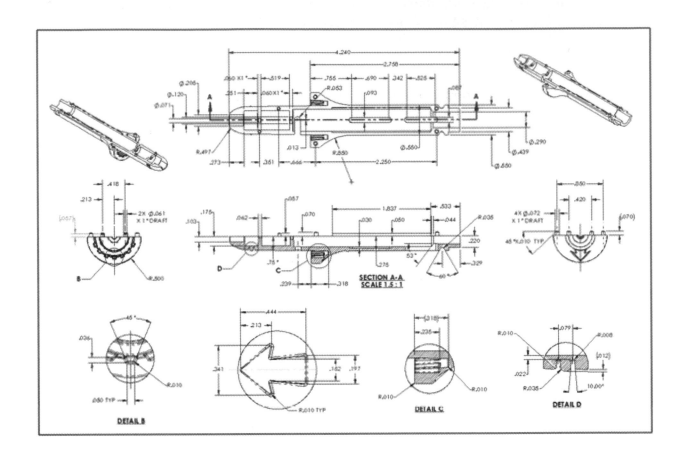

CHAPTER 15

Assembly Drawings

Assembly Drawings
Links Assembly

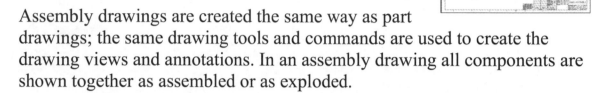

Assembly drawings are created the same way as part drawings; the same drawing tools and commands are used to create the drawing views and annotations. In an assembly drawing all components are shown together as assembled or as exploded.

The standard drawing views such as the Front, Top, Right, and Isometric views can be created with the same drawing tools, or they can be dragged and dropped from the View Palette.

When a Section view is created, it will be cross hatched automatically and the hatch pattern that represents the material of the part can be added and easily edited as well.

A parts list or a Bill of Materials (B.O.M.) is created to report the details of the components such as:

    * Item Number

    * Part Number

    * Description

    * Quantity, etc.

4	004-12345	Single Link	4
3	003-12345	Pin Head	6
2	002-12345	Alignment Pin	6
1	001-12345	Double Link	3
ITEM NO.	PART NUMBER	DESCRIPTION	QTY.

The components will then be labeled with Balloons (or Dash Numbers) for verification against the Bill of Materials.

This chapter discusses the basics of creating an assembly drawing using SOLIDWORKS 2023.

Assembly Drawings
Links Assembly

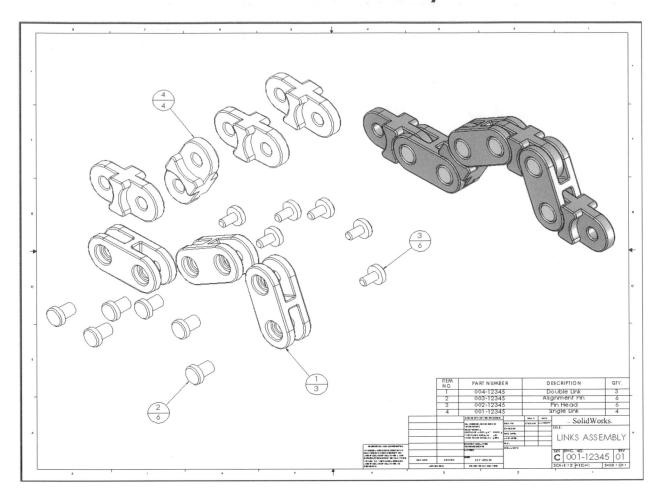

ITEM NO.	PART NUMBER	DESCRIPTION	QTY.
1	004-12345	Double Link	3
2	003-12345	Alignment Pin	6
3	002-12345	Pin Head	6
4	001-12345	Single Link	4

- SolidWorks-

TITLE: LINKS ASSEMBLY

C | DWG. NO. 001-12345 | REV 01

SCALE: 1:2 WEIGHT: SHEET 1 OF 1

Dimensioning Standards: **ANSI**	Third Angle Projection
Units: **INCHES** – 3 Decimals	

Tools Needed:

 New Drawing

 Model View

 Shaded View

 Balloon

 Bill of Materials

 Properties...

1. Creating a new drawing:

Select **File, New, Draw** (or Drawing) and click **OK**.

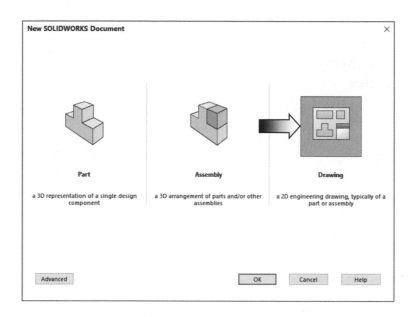

Under Standard Sheet Size, select **C-Landscape** for paper size.

A = 8.5"   x   11.00" (Landscape)
B = 17.00"   x   11.00" (Landscape)
C = 22.00"   x   17.00" (Landscape)
D = 34.00"   x   22.00" (Landscape)
E = 44.00"   x   34.00" (Landscape)

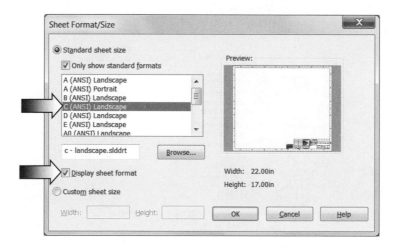

Enable **Display Sheet Format** checkbox to display the revision and title blocks.

Click **OK**. *(Right-click inside the drawing and select Properties to edit the settings.)*

The drawing template appears in the graphics area.

Right-click in the drawing and select **Properties**.

Set the Drawing Views **Scale** and **Angle of Projection** to:

Set **Scale** to **1:1** (full scale).

Set **Type of Projection** to: **Third Angle**

Select **C (ANSI) Landscape** for paper size.

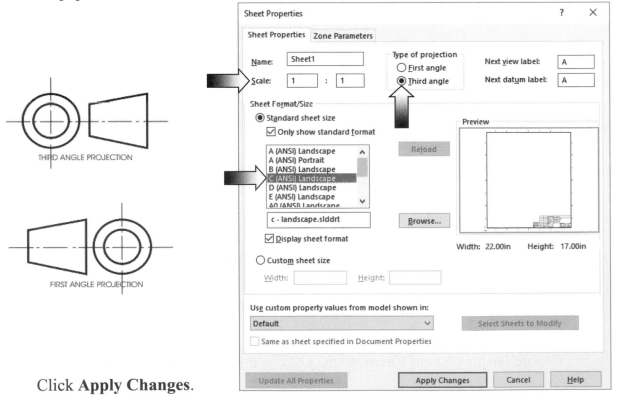

Click **Apply Changes**.

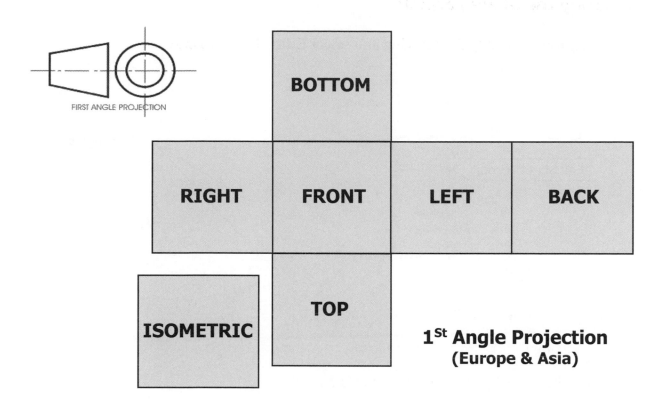

1St Angle Projection
(Europe & Asia)

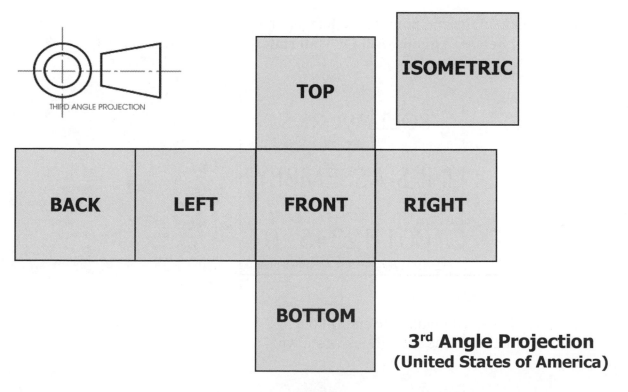

3rd Angle Projection
(United States of America)

2. Editing the Sheet Format:

Right-click inside the drawing and select **Edit-Sheet-Format**.

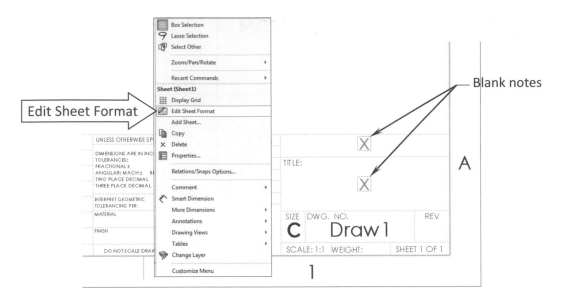

The **Format** layer is brought up to the front.

3. Setting up the B.O.M. anchor point:

Zoom in on the lower right side of the title block.

Right-click on the end point of the line (as shown) and select **Set-As-Anchor, Bill Of Materials.**

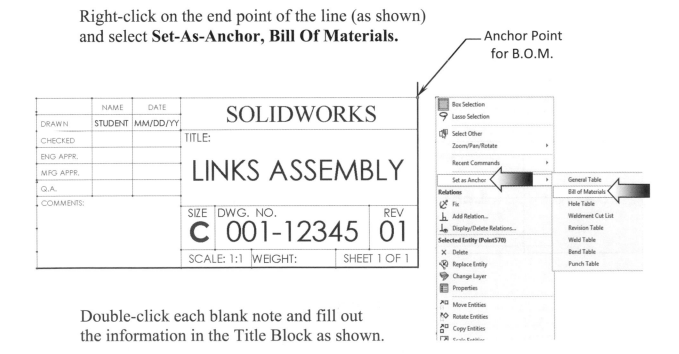

Double-click each blank note and fill out the information in the Title Block as shown.

4. Switching back to the Sheet layer:

Right-click inside the drawing and select **Edit-Sheet**. The sheet layer is now active.

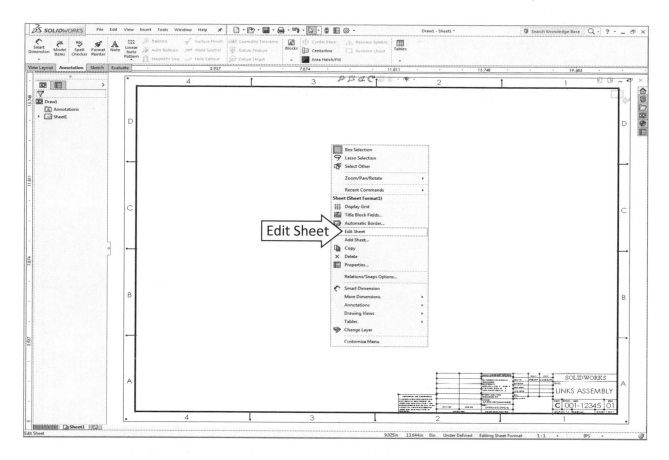

To Change the Drawing Paper's color:

Select **Tools, Options, System Options, Colors, Drawing Paper Colors, Edit**.

Select the **White** color and enable **Use Specified Color for Drawings Paper-Color** option.

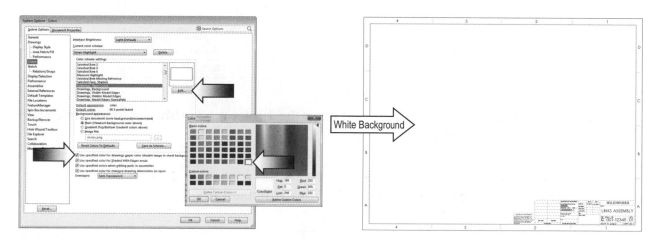

5. Opening an existing assembly document:

Click the **Model View** command from the **View Layout** tab.

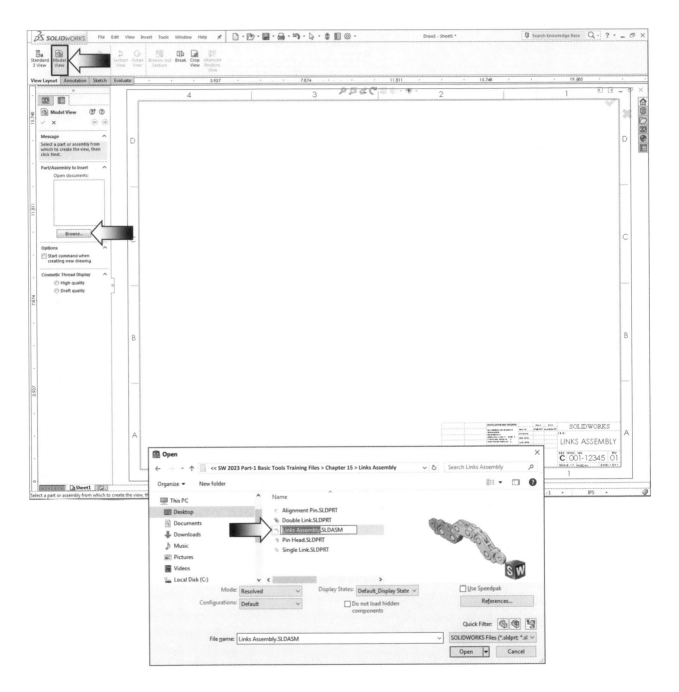

Click the **Browse** button; locate and open the **Links Assembly** document from the Training Files folder.

There are several different methods to create the drawing views; in this lesson, we will discuss the use of the **Model-View** command first.

Select the **Isometric** view button (arrow) and place the drawing view approximately as shown.

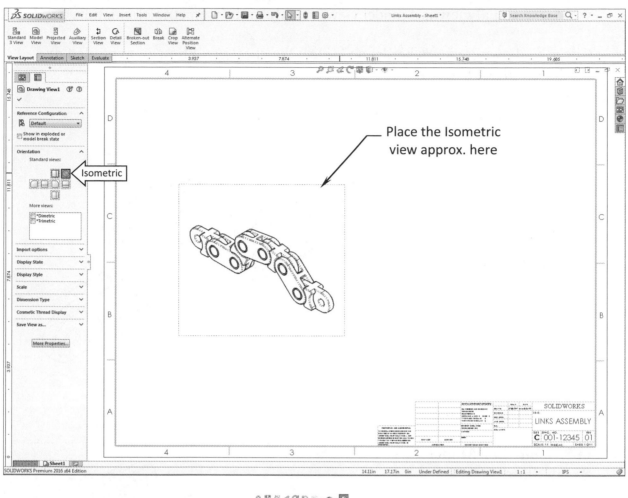

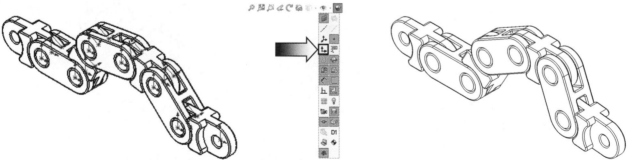

Click the drop-down arrow under the Hide/Show Items and <u>deselect</u> the **Origins**.

The isometric view is created based on the default orientations of the last saved assembly document. The scale will be changed in the next couple of steps.

NOTE: *Press **Ctrl + Tab** to toggle back and forth between the Drawing and the Assembly documents.*

6. Switching to the Exploded View state:

You can create an exploded drawing view from an existing exploded assembly view. The actual view is a model view, usually in the isometric orientation.

Click the drawing view's border and select the checkbox **Show in exploded or model break state** (arrow).

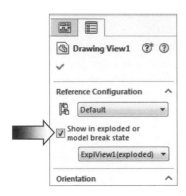

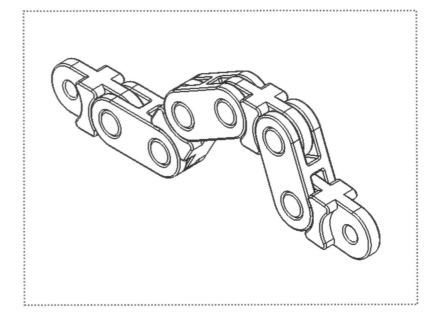

Alternatively, you can right-click the drawing view and select **Show-in Exploded or Model Break State**.

The exploded checkbox is only available if an exploded view has been created earlier in the assembly document.

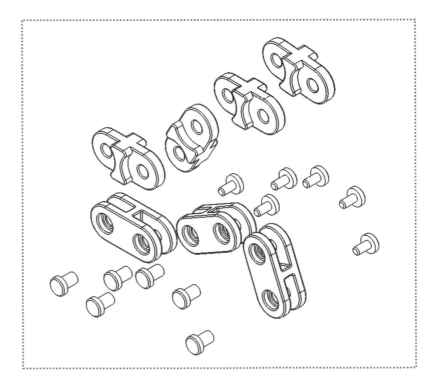

Click **OK**.

7. Changing the Line Style:

The tangent edges of the components should be changed to Phantom Line style.

Right-click the drawing view's border, and select **Tangent Edge, Tangent Edges With Font** (arrow).

The tangent edges are displayed as phantom lines.

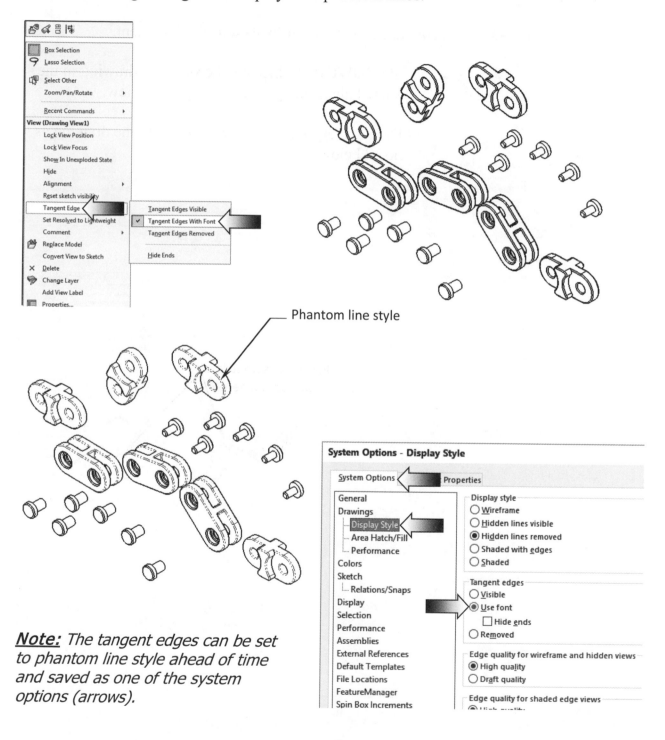

Phantom line style

**Note:** _The tangent edges can be set to phantom line style ahead of time and saved as one of the system options (arrows)._

8. Using the View Palette:

The View Palette allows a quick way to add drawing views to the drawing sheet by dragging and dropping. Each view is created as a model view. The orientations are based on the eight standard orientations (Front, Right, Top, Back, Left, Bottom, Current, and Isometric) and any custom views in the part or assembly.

Click the **View Palette** tab on the right side of the screen (arrow).

Click the **drop-down arrow** and select the **Links Assembly** document.

SOLIDWORKS displays the standard drawing in the View Palette along with the exploded view.

Drag and drop the Isometric view into the drawing as noted below.

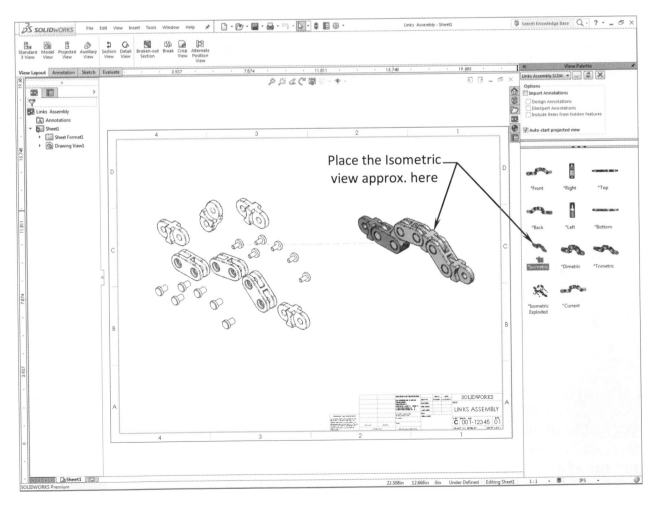

Place the Isometric view approx. here

9. Switching to Shaded view:

Sometimes for clarity, a drawing view is changed to Shaded instead of Wireframe or Hidden Lines Removed.

The shaded view will be printed as shaded. To change the shading of more than one view at the same time, hold the control key, select the dotted borders of the views and then change the shading.

Select the Isometric view's border.

Click **Shaded With Edges** in the Display Style section (arrow).

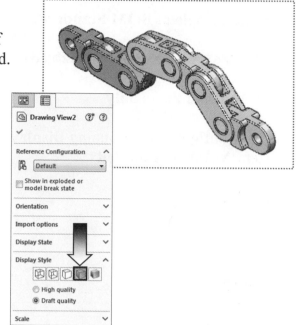

The Isometric view is now shown as Shaded with Edges.

10. Adding the Bill of Materials (B.O.M.) to the drawing:

Click the Isometric view's border and switch to the **Annotation** tool tab.

Click **Tables, Bill of Materials** – OR – select **Insert / Tables / Bill of Materials**.

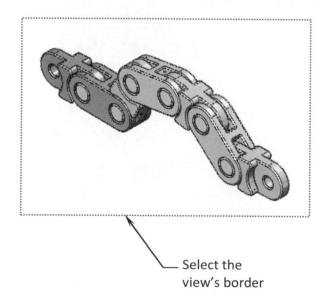

Select the view's border

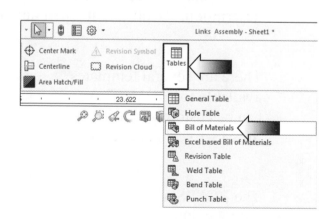

NOTE: A BOM can be anchored, moved, edited, and split into sections.

11. Selecting the B.O.M. options:

Table Template: **BOM Standard.**

Attach to Anchor Point*: **Enabled**.

BOM Type: **Parts only**.

Part Configuration Grouping **Display As One Item Number: Enabled**.

Click **OK**.

ITEM NO.	PART NUMBER	DESCRIPTION	QTY.
1	Double Link		3
2	Single Link		4
3	Alignment Pin		6
4	Pin Head		6

SOLIDWORKS

TITLE:
LINKS ASSEMBLY

SIZE DWG. NO.   REV
C   001-12345   01

SCALE: 1:1 WEIGHT: SHEET 1 OF 1

A BOM is created automatically and the information in the assembly document is populated with item numbers, description, quantities, and part numbers.

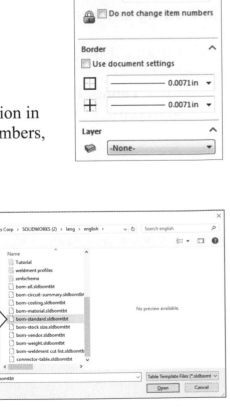

There are several templates available for BOM. They can be customized using the MS-Excel application.

The BOM Standard comes with 4 columns:
Item Number, **Part Number**, **Description**, and **Quantity**.

The Bill of Materials is created and anchored to the lower right corner of the title block, where the anchor point was set earlier.

* The Anchor point is set and saved in the Sheet Format layer. To change the location of the anchor point, first right-click anywhere in the drawing and select Edit Sheet Format then right-click on one of the end points of a line and select Set-As-Anchor / Bill of Material.

To change the anchor corner, click anywhere in the B.O.M, click the 4-way cursor on the upper left corner and select the stationary corner where you want to anchor the table (use the Bottom Right option for this lesson).

ITEM NO.	PART NUMBER	DESCRIPTION	QTY.
1	Double Link		3
2	Single Link		4
3	Alignment Pin		6
4	Pin Head		6

12. Creating the custom properties:

By default the **Part Number** column in the BOM is tied to the configuration name or file name.

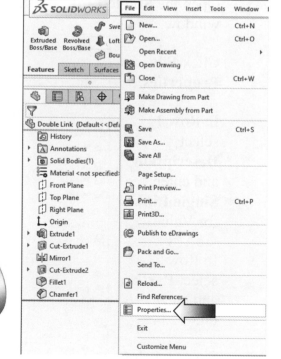

The next step is to open each component and enter the Part Number and Description into the Custom tab.

This information will automatically populate to the BOM when it is added to the drawing.

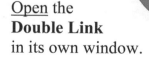

Open the **Double Link** in its own window.

Click **File, Properties**.

Click the drop-down arrow under the **Property Name** and select:
> **Description**

In the **Value / Text Expression** column, enter:
> **Double Link**

For Row 2, expand the drop-down arrow and select:
> **PartNo**

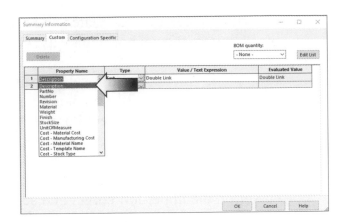

For Value / Text Expression, enter:
> **004-12345**

Click **OK**.

Save and close the Double Link.

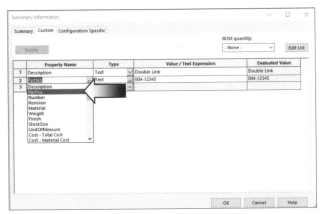

Open the **Single Link** in its own window.

Click **File**, **Properties**.

In Row 1, select **Description** and enter **Single Link** for Value.

In Row 2, select **PartNo** and enter **003-12345**

Save and close the document.

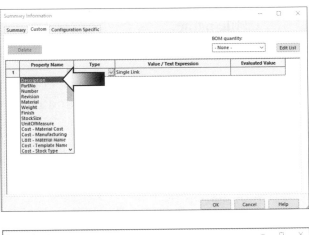

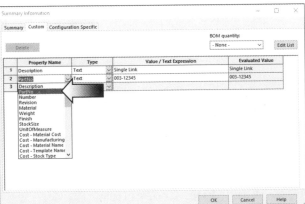

Open the component **Alignment Pin**.

Click **File, Properties**.

Add the following:

> Description: **Alignment Pin**
> PartNo: **002-12345**

Open the component **Pin Head** and add the following:

> Description: **Pin Head**
> PartNo: **001-1234**

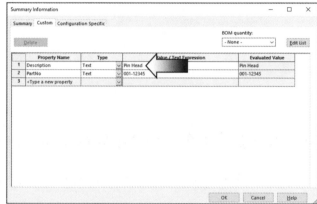

Save and close the documents and return to the drawing.

13. Modifying the B.O.M.:

As soon as the drawing is open, the Description column updates automatically to show the Value / Text that were entered in the part mode (click **Rebuild** if manual update is needed).

The Part Number column is tied to the configuration or file name, we will change it to the PartNo instead.

Double-click the **Header B**. Change the Column Type to **Custom Property**. Expand the Property Name and select **PartNo**.

Click outside the BOM to update the table. The part numbers that were entered in the models are populated to the BOM.

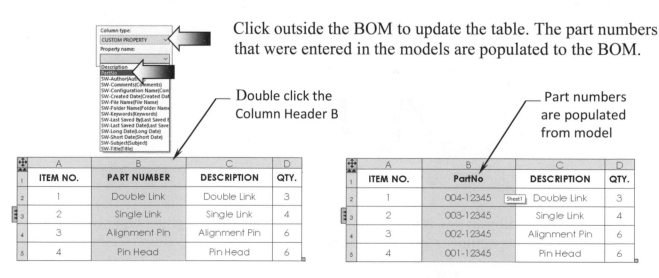

Double click the
Column Header B

Part numbers
are populated
from model

	A	B	C	D
1	ITEM NO.	PART NUMBER	DESCRIPTION	QTY.
2	1	Double Link	Double Link	3
3	2	Single Link	Single Link	4
4	3	Alignment Pin	Alignment Pin	6
5	4	Pin Head	Pin Head	6

	A	B	C	D
1	ITEM NO.	PartNo	DESCRIPTION	QTY.
2	1	004-12345	Double Link	3
3	2	003-12345	Single Link	4
4	3	002-12345	Alignment Pin	6
5	4	001-12345	Pin Head	6

The PartNo column is updated, displaying the values entered in each part's properties.

ITEM NO.	PartNo	DESCRIPTION	QTY.
1	004-12345	Double Link	3
2	003-12345	Single Link	4
3	002-12345	Alignment Pin	6
4	001-12345	Pin Head	6

A

UNLESS OTHERWISE SPECIFIED:		NAME	DATE	SOLIDWORKS
DIMENSIONS ARE IN INCHES	DRAWN	STUDENT	MM/DD/YY	TITLE:
TOLERANCES: FRACTIONAL±	CHECKED			
ANGULAR: MACH± BEND ±	ENG APPR.			LINKS ASSEMBLY
TWO PLACE DECIMAL ± THREE PLACE DECIMAL ±	MFG APPR.			
INTERPRET GEOMETRIC TOLERANCING PER:	Q.A.			
MATERIAL	COMMENTS:			SIZE / DWG. NO. / REV
				B 001-12345 01
FINISH				SCALE: 1:2 WEIGHT: SHEET 1 OF 1

PROPRIETARY AND CONFIDENTIAL
THE INFORMATION CONTAINED IN THIS DRAWING IS THE SOLE PROPERTY OF <INSERT COMPANY NAME HERE>. ANY REPRODUCTION IN PART OR AS A WHOLE WITHOUT THE WRITTEN PERMISSION OF <INSERT COMPANY NAME HERE> IS PROHIBITED.

NEXT ASSY   USED ON

APPLICATION   DO NOT SCALE DRAWING

2   1

14. Using the Assembly Structure preview:

Click anywhere inside the B.O.M. table to access its **Properties**.

Click the **Assembly Structure** tab to see the preview of the components. The Formatting toolbar appears on top of the BOM.

Click anywhere in the BOM to access its Properties...

Assembly Structure and Preview pane

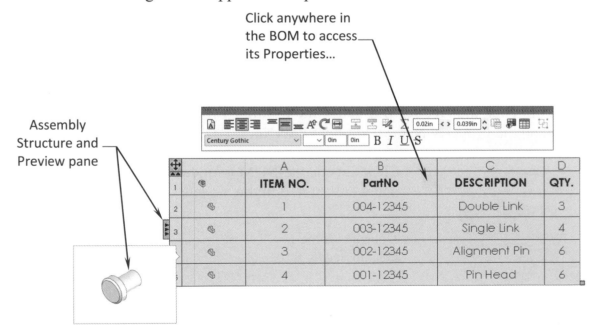

A thumbnail preview is available for each item in a BOM when you hover over its icon in the assembly structure column.

15. Reversing the Table Header:

The Column Headers can be positioned above or below the rows.

Click the **Table Header** button to reverse it as noted.

Click **OK**.

Table Header
Top/Bottom

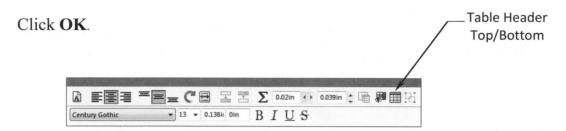

4	001-12345	Pin Head	6
3	002-12345	Alignment Pin	6
2	003-12345	Single Link	4
1	004-12345	Double Link	3
ITEM NO.	PartNo	DESCRIPTION	QTY.

ITEM NO.	PartNo	DESCRIPTION	QTY.
1	004-12345	Double Link	3
2	003-12345	Single Link	4
3	002-12345	Alignment Pin	6
4	001-12345	Pin Head	6

Table Header Bottom	Table Header Top

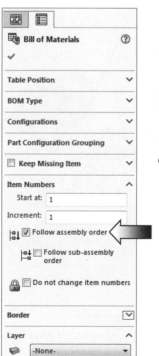

NOTE: Select **Follow Assembly Order** 📊 to reorder items in a drawing to follow the assembly from which it is created. The Bill of Materials automatically updates when a component is reordered, or a new component is added to the assembly document.

For components with multiple instances in the Bill of Materials, the first instance appears in the order in which it appears in the FeatureManager design tree. Subsequent instances increment the quantity.

16. Adding Balloon callouts:

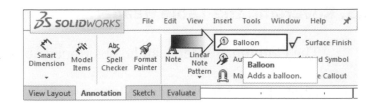

Balloons are added to identify and label the components. They can be added manually or automatically. The item number in the balloon is associated to the assembly order. If the components are reordered in the assembly, the item number will be updated to reflect the change.

Click [icon] or select **Insert, Annotations, Balloon**.

Click on the **edge** of the Single-Link. The system places a balloon on the part.

The Item Number matches the one in the B.O.M. automatically.

Click on an edge of the Double Link, the Alignment Pins, and Pin Heads to attach a balloon to each one of them.

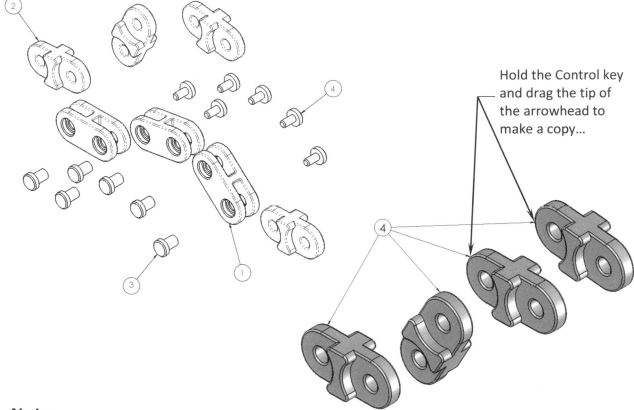

Hold the Control key and drag the tip of the arrowhead to make a copy...

Note:

Balloons are like notes; they can have multiple leaders or attachment points.
To copy the leader line, hold down the CONTROL key and drag the tip of one of the arrows.

17. Changing the balloon style:

Hold down the CONTROL key and select all balloons; the balloon Properties tree appears on the left side of the screen.

Select **Circular Split Line** under the Style menu.

Select **2 Character** size.

Click **OK**.

> 💡 **Circular** ①/②
> ─────
> **Split Line**
>
> The Upper Number represents the Item-Number.
> The Lower Number represents the Quantity.

Settings

| Circular Split Line | ▾ |
| 2 Characters | ▾ |

Padding:
0.000in

User defined:
0.400in

Balloon text:
Item Number ▾

Lower text:
Quantity ▾

Item Number

Quantity

4X 2/4

NOTE: To switch from Circular Split Lines to number of instances (X), activate the Quantity checkbox, then select the Placement and the X denotation options.

☑ **Quantity**

Placement:

Denotation:
X

Distance:
0.039in

Quantity value:

☐ Override value:

18. Saving your work:

Select **File / Save As /
Links-Assembly / Save**.

Questions for Review

1. The drawing's paper size can be changed at any time.
   a. True
   b. False

2. The First or Third angle projection can be selected in the Drawing Sheet Properties.
   a. True
   b. False

3. To access the sheet format and edit the information in the title block:
   a. Right click in the drawing and select Edit Sheet Format
   b. Right click the Sheet Format icon from the Feature Tree and select: Edit Sheet Format.
   c. All the above.

4. The BOM anchor point should be set on the Sheet Format not on the drawing sheet.
   a. True
   b. False

5. The Model View command can also be used to create the Isometric drawing view.
   a. True
   b. False

6. The Drawing views scale can be changed individually or all at the same time.
   a. True
   b. False

7. A Bill of Materials (BOM) can automatically be generated using Excel embedded features.
   a. True
   b. False

8. The Balloon callouts are linked to the Bill of Materials and driven by the order of the Feature Manager Tree.
   a. True
   b. False

8. TRUE   9. TRUE
5. TRUE   7. TRUE
3. C   4. TRUE
1. TRUE   2. TRUE

Exercise: Assembly Drawings

File location: Training Files folder,
Mini Vise folder.

1. Create an Assembly Drawing
   using the components provided.

2. Create a Bill of Materials and
   modify the Part Number and
   the Description columns as shown.

ITEM NO.	PART NUMBER	DESCRIPTION	QTY.
1	001-12345	Base	1
2	002-23456	Slide Jaw	1
3	003-34567	Lead Screw	1
4	004-45678	Crank Handle	1
5	005-56789	Crank Handle Knob	1

3. Save your work as **Mini Vise Assembly.slddrw**

Exercise: Assembly Drawings

File Location:
Training Files folder
Egg Beater.sldasm

ITEM NO.	PART NUMBER	DESCRIPTION	QTY.
1	010-8980	Egg Beater Handle	1
2	010-8981	Main Gear	1
3	010-8982	Support Rod	1
4	010-8983	Right Spinner	1
5	010-8984	Left Spinner	1
6	010-8985	Crank Handle	1
7	010-8986	Small Gear (RIGHT)	1
8	010-8987	Small Gear (LEFT)	1

1. Create an Assembly Drawing with 5 views.

2. Add Balloons and a Bill of Materials.

3. Save your work as **Egg Beater Assembly**.

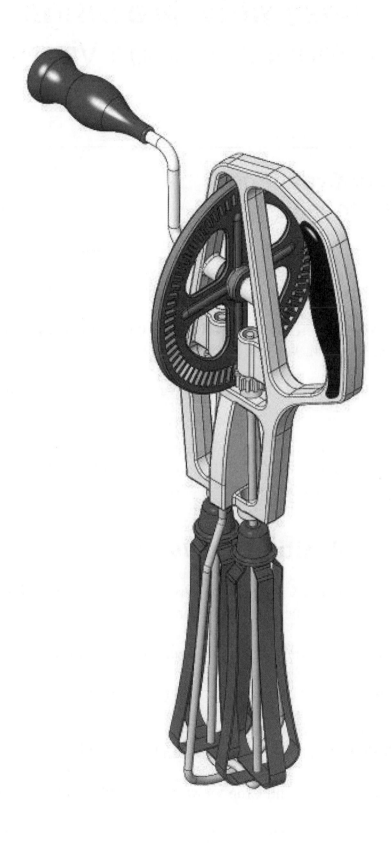

Assembly Drawings
Alternate Position Views

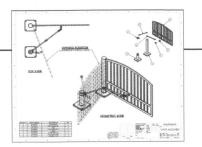

Assembly Drawings
Alternate Position Views

Assembly drawings are created in the same way as part drawings, except for all components in the assembly are shown together, as assembled or exploded.

A Bill of Materials is created to specify the details of the components, such as Part Number, Material, Weight, Vendor, etc.

The BOM template can be modified to have additional columns, rows, or different headers.

Balloons are also created on the same drawing to help identify the parts from its list (B.O.M). These balloons are linked to the Bill of Materials parametrically, and the order of the balloon numbers is driven by the Assembly's Feature tree.

Alternate Position Views

The Alternate Position View allows users to superimpose one drawing view precisely on another (open/close position for an example).

The alternate position view(s) is shown with phantom lines.

Dimension between the primary view and the Alternate Position View can be added to the same drawing view.

More than one Alternate Position View can be created in a drawing.

Section, Detail, Broken, and Crop views are currently not supported.

Assembly Drawings
Alternate Position Views

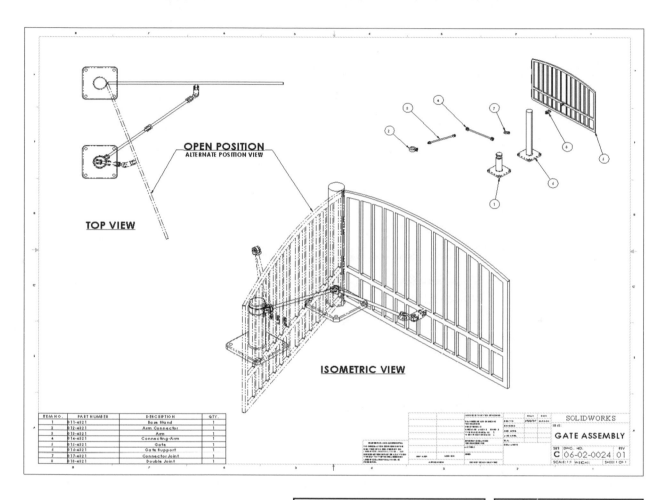

OPEN POSITION
ALTERNATE POSITION VIEW

TOP VIEW

ISOMETRIC VIEW

ITEM NO.	PART NUMBER	DESCRIPTION	QTY.
1	011-4221	Base Stand	1
2	012-4221	Arm Connector	1
3	012-4221	Arm	1
4	014-4221	Connecting-Arm	1
5	016-4221	Gate	1
6	016-4221	Gate Support	1
7	017-4221	Connector Joint	1
8	018-4221	Double Joint	1

SOLIDWORKS

GATE ASSEMBLY

SIZE **C** | DWG. NO. 06-02-0024 | REV 01

Dimensioning Standards: **ANSI** Units: **INCHES** – 3 Decimals	Third Angle Projection

Tools Needed:

 Alternate Position

 Model View

 Note

 Auto Balloon

 Bill of Materials

☑ Show in exploded or model break state

1. Creating a new drawing:

Select **File / New / Drawing / OK**.

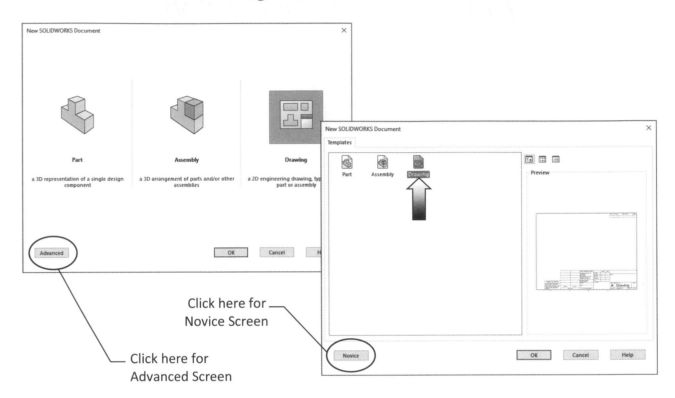

Click here for
Novice Screen

Click here for
Advanced Screen

For Standard Sheet Size, select **C-Landscape**.

Enable the **Display Sheet Format** check box.

Click **OK**.

Right-click in the drawing and select **Properties.**

Set **Scale** to **1:1** (full scale).

Set **Type of Projection** to **Third Angle** .

Click **OK**.

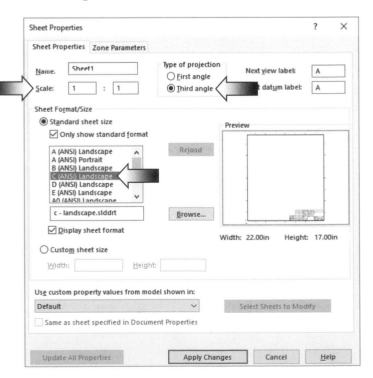

2. Creating the Isometric Drawing View:

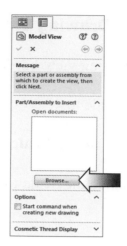

Switch to the **View Layout** tab.

Select the **Model View** command.

Click **Browse** and open the **Gate Assembly** document from the Training Files folder.

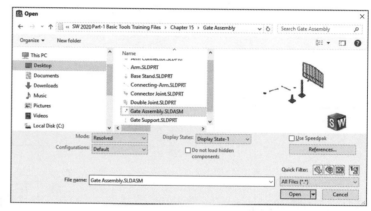

Select the **Isometric** view from the Orientation section (arrow).

Place the first drawing view (Isometric) approximately in the center of the sheet.

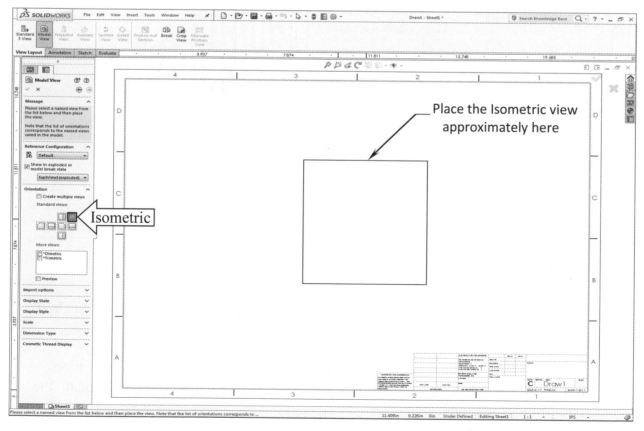

3. Changing the Drawing View Scale:

Click the drawing view border to display its options on the tree.

Change the Scale to:

* **Use Custom Scale**.

* Change the Scale to **1:3**.

Click **OK**.

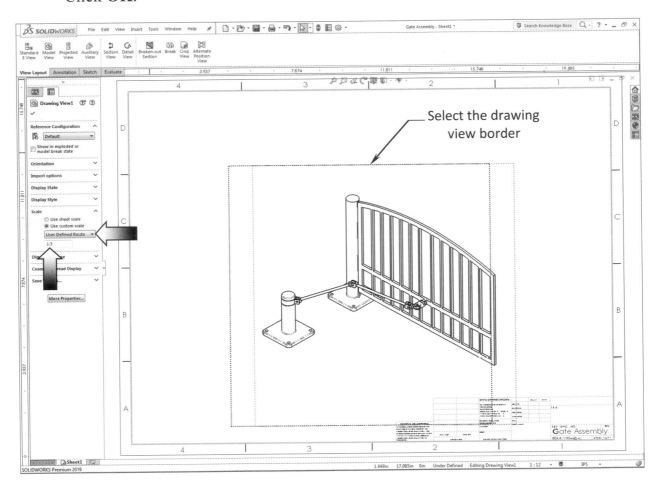

The Isometric View is scaled to (1:3) or 1/3 the actual size.

When selecting the User Defined Route option, you can enter your own scale if the scale that you want is not on the list.

The scale in the Sheet properties is linked to the scale callout in the title block (page 14-25). Changing the scale in the Properties will automatically update the one in the title block.

4. Creating an Alternate Position drawing view:

Use Alternate Position Views to show different positions of the components in an assembly.

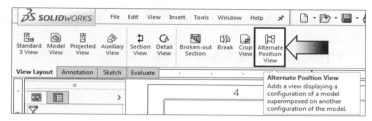

From the **View Layout** tool tab click the **Alternate Position View** command.

In the Configuration section, select the **Existing Configuration** option.

Expand the drop-down list and select the **Gate Open** configuration.

Click **OK**.

The new drawing view is shown with Phantom lines and is superimposed over the original view.

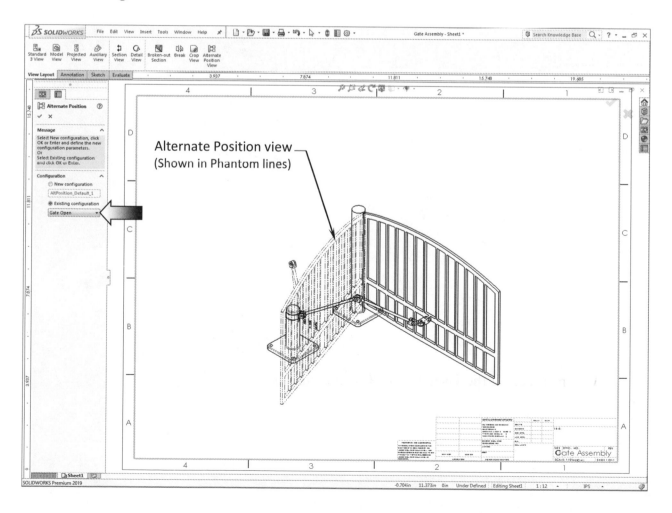

Alternate Position view
(Shown in Phantom lines)

5. Adding the Top Drawing view:

From the **View Layout** tab, click **Model View** command .

Click the **NEXT** arrow .

Select the **TOP** view from the Orientation dialog.

Place the Top drawing view approximately as shown below.

Click **OK**.

For clarity, change the view display to **Tangent-Edges With Font**. (Right-click the view's border, select Tangent Edge, and click Tangent Edges with Font).

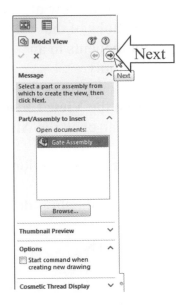

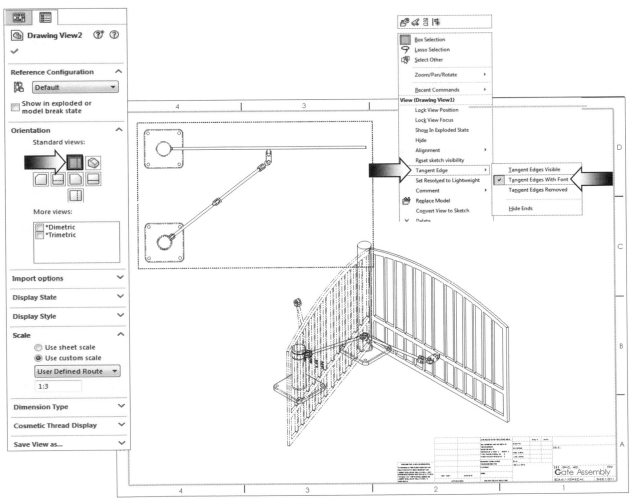

6. Creating the 2nd Alternate Position View:

Click the top view drawing's border to activate it.

Click **Alternate Position View** 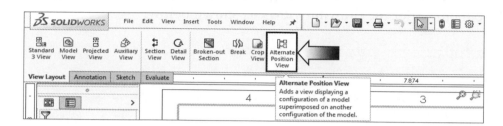 on the **View Layout** tab.

In the Configuration section, click the **Existing Configuration** option.

Select the **Gate Open** configuration from the drop-down list.

Click **OK**.

The new drawing view is shown in Phantom lines and is superimposed over the original view.

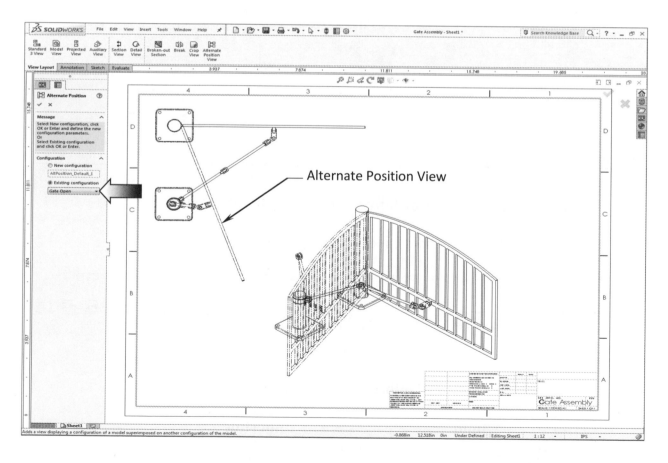

7. Adding Text / Annotations:

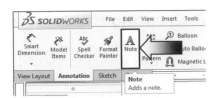

Select the top view drawing's border and switch to the **Annotation** tab.

Click the **Note** [A] command and create 2 notes **Top View** and **Isometric View** and place them under the drawing views as shown below.

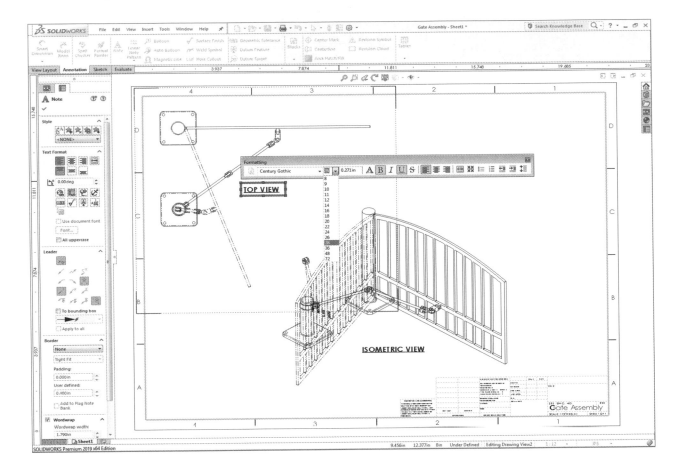

Use the options in the Formatting toolbar to modify the text.

Highlight the text and change it to match the settings below:

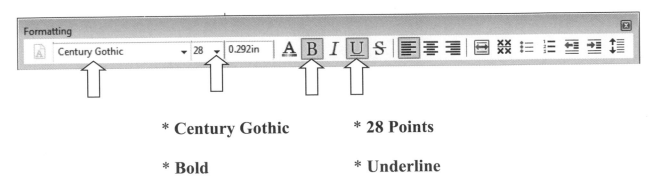

* **Century Gothic**   * **28 Points**

* **Bold**   * **Underline**

8. Creating an Exploded Isometric view:

Click the **Model View** command on the **View Layout** tab.

Click the **NEXT** arrow ⬅.

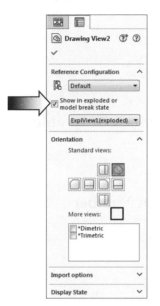

Select the **Isometric** view button from the Orientation dialog.

Place the Isometric view on the upper right side of the sheet.

On the upper left corner of the Properties tree, enable the **Show-In Exploded or Model Break State** checkbox (arrow).

Set the scale of the view to **1:8**.

NOTE:
The exploded view must be already created in the assembly for this option to be available in the drawing.

Click **OK**.

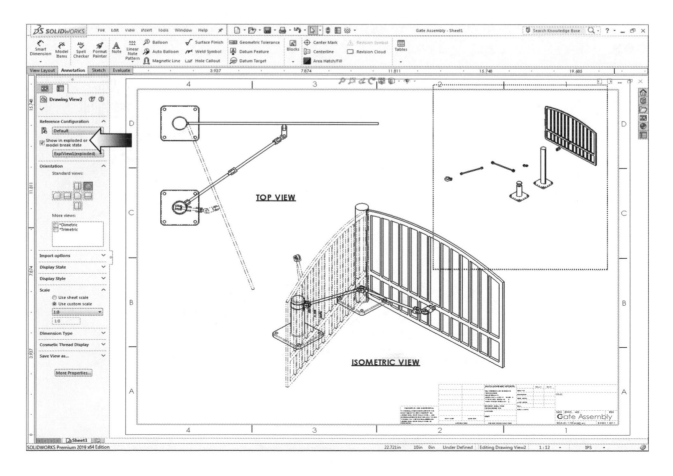

9. Adding Auto-Balloons to the Exploded view:

Switch to the **Annotation** tab.

Select the Isometric drawing view's border.

Click 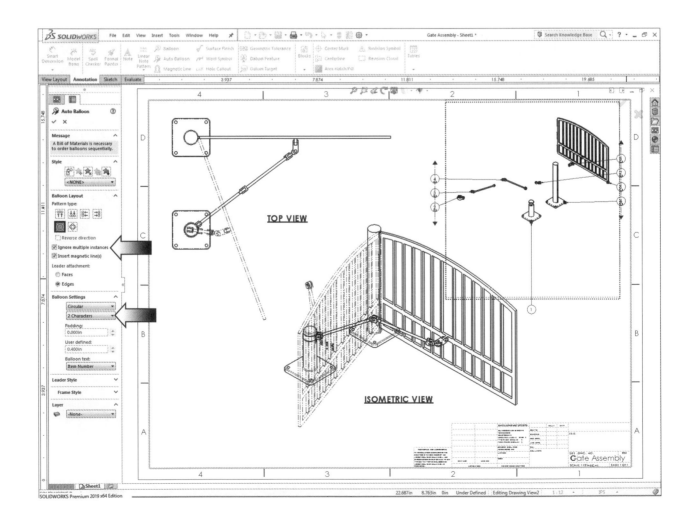 or select **Insert, Annotations, Auto Balloon**.

In the Auto Balloon properties tree, set the following:

* Balloon Layout: **Square**   * Ignore Multiple Instances: **Enable**

* Insert Magnetic Line(s): **Enable**   * Style: **Circular**

* Size: **2 Character**   * Balloon Text: **Item Number**

Click **OK**.

10. Adding the Bill of Materials:

Click **Bill of Materials** on the Annotation tab.

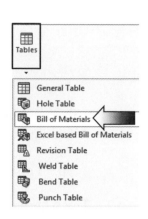

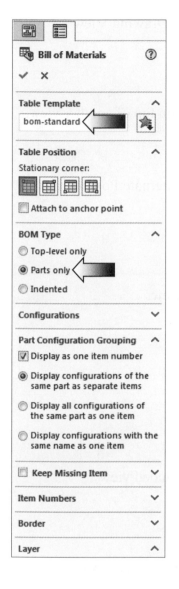

For Table Templates select **Bom-Standard**.

Uncheck the Attached to Anchor box.

For BOM Type select **Parts Only**.

Click **OK**.

	A	B	C	D
1	ITEM NO.	PART NUMBER	DESCRIPTION	QTY.
2	1	011-4321	Base Stand	1
3	2	012-4321	Arm Connector	1
4	3	013-4321	Arm	1
5	4	014-4321	Connecting-Arm	1
6	5	015-4321	Gate	1
7	6	016-4321	Gate Support	1
8	7	017-4321	Connector Joint	1
9	8	018-4321	Double Joint	1

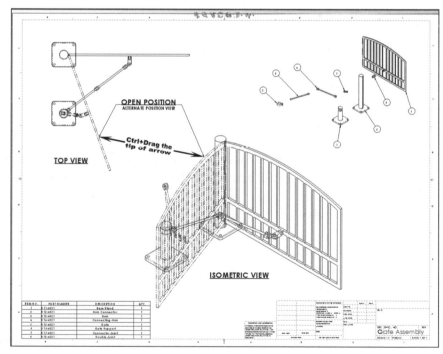

Modify the BOM as shown above to add the Part Numbers and Descriptions.

11. Saving a copy of your work:

Click **File / Save As.**

Enter **Gate Assembly.slddrw** for the file name and click **Save**.

Questions for Review

1. The Alternate Position command can also be selected from Insert / Drawing View / Alternate Position.
    a. True
    b. False

2. The Alternate Position command precisely places a drawing view on top of another view.
    a. True
    b. False

3. The types of views that are not currently supported to work with Alternate Position are:
    a. Broken and Crop Views
    b. Detail View
    c. Section View
    d. All the above.

4. Dimensions or annotations cannot be added to the superimposed view.
    a. True
    b. False

5. The Line Style for use in the Alternate Position view is:
    a. Solid line
    b. Dashed line
    c. Phantom line
    d. Centerline

6. In order to show the assembly exploded view on a drawing, an exploded view configuration must be created first in the main assembly.
    a. True
    b. False

7. Balloons and Auto Balloons can also be selected from Insert / Annotations / (Auto) Balloon.
    a. True
    b. False

8. The Bill of Materials contents can be modified to include the Material Column.
    a. True
    b. False

7. TRUE	8. TRUE
5. C	6. TRUE
3. D	4. FALSE
1. TRUE	2. TRUE

CHAPTER 16

Drawing Views

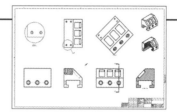

Drawing Views
Machined Block

When creating an engineering drawing, one of the first things to do is to layout the drawing views such as:

      * The standard views Front, Top, Right, and Isometric.

      * Other drawing views such as Detail, Cross section, Auxiliary views, etc., can be created or projected from the 4 standard views.

The Orthographic drawing views in SOLIDWORKS are automatically aligned with one another (default alignments); each view can only be moved along the direction that was defined for that particular view (vertical, horizontal, or at some projected angle).

Secondly, dimensions and annotations will then be added to the drawing views. The dimensions created in the model will be inserted into the drawing views, so that the association between the model and the drawing views can be maintained. Changes done to these dimensions will update all drawing views and the solid model as well.

And lastly, tolerances and annotations such as hole callouts, number of instances, general notes, and the title block information are added at this time.

Configurations are also used in this lesson to create some specific views.

This chapter will guide you through the creation of some of the most commonly used drawing views in an engineering drawing, such as 3 Standard views, Section views, Detail views, Projected views, Auxiliary views, Broken Out Section views, and Cross Hatch patterns.

Drawing View Creation
Machined Block

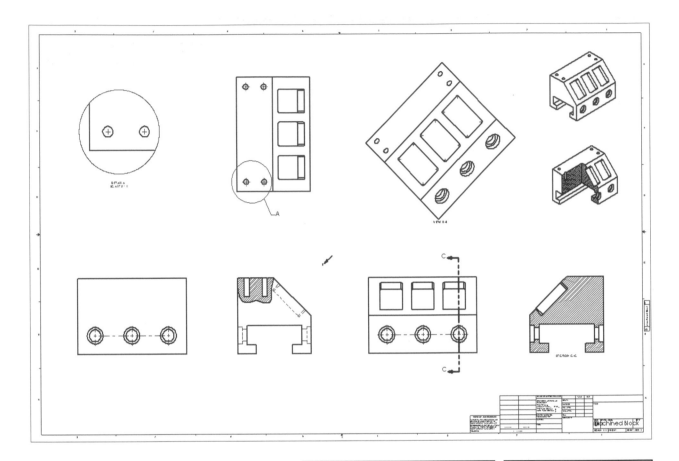

Dimensioning Standards: **ANSI**	Third Angle Projection
Units: **INCHES** – 3 Decimals	

Tools Needed:

 New Drawing   Model View   Projected View

 Auxiliary View   Detail View   Section View

 Broken Out Section   Area Hatch Fill   Note

1. Creating a new drawing:

Select **File / New / Drawing** template.

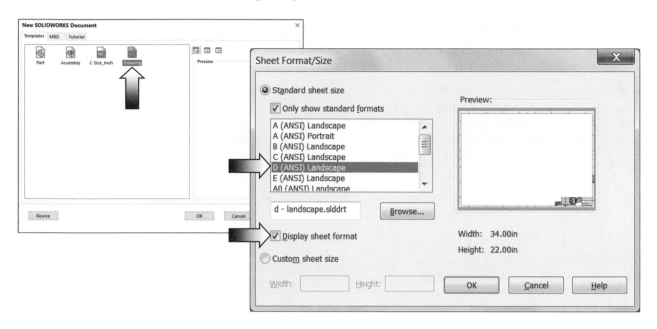

Under Standard Sheet Size, select **D-Landscape***. (34.00in. X 22.00in.)

Enable the **Display Sheet Format** check box to include the title block.

Click **OK**.

* If the options above are not available, **right click** inside the drawing and select **Properties**.

* Select the **Third Angle** option under Type of Projection.

* Set the default view scale to **1:2**.

* The Units can be changed later on.

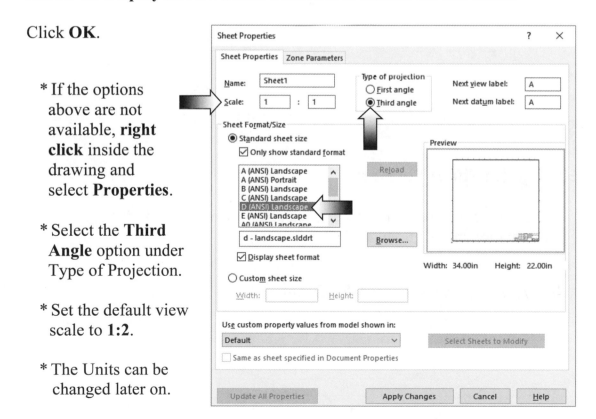

The drawing template comes with 2 default "layers":

* The "Front layer" is called the **Sheet** layer, where drawings are created.

* The "Back layer" is called the **Sheet Format,** where the title block and the revision information is stored.

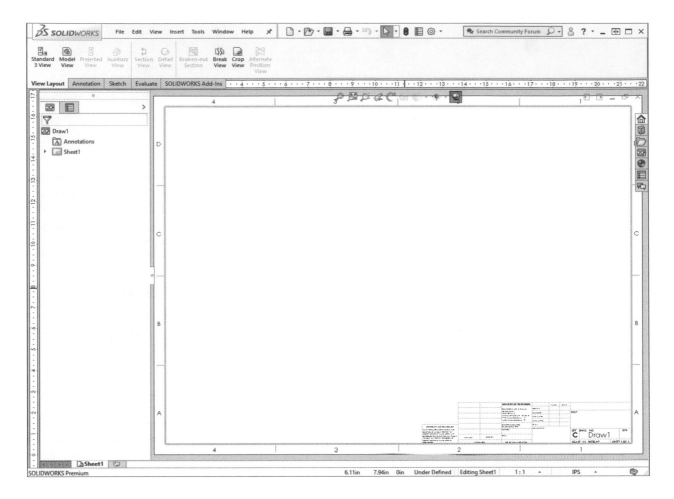

2. Editing the Sheet Format:

Zoom in closer to the Title Block area. We are going to fill out the information in the title block in the next step.

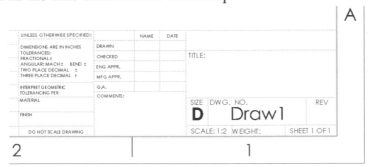

Right-click in the drawing area and select **Edit Sheet Format** (arrow).

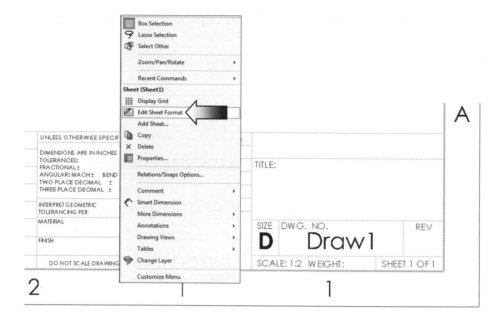

The Sheet Format is brought up to the front.

New annotation and sketch lines can now be added to customize the title block.

Any existing text or lines can also be modified at this time.

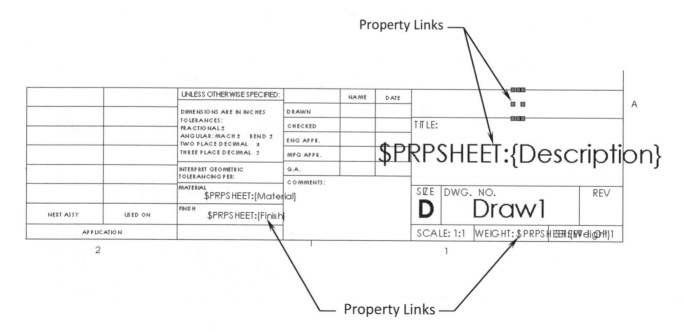

Notice the link to properties text strings: "$PRPS". Some of the annotations have already been linked to the part's properties.

3. Modifying the existing text:

Double-click on the "Company Name" and enter the name of your company.

Click **OK** when finished.

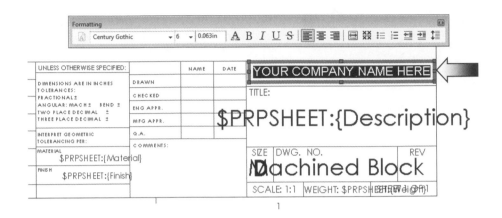

4. Adding the Title of the drawing:

Double-click on the Property-Link in the Title area ($PRPSHEET:{Description}).

Enter **Machined Block** for the title of the drawing (use uppercase letters).

Fill out the information as shown in the title block for the other areas.

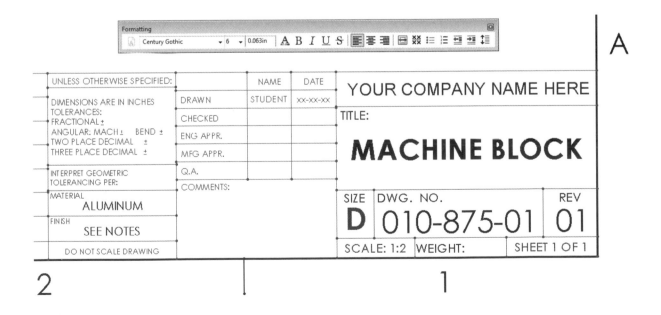

Click **OK**.

5. Switching back to the drawing Sheet:

Right-click in the drawing and select **Edit Sheet** (arrow).

The drawing Sheet "layer" is brought back up on top.

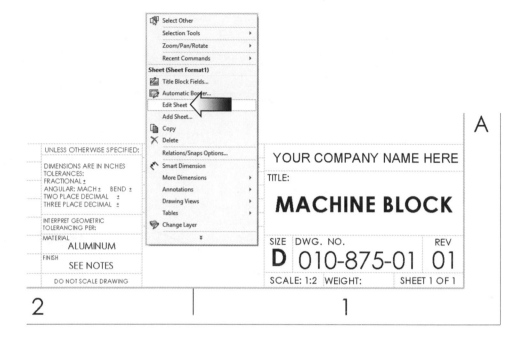

6. Using the View-Palette:

The **View Palette** is located on the right side of the screen in the Task Pane area; click on its icon to access it.

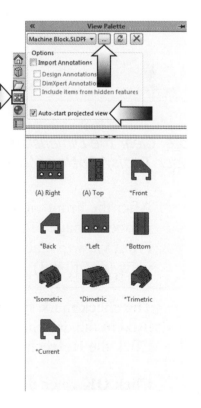

Note: If the drawing views are not visible in the View Palette window, click the Browse button then locate and open the part Machined Block.

The standard drawing views like the Front, Right, Top, Back, Left, Bottom, Current, Isometric, and Sheet Metal Flat-Pattern views are contained within the View Palette window.

These drawing views can be dragged into the drawing sheet to create the drawing views.

The **Auto Start Projected View** checkbox should be selected by default.

7. Using the View Palette:

Click/Hold/Drag the Front view from the View-Palette (arrow) and drop it on the drawing sheet approximately as shown. Make sure the Auto-Start projected view checkbox is enabled (arrow).

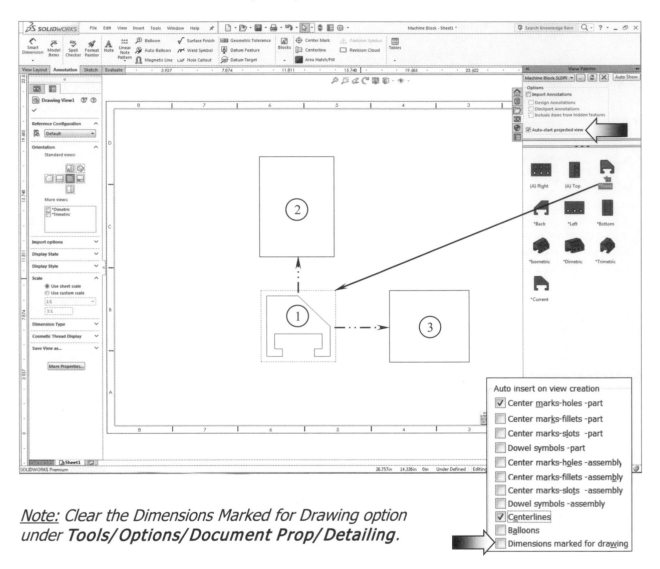

Note: _Clear the Dimensions Marked for Drawing option under **Tools/Options/Document Prop/Detailing**._

After the Front view is placed (position 1), move the mouse cursor upward, a preview image of the Top view appears, click in an approximate spot above the Front view to place the Top view (position 2); repeat the same step for the Right view (position 3).

The check mark symbol appears next to the drawing views that have been used in the drawing.
Click the **Refresh** button to update the View Palette.

Click **OK** when you are done with creating the first three views.

8. Adding an Isometric view:

The Isometric view can also be inserted from the View Palette but we will look at another option to create it instead.

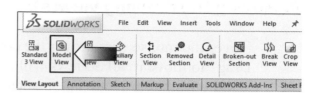

Switch to the **View Layout** tab and select the **Model View** command .

Click the **Next** arrow .

Select **Isometric** (arrow) from the Model View properties tree and place it approximately as shown below.

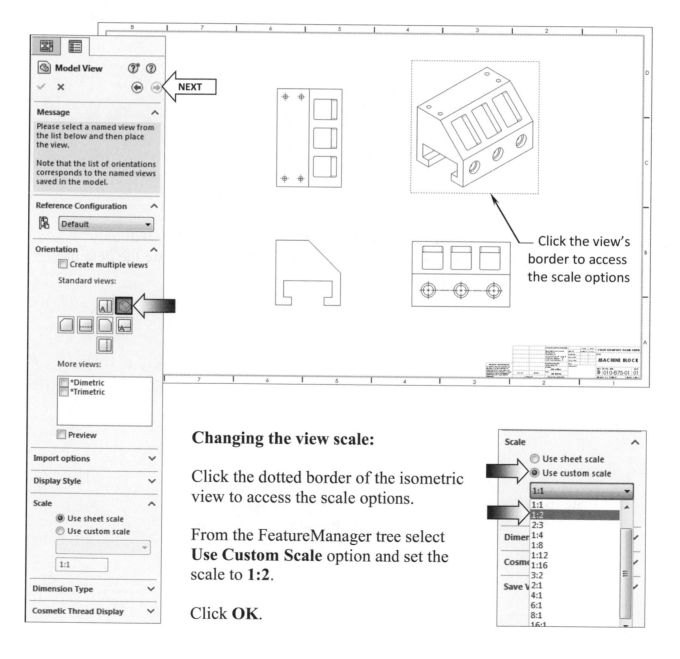

Click the view's border to access the scale options

Changing the view scale:

Click the dotted border of the isometric view to access the scale options.

From the FeatureManager tree select **Use Custom Scale** option and set the scale to **1:2**.

Click **OK**.

9. Moving the drawing view(s):

Hover the mouse cursor over the dotted border of the Front view and "Click/Hold/Drag" to move it; all 3 views will move at the same time. By default, the Top and the Right views are automatically aligned with the Front view.

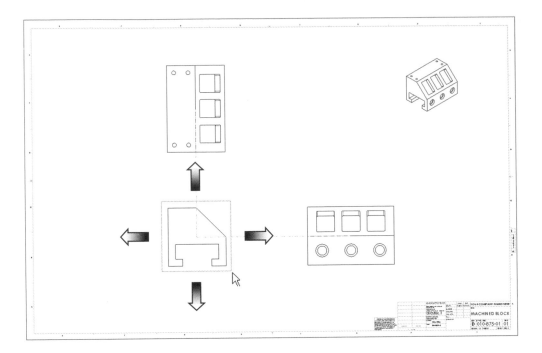

When moving either the Right or the Top view, notice how they will only move along the default vertical or horizontal directions. These are the default alignments for the standard drawing views.

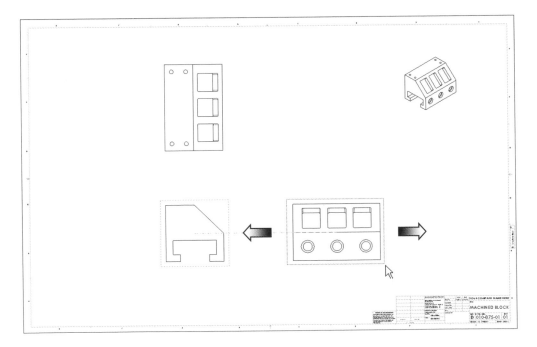

10. Breaking the alignments between the views:

Right-click inside the Top drawing view (or on its dotted border) and select: **Alignment / Break-Alignment*** (arrow).

The Top view is no longer locked to the default direction; it can now be moved freely.

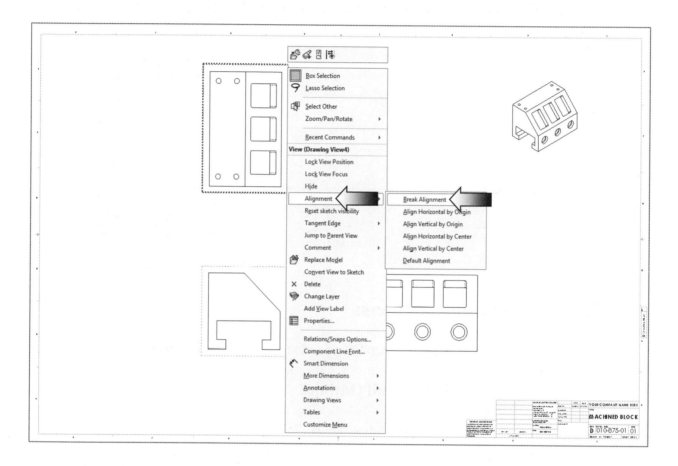

Dependent views like Projected Views, Auxiliary Views, Section views, etc. are aligned automatically with the views they were created from. Their alignments can be broken or also reverted back to their default alignments.

Independent views can also be aligned with other drawing views by using the same alignment options as shown above.

* To re-align a drawing view:

Right-click the drawing view's border and select **Alignment / Default-Alignment**.

11. Creating a Detail View:

Click  or select **Insert / Drawing View / Detail**.

Sketch a **circle** approximately as shown below, over the top drawing view.

A Detail View is created automatically, place it on the left side of the top view.

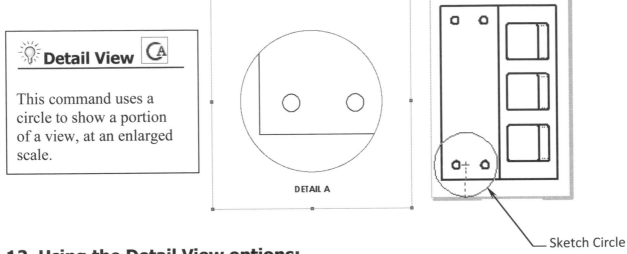

> 💡 **Detail View** Ⓐ
>
> This command uses a circle to show a portion of a view, at an enlarged scale.

Sketch Circle

12. Using the Detail View options:

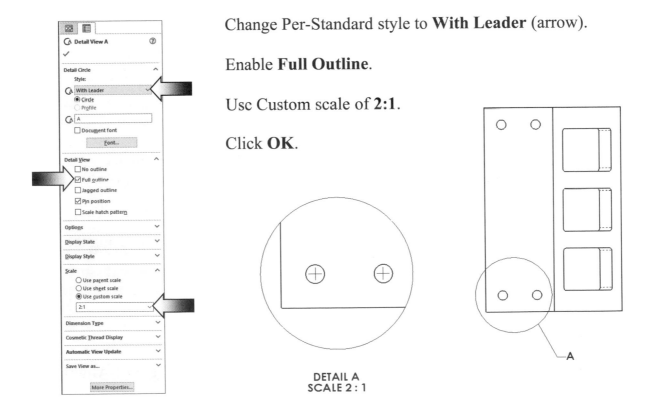

Change Per-Standard style to **With Leader** (arrow).

Enable **Full Outline**.

Usc Custom scale of **2:1**.

Click **OK**.

DETAIL A
SCALE 2 : 1

13. Creating a Projected View:

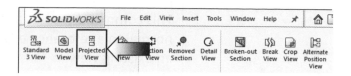

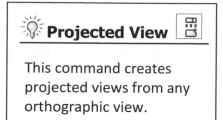

Projected View

This command creates projected views from any orthographic view.

Click the **Front view** dotted border to activate it.

Click 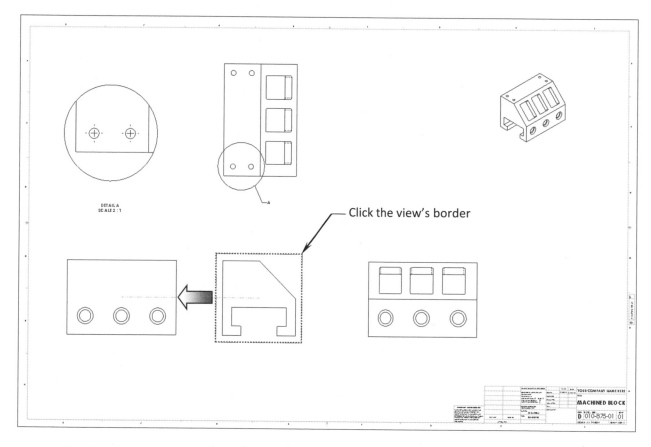 or select **Insert / Drawing View / Projected**.

The preview of the **Left view** appears, place it on the left side of the Front view.

Click the view's border

Notice the projected view is also aligned automatically to the horizontal axis of the Front view. It can only be moved from left to right.

For the purpose of this lesson, we will keep all the default alignments the way SOLIDWORKS creates them. (To break these default alignments at any time, simply right-click on their dotted borders and select Break-Alignment.)

Note: The Left side view can also be dragged and dropped from the View Palette.

14. Creating an Auxiliary View:

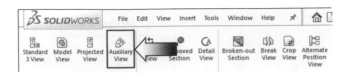

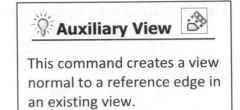

Auxiliary View

This command creates a view normal to a reference edge in an existing view.

Select the **angled-edge** as indicated and click the **Auxiliary** command.

Place the Auxiliary view approximately as shown below.

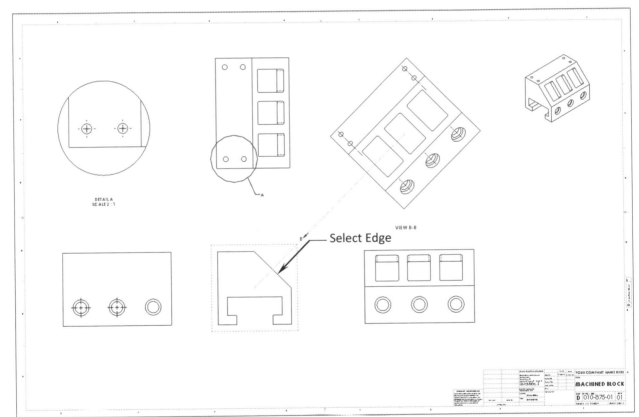

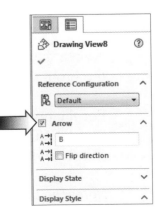

The Auxiliary view is also aligned to the edge from which it was originally projected.

The projection arrow can be toggled on and off from the PropertiesManager tree (arrow).

Use the Break-Alignment option if needed, to re-arrange the drawing views to different locations.

15. Creating a Section View:

Zoom in on the **Right view**.

Click 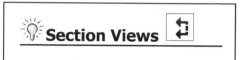 or select **Insert / Drawing View / Section**.

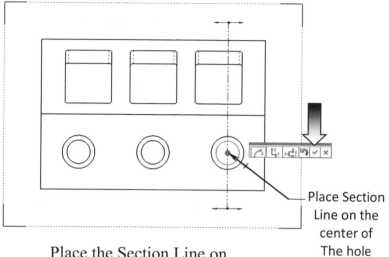

Place Section Line on the center of The hole

Section Views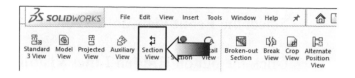

The Section View command creates a cut through a view, using single or multiple lines to show the interior details.
The sectioned surfaces are fully crosshatched automatically.

Place the Section Line on the center of the hole on the right.

The Section View pop-up appears. Click **OK** on the pop-up and place the section view on the right side.

The Section view is aligned horizontally with the view from which it was created from and the crosshatches are added automatically.

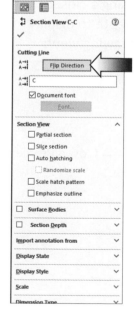

To change the direction of the cut:

* Double-Click on the section line and click Rebuild – OR –
* Enable the Flip Direction checkbox in the Feature tree (arrow).

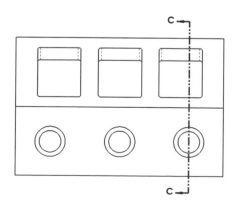

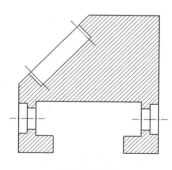

SECTION C-C

16. Showing the hidden lines in a drawing view:

Select the **Front view** dotted border.

Click the button **Hidden Lines Visible** under the Display Style section.

The hidden lines are now visible.

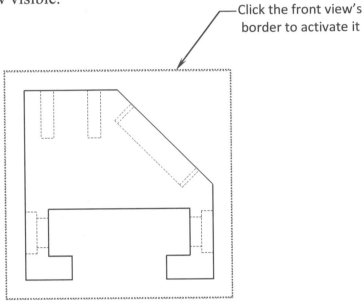

Click the front view's border to activate it

17. Creating a Broken-Out-Section:

Switch to the Sketch tab and sketch a <u>closed, free-form shape</u> over the view, around the upper area as shown. (Use either the Line or the Spline command to create the profile.)

Select the entire <u>closed profile</u> and switch to the **View Layout** tab.

Click **Broken-Out-Section** .

☼ **Broken Out Section View**

Creates a partial cut, using a closed profile, at a specified depth to display the inner details of a drawing view.

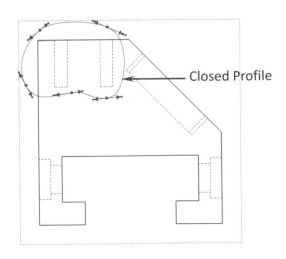

Closed Profile

Enable the **Preview** check box on the Properties tree.

Enter **.500** in. for Depth.

Click **OK**.

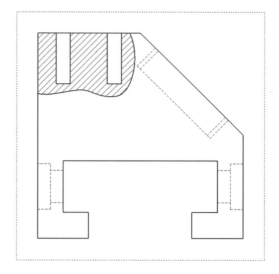

18. Adding a Cutaway view: (Previously created as a Configuration in the model.)

Click the **Model View** command .

Click **Next** ⮕ and select the **Isometric** under Orientation window, select the **Custom Scale, 1:2**.

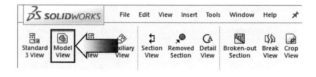

Place the 2nd Isometric view <u>below</u> the 1st isometric view.

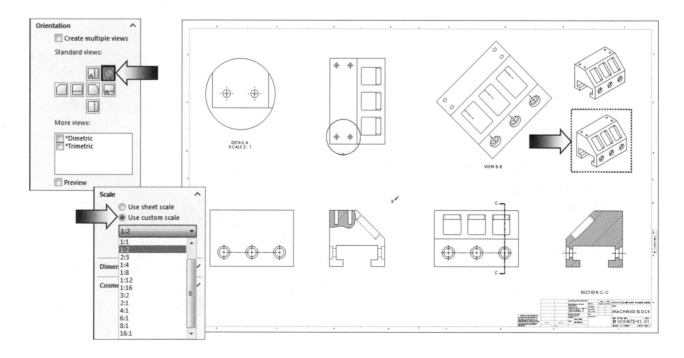

19. Changing Configurations: (From Default to Cutaway View).

Click the new **Isometric view** dotted border.

On the Properties tree, select the **Cutaway View** configuration from the list.

Click **OK**.

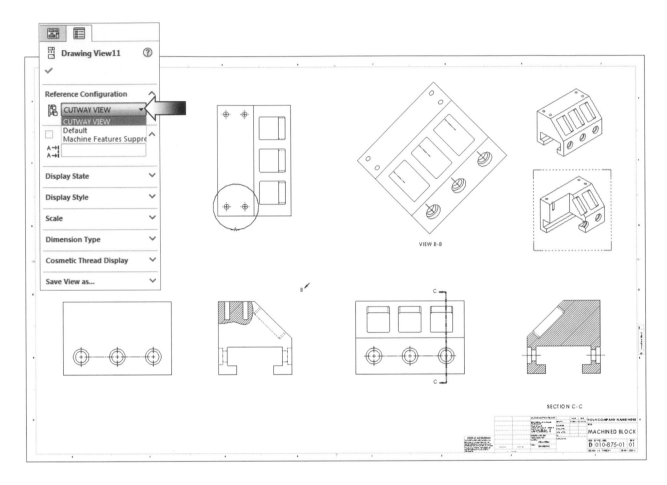

The **Cutaway View** configuration is activated and the extruded-cut feature that was created earlier in the model is now shown here.

Since the cutout was created in the model, there is no crosshatch on any of the sectioned faces.

Crosshatch is added at the drawing level and will not appear in the model.

The hatch lines are added next...

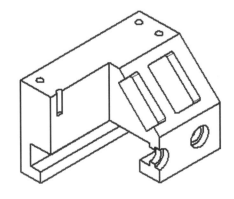

20. Adding crosshatch to the sectioned surfaces:

Crosshatch is added automatically when section views are made in the drawing. Since the cut was created in the model, we will have to add the hatch manually.

Hold the Control key and select the **3 faces** as indicated.

Click the **Area Hatch/Fill** on the **Annotation** tab.

Set the parameters indicated in the dialog box.

Select 3 faces

The Cutaway View is crosshatched.

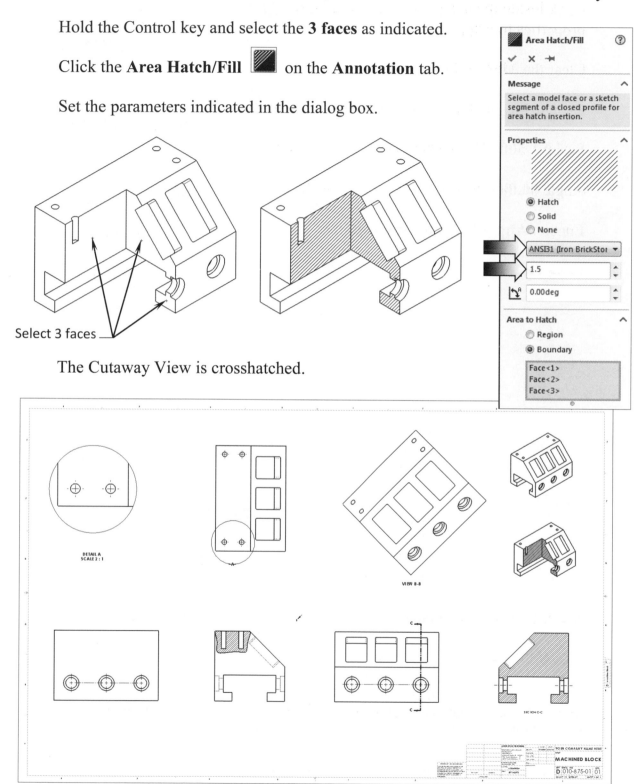

21. Modifying the crosshatch properties:

Zoom in on **Section C-C**.

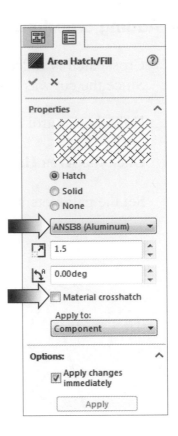

Click inside the hatch area; the **Area Hatch / Fill** properties tree appears.

Clear the **Material Crosshatch** checkbox.

Change the hatch pattern to **ANSI38** (Aluminum).

Scale: **1.500** (The spacing between the hatch lines).

Angle **00.00 deg**. (Sets the angle of the hatch lines).

Enable: **Apply Changes Immediately**.

Click **OK**.

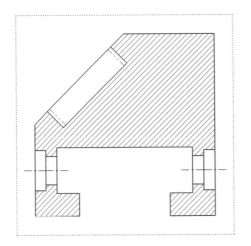

ANSI31 (Iron Brick Stone)

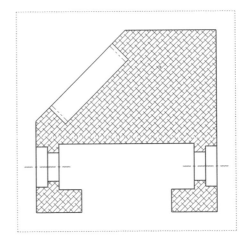

ANSI38 (Aluminum)

When the hatch pattern is applied locally to the selected view, it does not override the global settings in the system options.

Use the same hatch pattern for any views that have the crosshatch in them.

22. Saving your work:

Select **File / Save As**.

Enter **Machine Block-Drawing Views** for file name and click **Save**.

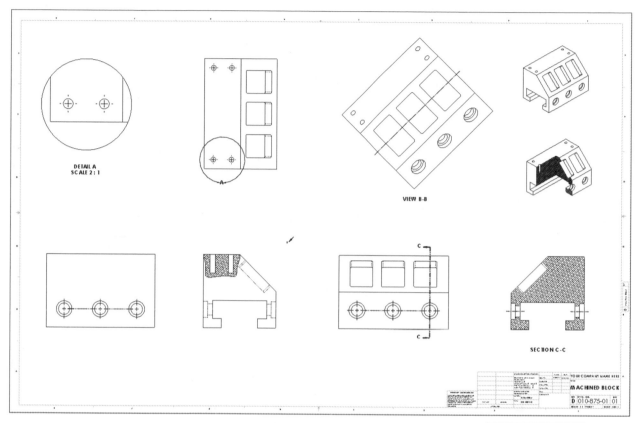

The drawing views can also be changed from Wireframe to **Shaded**.

The color of the drawing views is driven by the part's color. What you see in the drawing is how it is going to look when printed (in color).

Changing the part's color will update the color of the drawing views automatically.

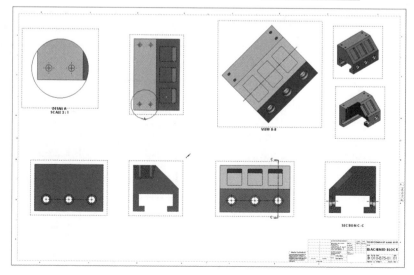

Geometric Tolerance & Flag Notes

Geometric Tolerance Symbols GCS are used in a Feature Control Frame to add Geometric Tolerances to the parts and drawings.

∠	Angularity	⊕	Position
↔	Between	⌓	Profile of Any Surface
◯	Circularity (Roundness)	↗	Simple Runout
◎	Concentricity and Coaxially	↗	Simple Runout (open)
⌭	Cylindricity	—	Straightness
▱	Flatness	≡	Symmetry
⌒	Profile of Any Line	↗↗	Total Runout
//	Parallelism	↗↗	Total Runout (open)
⊥	Perpendicularity	To access the symbol libraries, select **Insert / Annotations / Note**

Modifying Symbols Library

Circle 1-99	Square 1-99	Square/Circle 1-99	Triangle 1-99
①. ①	①. ①	①. ①	①. ①
Circle A-Z	Square A-Z	Square/Circle A-Z	Triangle A-Z
Ⓐ. Ⓐ	Ⓐ. Ⓐ	Ⓐ. Ⓐ	Ⓐ. Ⓐ

Modifying & Hole Symbols

Modifying Symbols MC

℄	Centerline	Ⓣ	Tangent Plane
°	Degree		Slope (Up)
⌀	Diameter		Slope (Down)
S⌀	Spherical Diameter		Slope (Inverted Up)
±	Plus/Minus		Slope (Inverted Down)
Ⓢ	Regardless of Feature Size	□	Square
Ⓕ	Free State	⊠	Square (BS)
Ⓛ	Least Material Condition	⟨ST⟩	Statistical
Ⓜ	Maximum Material Condition		Flattened Length
Ⓟ	Projected Tolerance Zone	ᴘ∟	Parting Line
Ⓔ	Encompassing	SOLIDWORKS supports **ASME-ANSI Y14.5** Geometric & True Position Tolerancing	

⌴ Counterbore (Spot face)   ∨ Countersunk   ▽ Depth/Deep   ⌀ Diameter

ASME (American Society for Mechanical Engineering)
ANSI Y14.5M (American National Standards Institute)

SYMBOL DESCRIPTIONS

ANGULARITY:

The condition of a surface or line, which is at a specified angle (other than 90°) from the datum plane or axis.

BASIC DIMENSION: 1.00

A dimension specified on a drawing as BASIC is a theoretical value used to describe the exact size, shape, or location of a feature. It is used as a basis from which permissible variations are established by tolerance on other dimensions or in notes. A basic dimension can be identified by the abbreviation BSC or more readily by boxing in the dimension.

CIRCULARITY (ROUNDNESS):

A tolerance zone bounded by two concentric circles within which each circular element of the surface must lie.

CONCENTRICITY:

The condition in which the axis of all cross-sectional elements of a feature's surface of revolution are common.

CYLINDRICITY:

The condition of a surface of revolution in which all points of the surface are equidistant from a common axis or for a perfect cylinder.

DATUM:

A point, line, plane, cylinder, etc., assumed to be exact for purposes of computation from which the location or geometric relationship of other features of a part may be established.
A datum identification symbol contains a letter (except I, C, and Q) placed inside a rectangular box.

DATUM TARGET:

The datum target symbol is a circle divided into four quadrants. The letter placed in the upper left quadrant identifies it as associated datum feature. The numeral placed in the lower right quadrant identifies the target; the dashed leader line indicates the target on far side.

FLATNESS:

The condition of a surface having all elements in one plane. A flatness tolerance specifies a tolerance zone confined by two parallel planes within which the surface must lie.

MAXIMUM MATERIAL CONDITION: Ⓜ
The condition of a part feature when it contains the maximum amount of material.

LEAST MATERIAL CONDITION: Ⓛ
The condition of a part feature when it contains the least amount of material. The term is opposite from maximum material condition.

PARALLELISM: //
The condition of a surface or axis which is equidistant at all points from a datum plane or axis.

PERPENDICULARITY: ⊥
The condition of a surface, line, or axis, which is at a right angle (90°) from a datum plane or datum axis.

PROFILE OF ANY LINE: ⌒
The condition limiting the amount of profile variation along a line element of a feature.

PROFILE OF ANY SURFACE: ⌓
Similar to profile of any line, but this condition relates to the entire surface.

PROJECTED TOLERANCE ZONE: Ⓟ
A zone applied to a hole in which a pin, stud, screw, etc. is to be inserted. It controls the perpendicularity of any hole, which controls the fastener's position; this will allow the adjoining parts to be assembled.

REGARDLESS OF FEATURE SIZE: Ⓢ
A condition in which the tolerance of form or condition must be met, regardless of where the feature is within its size tolerance.

RUNOUT: ↗
The maximum permissible surface variation during one complete revolution of the part about the datum axis. This is usually detected with a dial indicator.

STRAIGHTNESS: —
The condition in which a feature of a part must be a straight line.

SYMMETRY: ⩵
A condition wherein a part or feature has the same contour and sides of a central plane.

TRUE POSITION: ⊕
This term denotes the theoretically exact position of a feature.

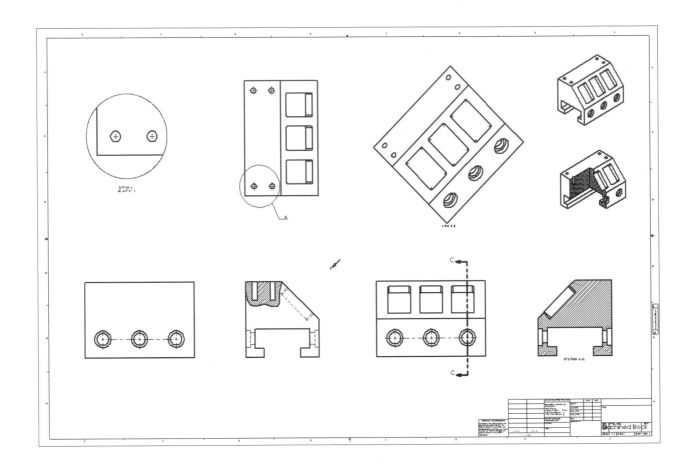

CHAPTER 17

Detailing

Detailing
Machined Block

After the drawing views are all laid out, they will be detailed with dimensions, tolerances, datums, surface finishes, notes, etc.

To fully maintain the associations between the model dimensions and the drawing dimensions, the Model Items options should be used. If a dimension is changed in either mode (from the solid model or in the drawing) they will both be updated automatically.

When a dimension appears in gray color it means that the dimension is being added in the drawing and it does not exist in the model. This dimension is called a Reference Dimension.

SOLIDWORKS uses different colors for different types of dimensions such as:
* Black color dimensions = Sketch dimensions (driving).
* Gray color dimensions = Reference dimensions (driven).
* Blue color dimensions = Feature dimensions (driving)
* Magenta color dimensions = Dimensions linked to Design Tables (driving).

For certain types of holes, the Hole-Callout option should be used to accurately call out the hole type, depth, diameter, etc...

The Geometric Tolerance option is used to control the accuracy and the precision of the features. Both Tolerance and Precision options can be easily created and controlled from the FeatureManager tree.

This chapter discusses most of the tools used in detailing an engineering drawing, including importing dimensions from the model, and adding the GD&T to the drawing (Geometric Dimensions and Tolerancing).

Detailing
Machined Block

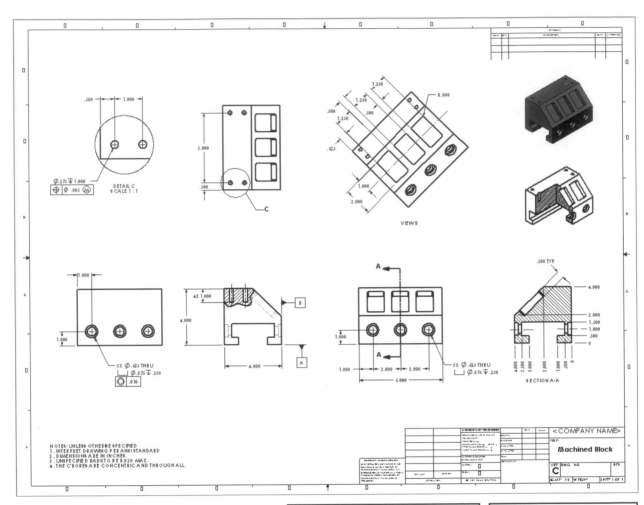

Dimensioning Standards: **ANSI**	Third Angle Projection
Units: **INCHES** – 3 Decimals	

Tools Needed:

 Model Dimensions

 Geometric Tolerance

 Datum Feature Symbol

 Surface Finish

 Hole-Callout

 Note

1. Opening a drawing document:

Select **File / Open** and open the previous drawing document: **Machined Block**.

2. Inserting dimensions from the model:

The dimensions previously created in the model will be inserted into the drawing.

Click the **Right** drawing view dotted border and select **Model Items** on the **Annotation** tab.

For Source/Destination, select **Entire Model**.

For Dimensions, enable **Marked for Drawings**, **Hole Wizard Locations** and **Eliminate Duplicates**.

Under Options, select **Use Dimension Placement in Sketch** and click **OK**.

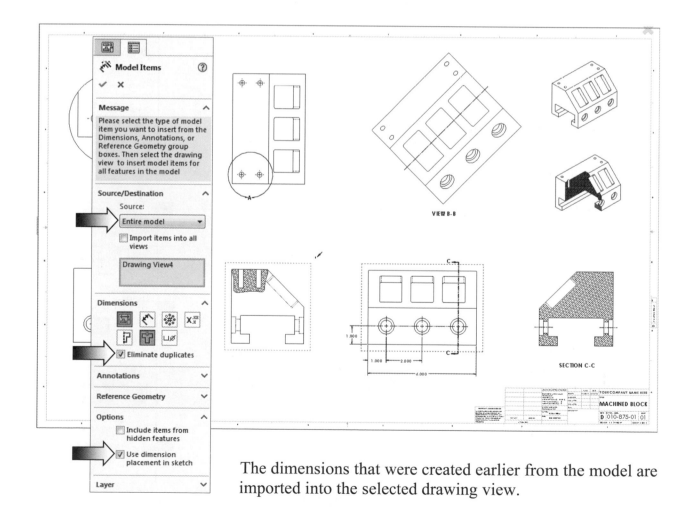

The dimensions that were created earlier from the model are imported into the selected drawing view.

3. Re-arranging the new dimensions: Zoom in 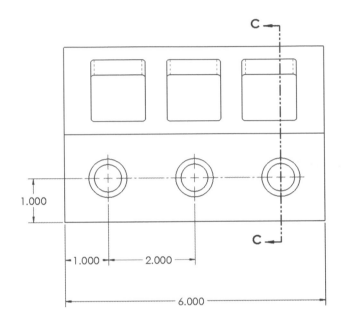 on the **Right** drawing view.

Keep only the hole location dimensions and delete the others. The counterbore hole callouts will be added later.

Use the **Smart Dimension** tool and add any missing dimensions manually.

Center Marks

Center Marks are added automatically unless its checkbox is cleared under: Tools, Options, Document Properties, Detailing.

4. Inserting dimensions to the Section view:

Select the **Section View** dotted border.

Click or select **Insert / Model Items**.

The previous settings (arrow) should still be selected.

Click **OK**.

The dimensions from the model are imported into the Section view.

Some of the dimensions are missing due to the use of the relations in the model.

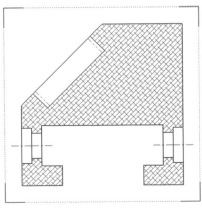

SECTION C-C

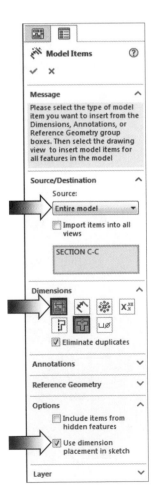

Use the **Smart Dimension** tool 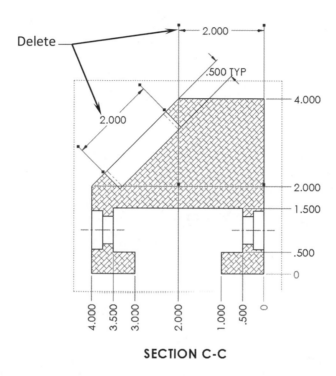 to add any missing dimensions.

If any of these dimensions are changed, both the model and the drawing views will also change accordingly.

Delete the dimensions indicated; they will be added to the auxiliary view later.

SECTION C-C

5. Repeating step 4:

Insert the model's dimensions into the **Top** drawing view.

Re-arrange the dimensions and make them evenly spaced either manually or by using the options in the Align toolbar (View / Toolbars / Align).

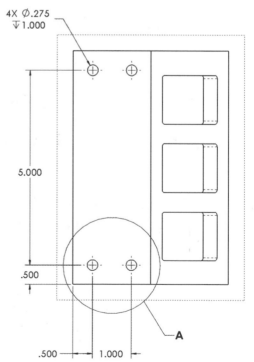

6. Adding dimensions to the Auxiliary view:

Add any missing dimensions to the drawing view using the **Smart Dimension** command.

<u>*NOTE:*</u> *To add the Centerline symbol, select the Note command from the Annotation tab, click the Add Symbol button, then select the Centerline option from the list.*

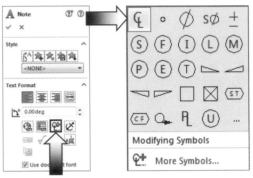

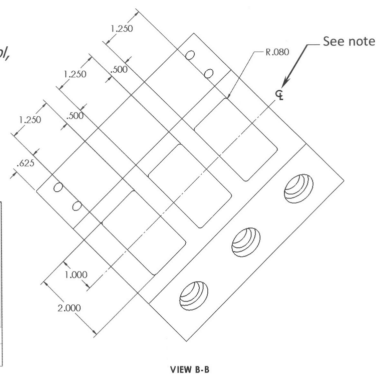

VIEW B-B

7. Adding the Center Marks:

(Skip this step if the center marks are already added.)

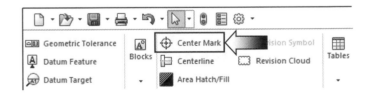

Zoom in 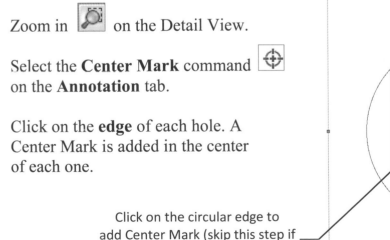 on the Detail View.

Select the **Center Mark** command on the **Annotation** tab.

Click on the **edge** of each hole. A Center Mark is added in the center of each one.

Click on the circular edge to add Center Mark (skip this step if the center marks are already added)

DETAIL A
SCALE 2 : 1

8. Adding center marks to other holes:

Click the **Center Mark** command once again.

Add a center mark to other holes by clicking on their circular edges.

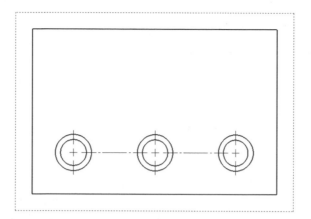

9. Adding the Datum Feature Symbols:

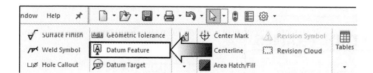

Click or select **Insert / Annotations / Datum Feature Symbol**.

Select **edge 1** and place **Datum A** as shown.

Select **edge 2** and place **Datum B** as shown.

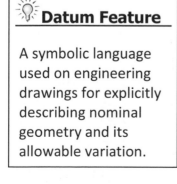

> 💡 **Datum Feature**
>
> A symbolic language used on engineering drawings for explicitly describing nominal geometry and its allowable variation.

1982 Datum symbol   − A −

1994 Datum symbol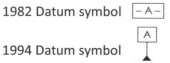

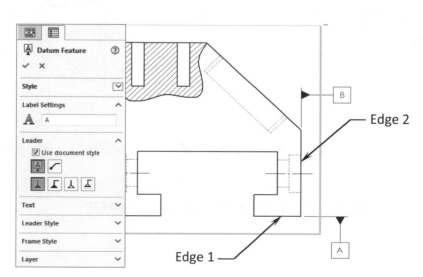

Edge 2

Edge 1

Datum Reference & Geometric Tolerance Examples

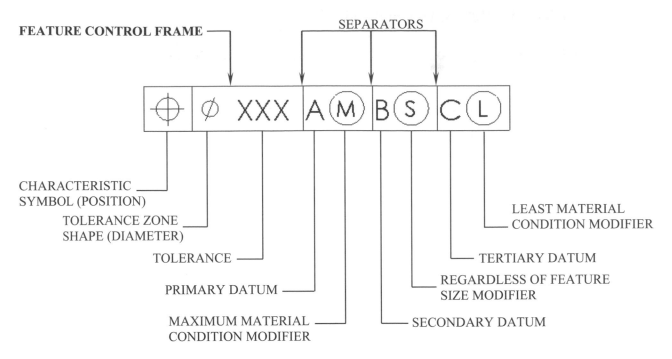

FEATURE CONTROL FRAME

SEPARATORS

CHARACTERISTIC SYMBOL (POSITION)

TOLERANCE ZONE SHAPE (DIAMETER)

TOLERANCE

PRIMARY DATUM

MAXIMUM MATERIAL CONDITION MODIFIER

LEAST MATERIAL CONDITION MODIFIER

TERTIARY DATUM

REGARDLESS OF FEATURE SIZE MODIFIER

SECONDARY DATUM

💡 FEATURE CONTROL FRAMES

A feature control frame symbolizes the tolerance requirements for a feature of a part. It can be added to a drawing note for a feature tolerance, or can be specified by running a leader line from the feature control frame directly to the feature. The box may be attached to an extension line from the feature or it can be placed on a dimension line. A feature can have more than one feature control frame, depending on its requirements.

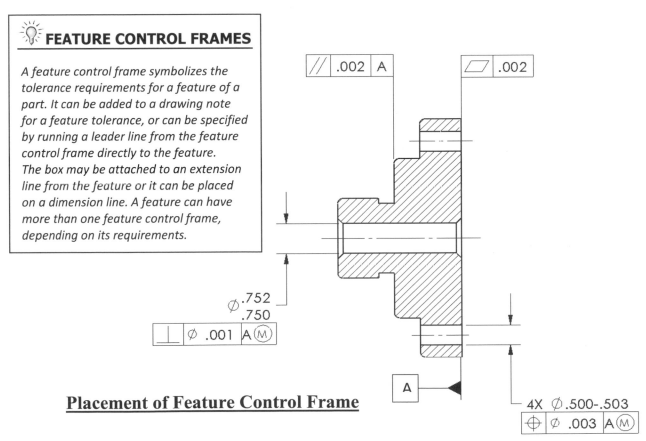

Placement of Feature Control Frame

10. Adding hole specifications using the Hole-Callout:

Click [⊔∅] or select **Insert / Annotations / Hole Callout**.

Select the <u>edge</u> of the **Counterbore** and place the callout below the hole.

A callout is added as a reference dimension (gray color); it includes the diameter and the depth of the hole and the Counterbore.

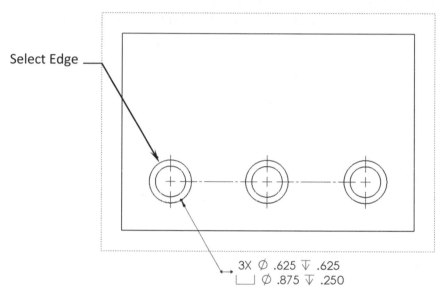

Select Edge

3X ∅ .625 ⯆ .625
⊔ ∅ .875 ⯆ .250

11. Adding Geometric Tolerances:

Geometric dimensioning and tolerancing (GD&T) is used to define the nominal (theoretically perfect) geometry of parts and assemblies to define the allowable variation in form and possibly size of individual features and to define the allowable variation between features. Dimensioning and tolerancing and geometric dimensioning and tolerancing specifications are used as follows:

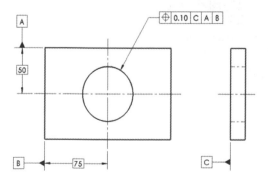

* Dimensioning specifications define the nominal, as modeled or as-intended geometry. One example is a basic dimension.

* Tolerancing specifications define the allowable variation for the form and possibly the size of individual features and the allowable variation in orientation and location between features. Two examples are linear dimensions and feature control frames using a datum reference.

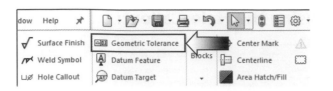

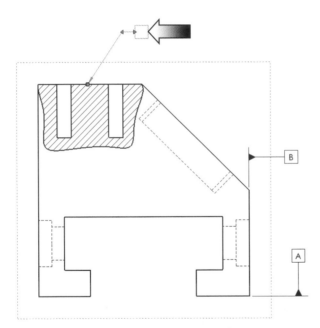

Zoom in on the **Front** drawing view.

Select the **upper edge** as shown.

Click or select **Insert / Annotations / Geometric Tolerance**.

A small <u>square</u> (arrow) and a leader line appear,
click to place the small square and access the **Tolerance Symbol** dialog box.

Select **Parallel** // from the list.

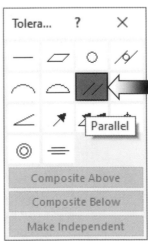

Enter **.010** in the Tolerance box.

Click **Add Datum** and enter **A** under Primary reference datum (arrow).

Click **Done** and **OK**.

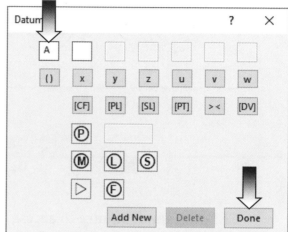

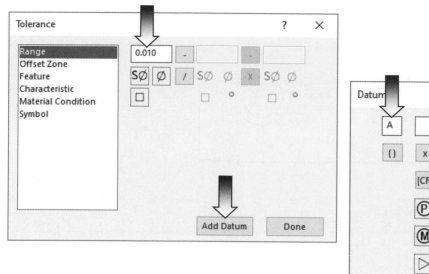

12. Align the Geometric Tolerance:

Drag the Control Frame to the left side until it snaps to the horizontal alignment with the upper edge, then release the mouse button.

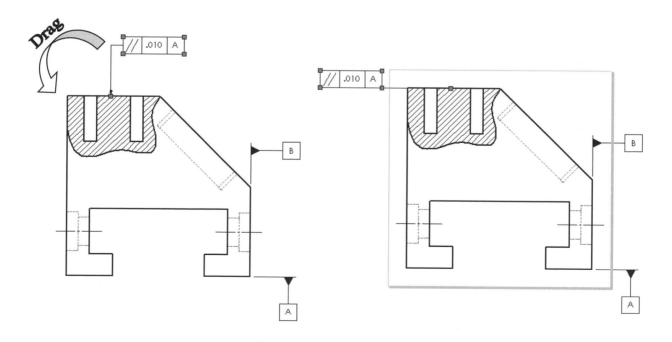

13. Attaching the Geometric Tolerance to the Driving dimension:

Select the **Counterbore dimension**. By pre-selecting a dimension, the geometric-tolerance will automatically be attached to this dimension.

Click 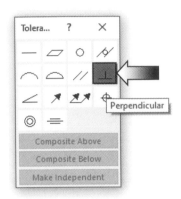 or select **Insert / Annotation / Geometric Tolerance**.

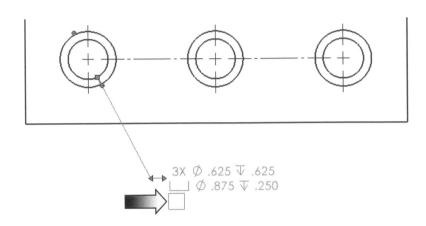

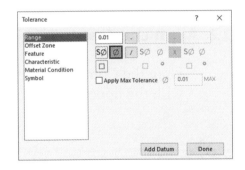

Click the **small square** to access the **Symbol Library**.

Select **Perpendicularity** from the list.

Click in the Tolerance box, select the **Diameter** symbol and enter **.005**.

Click **Add Datum** and enter **B** in the reference datum box.

Select the Ⓜ (Maximum Material Condition) to place it next to the Datum B.

Click **Done** and **OK**.

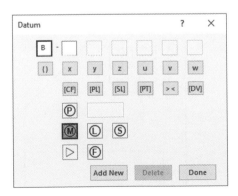

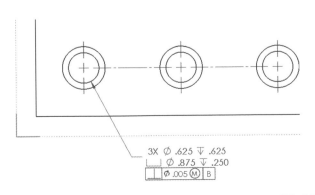

The Geometric Tolerance Control Frame is attached to the counterbore dimension.

14. Adding Tolerance/Precision to dimensions:

Select the dimension **.500** (circled).

The dimension properties tree pops up; select **Bilateral** under the Tolerance/ Precision section (arrow).

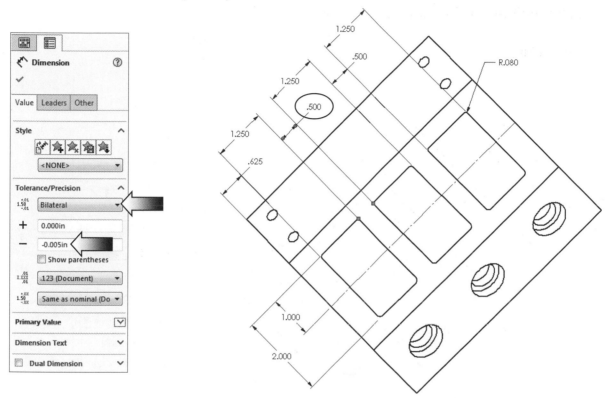

Enter **.000 in**. for Max variation.

Enter **.005 in**. for Min variation.

Click **OK**.

VIEW B

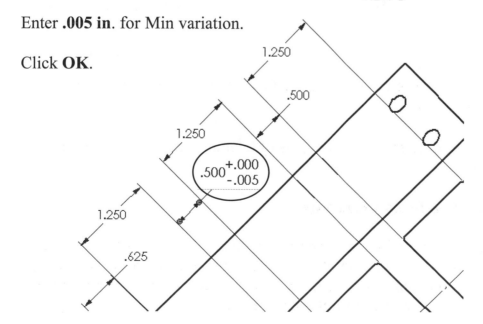

15. Adding Symmetric tolerance to a dimension:

Click the width dimension **1.250** (circled).

Select **Symmetric** under Tolerance/Precision list (arrow).

Enter **.003 in**. for Maximum Variation.

Click **OK**.

Repeat step 15 and add a Symmetric tolerance to the height dimension: (**2.000 ±.003**).

For practice purposes: Add other types of tolerance to some other dimensions such as Min–Max, Basic, Fit, etc.

16. Adding Surface Finish callouts:

Surface finish is an industrial process that alters the surface of a manufactured item to achieve a certain property. Finishing processes such as machining, tumbling, grinding, or buffing etc., may be applied to improve appearance and other surface flaws and control the surface friction.

Zoom in 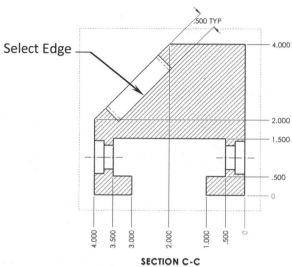 on the **Section C-C**.

Select the <u>edge</u> as indicated.

Click the **Surface Finish** command ⌐√.
on the **Annotation** tab.

Choose **Machining-Required** ⌐√ under Symbol.

Under **Maximum Roughness**,
enter **125**

For Leader, select **Bent leader.**

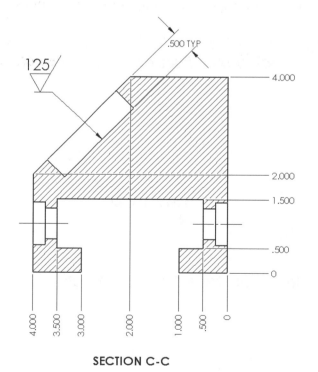

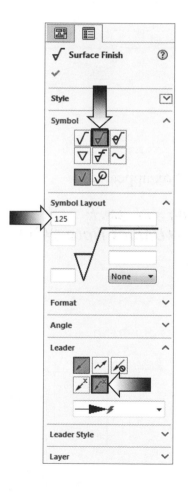

Click **OK**.

17. Adding non-parametric callouts:

Click **Smart Dimension** 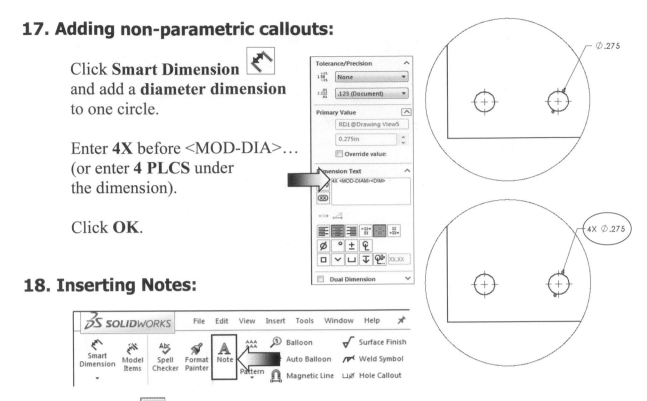 and add a **diameter dimension** to one circle.

Enter **4X** before <MOD-DIA>… (or enter **4 PLCS** under the dimension).

Click **OK**.

18. Inserting Notes:

Zoom in on the upper left side of the drawing.

Click Note **A** on the **Annotations** tool tab.

Click in upper left area, approximately as shown, a note box appears.

NOTE: To lock the note to the sheet, right-click in the drawing and select: **Lock-Sheet-Focus** (or select the note and click **Lock/Unlock Note**)

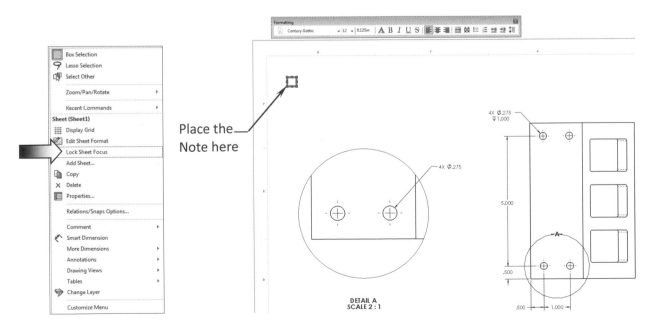

Enter the notes below:

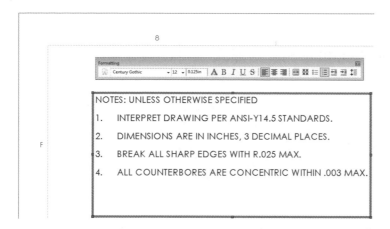

Click **OK** when you are finished typing the notes.

19. Changing document's font:

Double-click anywhere inside the note area to activate it.

Highlight the entire note and select the following:

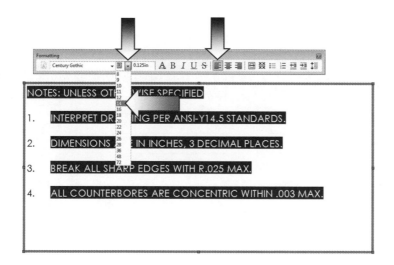

Font: **Century Gothic**.

Point size: **14**.

Alignment: **Left**.

Click **OK**.

20. Saving your work:

Select **File / Save As**.

Enter **Machined Block Detailing** for the name of the file.

Click **Save**.

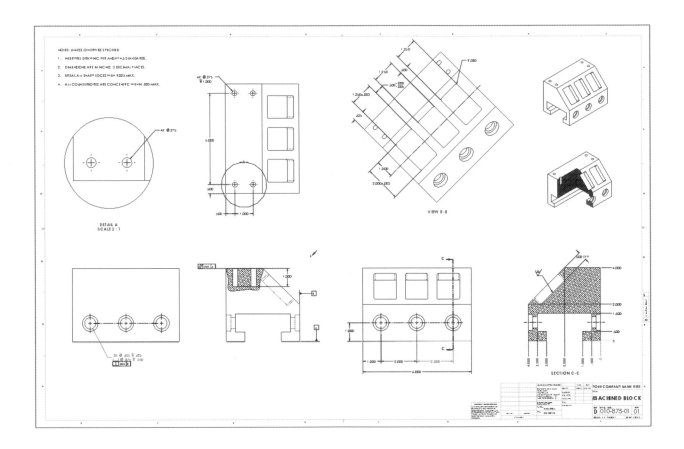

Questions for Review

1. Existing text in the title block can be edited when the Sheet-Format is active.
   a. True
   b. False

2. The standard drawing views can be created using the following method:
   a. Use Insert / Drawing views menu
   b. Use the Model-View command
   c. Drag and drop from an open window
   d. All the above

3. The default view alignments can be broken or re-aligned.
   a. True
   b. False

4. The Detail-View scale cannot be changed or controlled locally.
   a. True
   b. False

5. The Projected view command projects and creates the Top, Bottom, Left, and Right views from another view.
   a. True
   b. False

6. To create an Auxiliary view, an edge must be selected, not a face or a surface.
   a. True
   b. False

7. Only a single line can be used to create a Section view. The System doesn't support a multi-line section option.
   a. True
   b. False

8. Hidden lines in a drawing view can be turned ON / OFF locally and globally.
   a. True
   b. False

9. Configurations created in the model cannot be shown in the drawings.
   a. True
   b. False

9. FALSE
7. FALSE   8. TRUE
5. TRUE   6. TRUE
3. TRUE   4. FALSE
1. TRUE   2. D

Exercise: Detailing I

1. Create the **part** <u>and</u> the **drawing** as shown below.
2. The Counter-Bore dimensions are measured from the Top planar surface.
3. Dimensions are in inches, 3 decimal places.
4. The part is symmetrical about the Top reference plane.
   (To create the "Back-Isometric-View" from the model: hold the Shift key and push the Up-arrow key twice. This will rotate the part 180°, and then insert the view to the drawing using the **Current-View** option.)

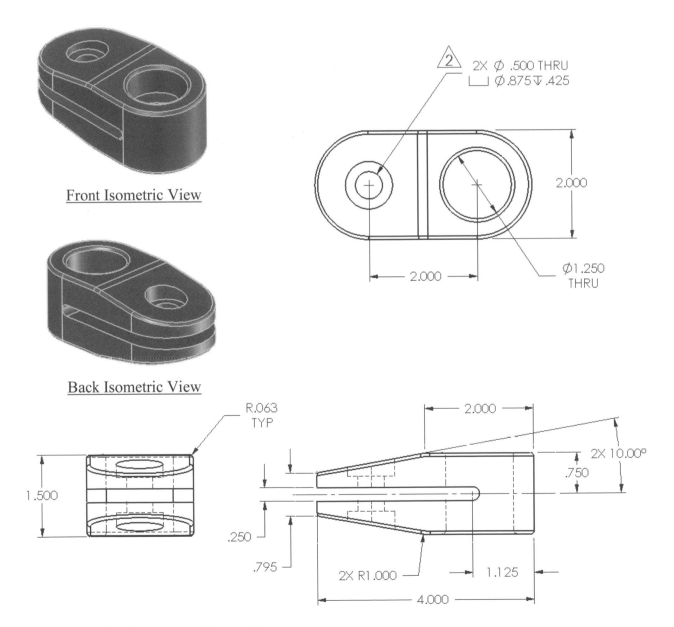

Front Isometric View

Back Isometric View

5. Save the drawing as **Clamp Block**.

Exercise: Detailing II

1. Open the part named **Base Mount Block** from the Training Files folder.
2. Create a drawing using the information provided below.
3. Add the Virtual-Sharps where needed prior to adding the Ordinate Dimensions.
   (To add the Virtual Sharps: Hold the Control key, select the 2 lines, and click the Sketch-Point command.)

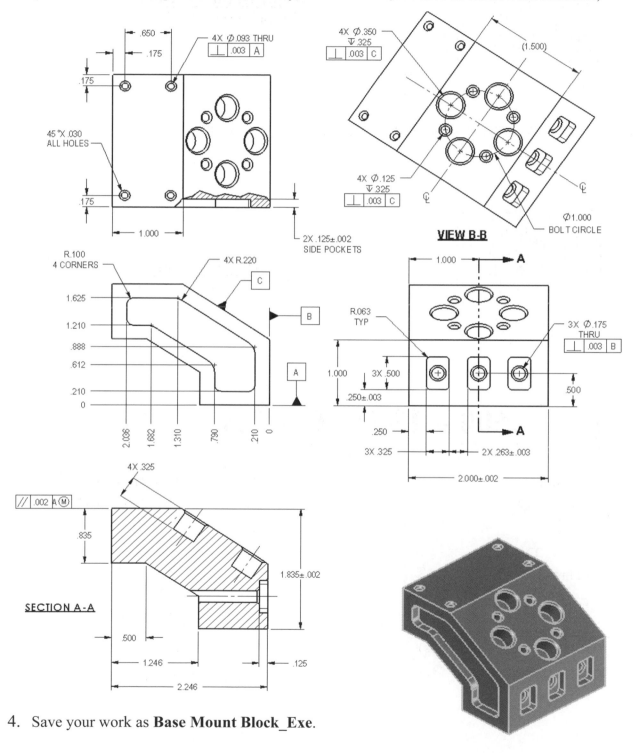

4. Save your work as **Base Mount Block_Exe**.

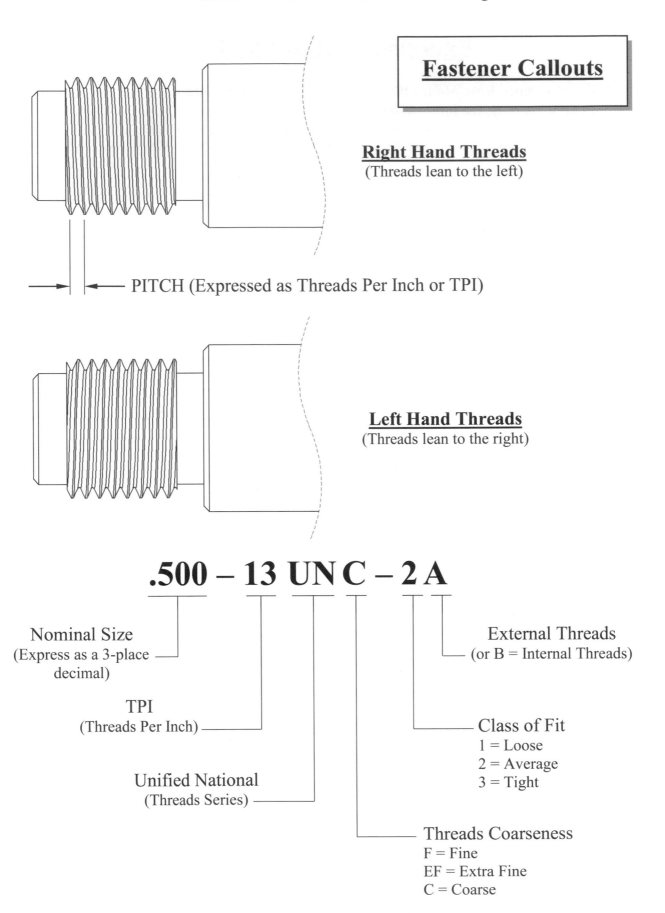

Fastener Callouts

Right Hand Threads
(Threads lean to the left)

PITCH (Expressed as Threads Per Inch or TPI)

Left Hand Threads
(Threads lean to the right)

.500 – 13 UN C – 2 A

Nominal Size
(Express as a 3-place decimal)

TPI
(Threads Per Inch)

Unified National
(Threads Series)

Threads Coarseness
F = Fine
EF = Extra Fine
C = Coarse

Class of Fit
1 = Loose
2 = Average
3 = Tight

External Threads
(or B = Internal Threads)

Thread Nomenclature

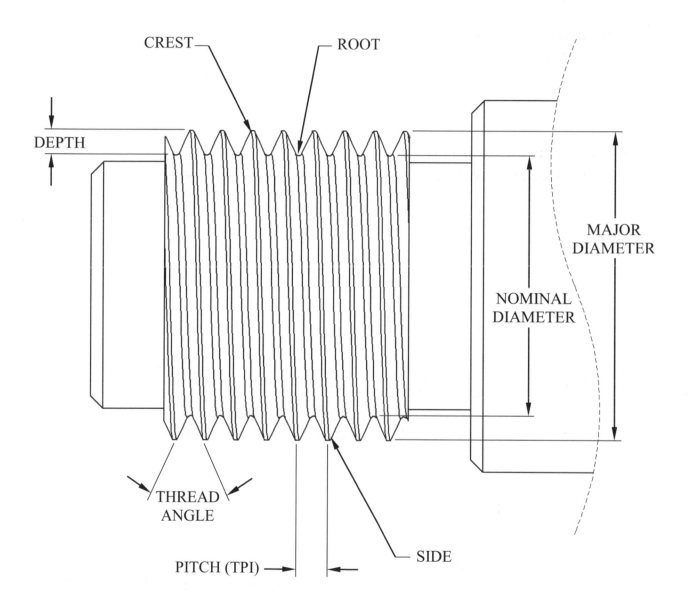

Attaching a note or a symbol to a dimension

1. Opening a drawing:

Open a drawing document named **Attaching Symbol to Dim.slddrw**.

2. Adding symbol to a dimension:

Select the **5.00** dimension (circled).

Expand the **Dimension Text** section on the property tree.

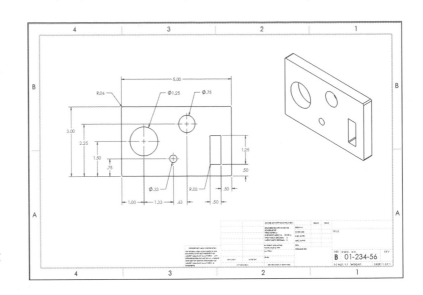

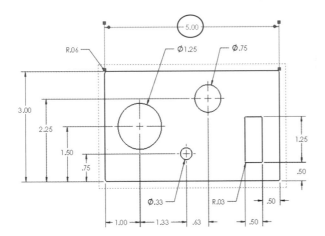

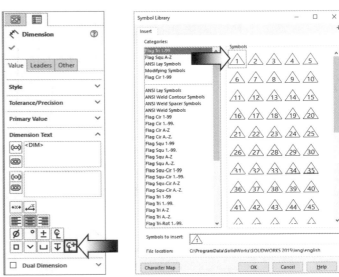

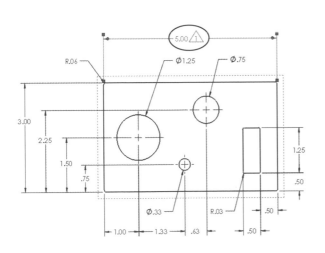

Click the **More Symbols** button (arrow).

From the Flag Tri 1-99 category, select the **Triangular 1** symbol (arrow).

Click **OK**.

CHAPTER 18

Sheet Metal Drawings

Sheet Metal Drawings
Post Cap

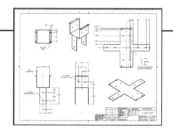

When a sheet metal drawing is made from a sheet metal part, SOLIDWORKS automatically creates a Flat Pattern view to use in conjunction with the Model-View command.

Any drawing views can be toggled to show the flattened stage, along with the bend lines. By accessing the Properties of the view, you can change from Default (Folded) to Flattened.

By default, the Bend Lines are visible in the Flat Pattern, but the Bend Regions are not. To show the bend regions, open the sheet metal part and right-click the **Flat-Pattern** (in the FeatureManager design tree) select **Edit Feature** and clear **Merge Faces**. You may have to rebuild the drawing to see the bend regions.

Both Folded and Flat Pattern drawing views can be shown on the same drawing sheet if needed, or they can be created using multiple sheets but within the same drawing document. The dimensions and annotations can then be added to define the drawing views.

Changes done to the Sheet Metal part <u>**will**</u> populate to the drawing views, and the drawing itself can also be changed to update the sheet metal part as well. However, unlike the machined parts, most sheet metal fabricators would prefer a 2D Flat-Pattern drawing of the part to re-calculate the setback allowances by themselves, rather than using the geometry directly from a 3D model.

This lesson will guide you through the basics of creating a sheet metal drawing and the use of the Default/Flat-Pattern configurations.

Sheet Metal Drawings
Post Cap

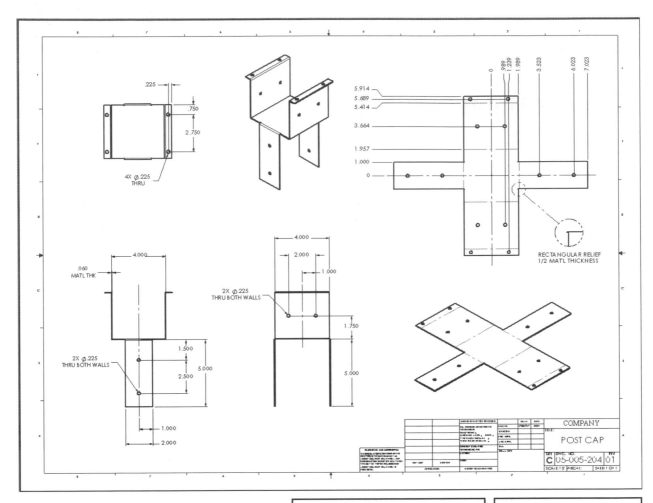

Dimensioning Standards: **ANSI**

Units: **INCHES** – 3 Decimals

Third Angle Projection

Tools Needed:

New Drawing

Model View

 Detail View

Model Items

Vertical Ordinate

Horizontal Ordinate

1. Starting a new drawing:

Select **File / New / Drawing** and click **OK**.

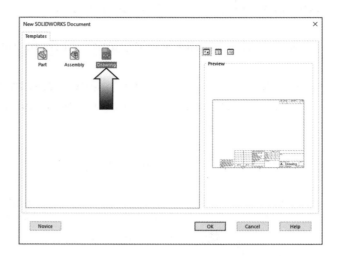

NOTE: *If you have already created and saved the template from the previous chapters, you can use that same template for this drawing.*

Click **Cancel** (arrow) to exit the Insert Part mode. The drawing paper size, drawing view scale, and the Projection angle must be set first.

If the Sheet Properties dialog does not appear, right-click in the drawing and select Properties.

Set **Scale** to **1:2** and set **Type of Projection** to **Third Angle**.

Select **C-Lanscape** sheet size.

Enable **Display Sheet Format** and click **OK** [OK].

Go to **Tools / Options** and change the **Units** to **IPS** (Inch / Pound / Second).

(The document units can also be changed at the bottom right corner of the screen.)

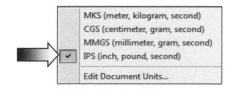

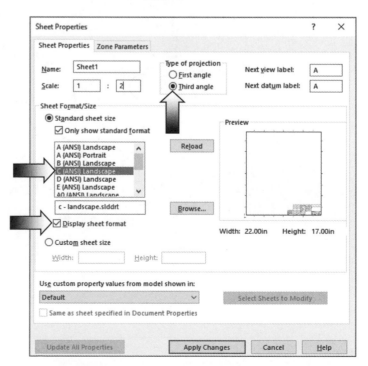

2. Creating the 3 Standard Views:

Select **the Model View** command from the **View Layout** tab.

Click the **Browse** [Browse...] button, and from the Training Files folder, locate and open the part named **Post Cap**.

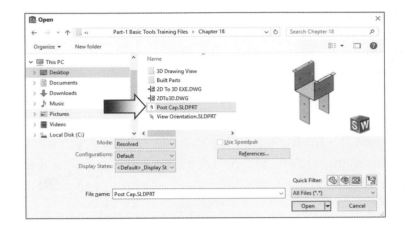
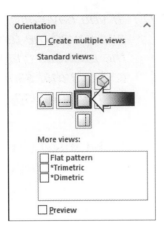

After a part document is selected SOLIDWORKS automatically switches to the Projected View mode.

Start by placing the **Front view** approximately as shown and then add the other **3 views** as labeled.

Click **OK** to stop the projection.

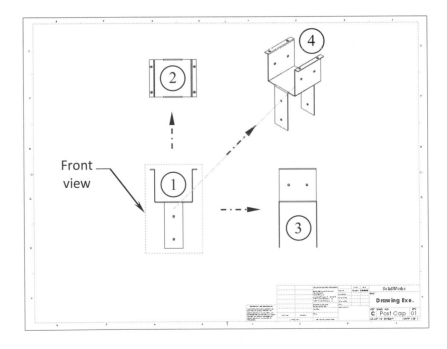

3. Rearranging the drawing views:

Rearrange the drawing views to the approx. positions by dragging their dotted borders. The flat-pattern view will be placed on the right side of the drawing.

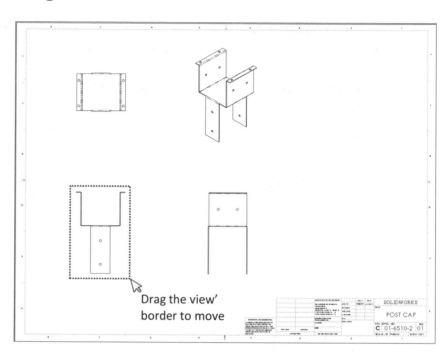

Drag the view' border to move

4. Creating the Flat-Pattern drawing view:

NOTE: Bend Notes should be turned off _prior_ to creating the Flat Pattern view: (**Tools / Option / Document-Properties / Sheet Metal**).

Select the **Model View** command ▦, click **Next** ➡, select the **Flat-Pattern** view (arrow) from the list and place it approximately as shown.

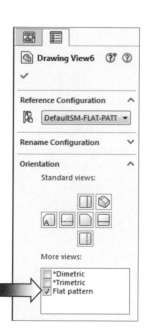

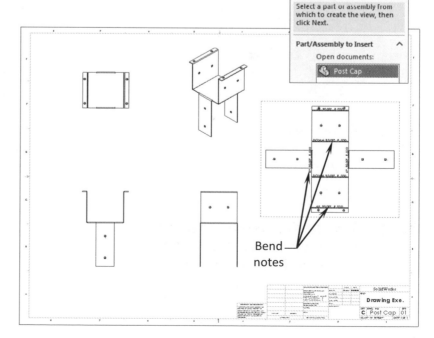

Bend notes

5. Creating a Detail view:

Click the **Detail View** command 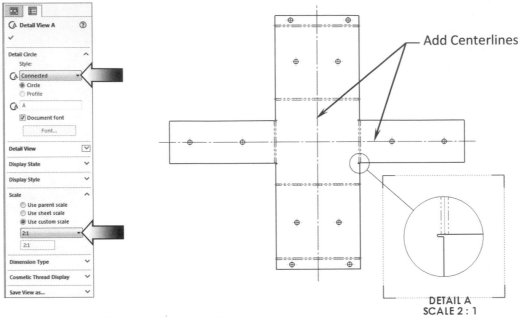 and sketch a **Circle** at the lower right corner of the flattened view, approximately as shown. Also add **2 centerlines**.

Use the **Connected** option and **Custom Scale** of **2:1** (arrows).

6. Adding the Ordinate dimensions:

Expand **Smart Dimension** to access its options.
Select **Horizontal Ordinate Dimension** (arrow).

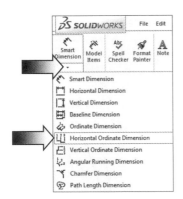

**NOTE:** Ordinate dimensions are a group of dimensions measured from a zero point. When adding them in a _drawing_, they become reference dimensions and their values cannot be changed.

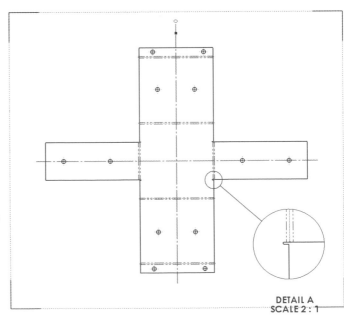

Starting at the
Vertical Centerline,
add the Horizontal
Ordinate dimensions
as shown here.

Click the drop-down
arrow under Smart
Dimension and
select **Vertical
Ordinate Dimension**

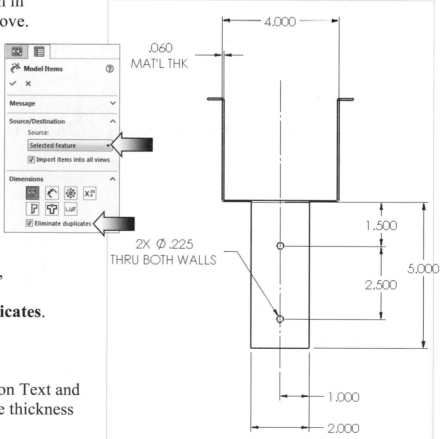

Start the Horizontal here

Start the
Vertical here

(BEND LINES AND HOLES
LOCATIONS DIMENSION)

RECTANGULAR RELIEF
1/2 MAT'L THICKNESS

Modify the label of the Detail View to say:
Rectangular Relief ½ Mat'l Thickness.

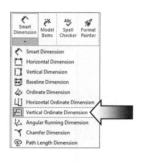

Add the Vertical Ordinate
dimensions as shown in
the drawing view above.

7. Adding the Model
dimensions:

Select the **Front**
view dotted border
to activate the view.

Switch to the
Annotation tab
and click the
Model Items button,
select **Entire Model**
and **Eliminate Duplicates**.

Click **OK**.

Modify the Dimension Text and
add the depth and the thickness
callouts as shown.

.060
MAT'L THK

4.000

2X Ø .225
THRU BOTH WALLS

1.500

2.500

5.000

1.000

2.000

Repeat step 6 and add the Model Dimensions 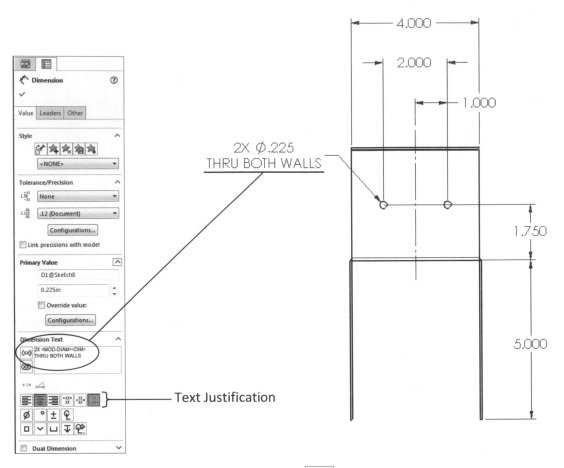 to the **Right** drawing view.

Add the annotations below the dimension text (circled).

Continue adding the Model Dimensions to the Top drawing view.

Modify the dimension
Ø.225 to include the
number of instances
and depth.

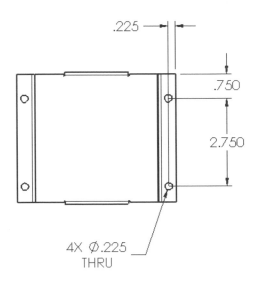

8. Creating the Isometric Flat Pattern view:

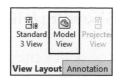

Click the **Model View** command from the **View Layout** tab.

Click the **Next button** ⊕, select the Isometric View from the menu and place it below the flat-pattern view.

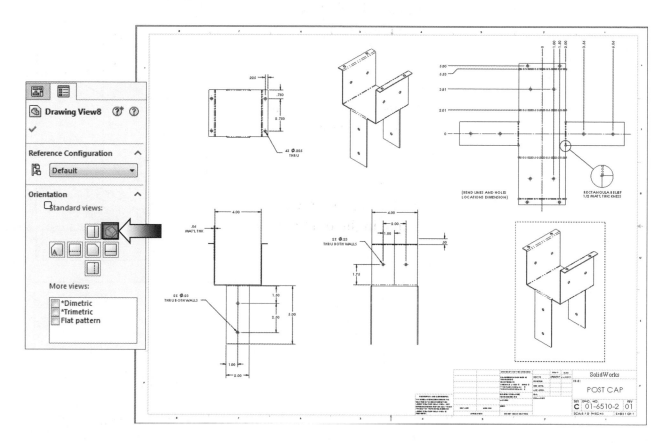

A drawing view can also be copied and pasted using Ctrl+C and Ctrl+V.

Under the Reference Configuration section, change the **Default** configuration to **SM-Flat-Pattern** configuration (arrow).

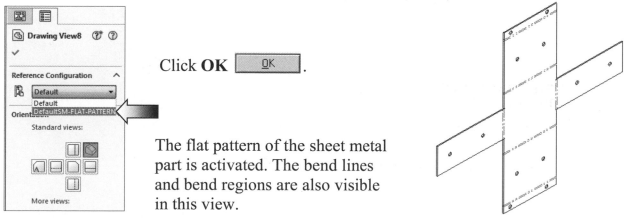

Click **OK**.

The flat pattern of the sheet metal part is activated. The bend lines and bend regions are also visible in this view.

9. Showing/Hiding the Bend Lines:

Click the **DrawingManager** tab (arrow), scroll down the tree and expand the last drawing view in the FeatureManager tree.

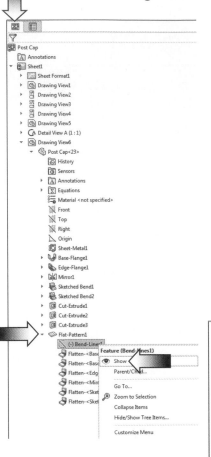

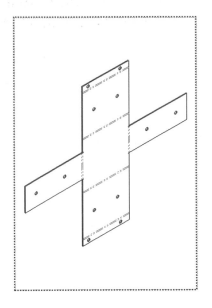

Expand the feature **Flat-Pattern1**.

Right-click Bend-Lines1 and select **Show**.

The Bend Lines are now visible in the drawing view.

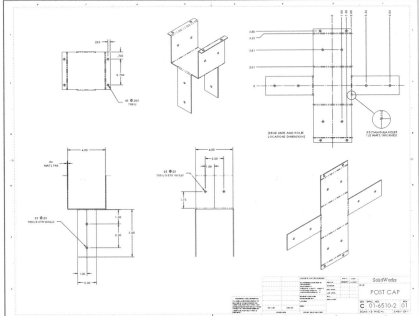

10. Saving your work:

Select **File / Save As**.

Enter **Post Cap.slddrw** for the name of the file.

Click **Save**.

SOLIDWORKS eDrawing & 3D Drawing View

eDrawing & 3D Drawing View
Soft-Lock Assembly

eDrawing:

eDrawing is one of the most convenient tools in SOLIDWORKS to create, share, and view your 2D or 3D designs.

The eDrawing Professional allows the user to create eDrawing markup files (***.markup**) that have markups, such as text comments and geometric elements.

With eDrawing 2023 and SOLIDWORKS® 2023, you can create an eDrawing from any CAD model or assembly from programs like AutoCAD®, Inventor, Creo®, and others.

The following types of eDrawing files are supported:
* 3D part files (***.eprt**)
* 3D assembly files (***.easm**)
* 2D drawing files (***.edrw**)

3D Drawing View:

The 3D drawing view mode lets you rotate a drawing view out of its plane so you can see components or edges obscured by other entities. When you rotate a drawing view in 3D drawing view mode, you can save the orientation as a new view orientation.

3D drawing view mode is particularly helpful when you want to select an obscured edge for the depth of a broken-out section view. Additionally, while in 3D drawing view mode, you can create a new orientation for another model view. 3D drawing view mode is not available for detail, broken, crop, empty, or detached views.

SOLIDWORKS e-Drawing & 3D Drawing View
Soft-Lock Assembly

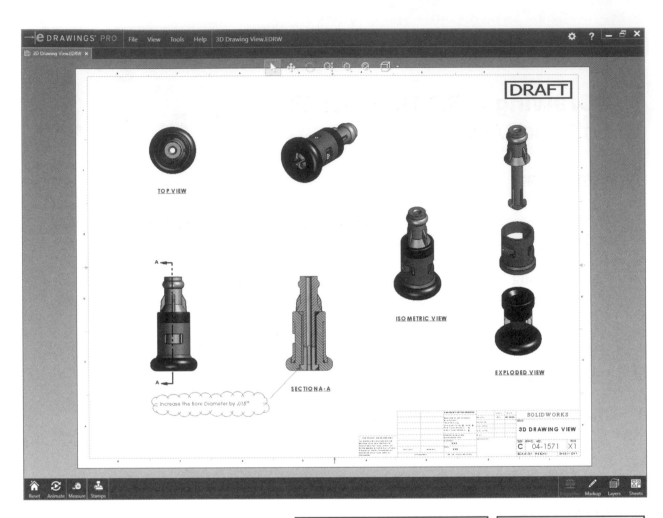

Dimensioning Standards: **ANSI**	Third Angle Projection
Units: **INCHES** – 3 Decimals	

Tools Needed:

 SOLIDWORKS e-drawing

 Play Animation

 Stop Animation & Returns to Full Screen

 Rotate View

 3D Drawing View

1. Opening an existing drawing:

Click **File / Open**.

Select **3D Drawing View.slddrw** from the Training Files folder and open it.

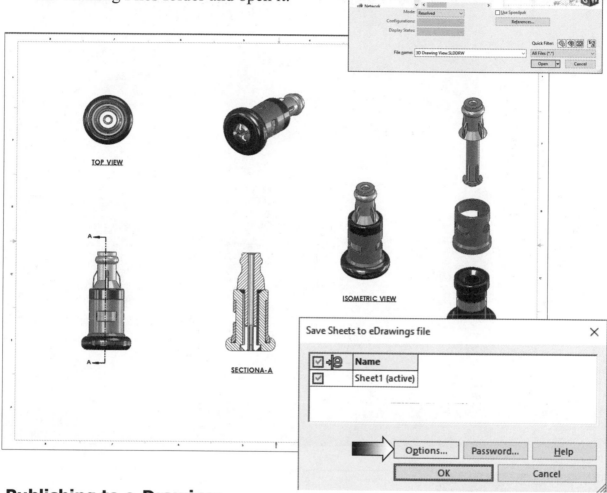

TOP VIEW

ISOMETRIC VIEW

SECTION A-A

2. Publishing to e-Drawing:

Select **File / Publish to eDrawing**.

Click the **Options** button.

Enable the **OK to Measure** and **Save Shaded Data in Drawings** checkboxes.

Click **OK** twice.

The eDrawing app is launched.

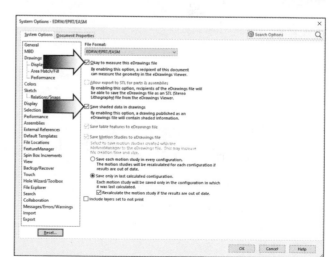

3. Working with eDrawing:

Open the **e-Drawing** program  if it did not open on its own from the last step.

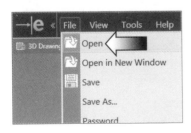

Click **Open**.

Select the document named: **3D Drawing View.edrw** and click **Open**.

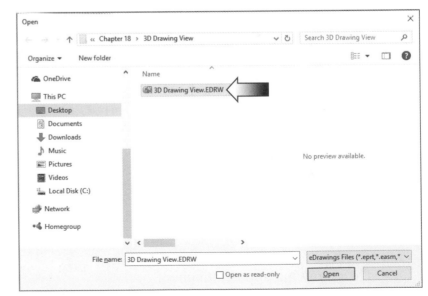

The drawing that was previously saved is opened in the eDrawing window. Click the eDrawing **Options** button to see all available options.

eDrawing Options

The eDrawing Options:

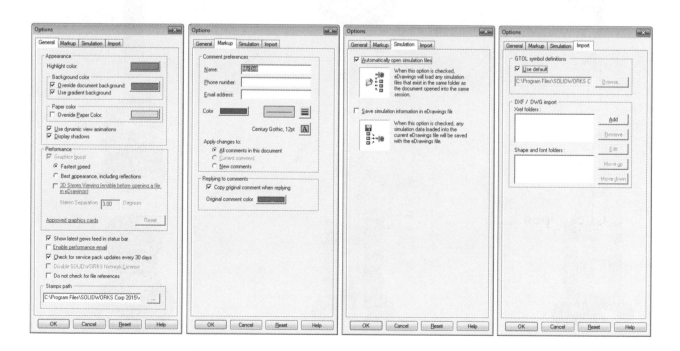

The Markup toolbar:

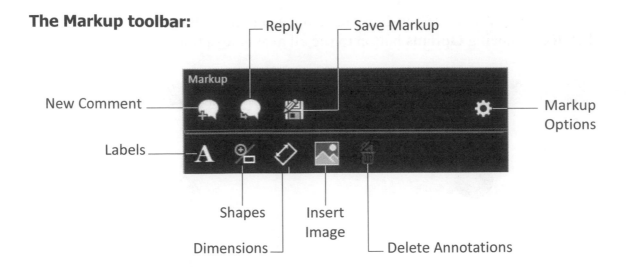

4. Playing the Animation:

Click **Animate** 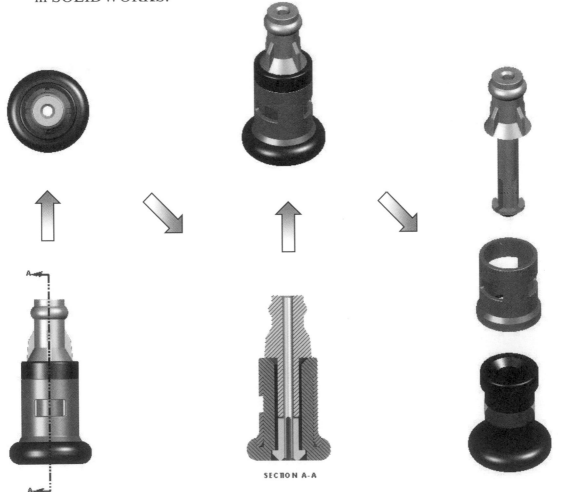 and **Play** (at the bottom left corner of the screen).

The eDrawing animates the drawing views based on the order that was created in SOLIDWORKS.

SECTION A-A

Click **Reset** 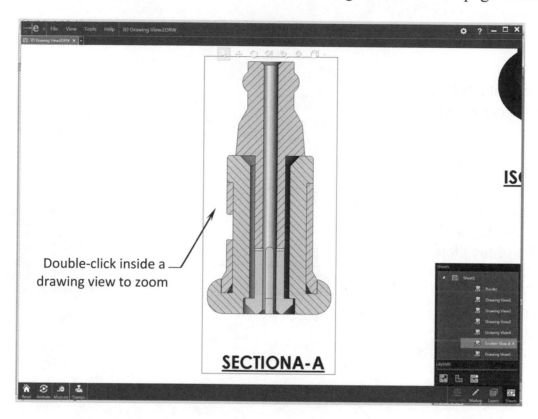 to stop the animation.

The Reset command also returns the eDrawing back to its full-page mode.

Double-click inside the Section view to zoom and fit it to the screen.

5. Adding the markup notes:

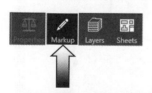

Click the **Labels** button and select the **Cloud With-Leader** command (2nd arrow).

Click on the **Edge** of the bottom bore hole and place the blank note on the lower left side.

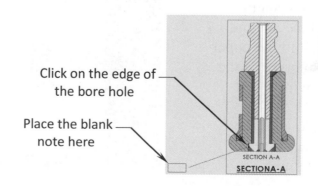

Click on the edge of the bore hole

Place the blank note here

Enter the note:
Increase the Bore Diameter by .015".

Click **OK** .

Zoom out a little to see the entire note.

Click inside the cloud; there are 4 handle points to move or adjust the width of the cloud.

Adjust the cloud by dragging one of the handle points.

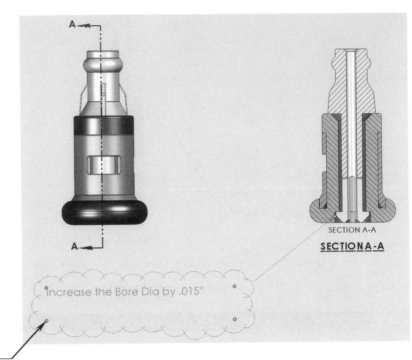

Drag handle point

6. Adding a "Stamp":

Click the **Stamp** button at the bottom left corner.

Locate the **DRAFT** stamp from the list and drag it to the upper right corner of the eDrawing.

Drag one of the handles to resize the stamp.

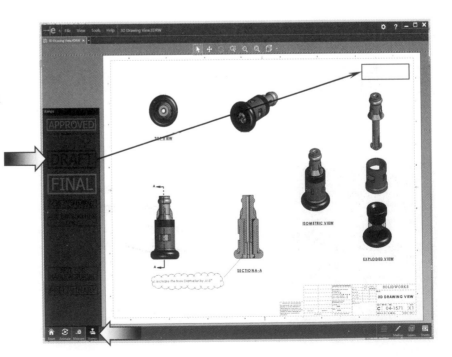

Click the **Reset** button 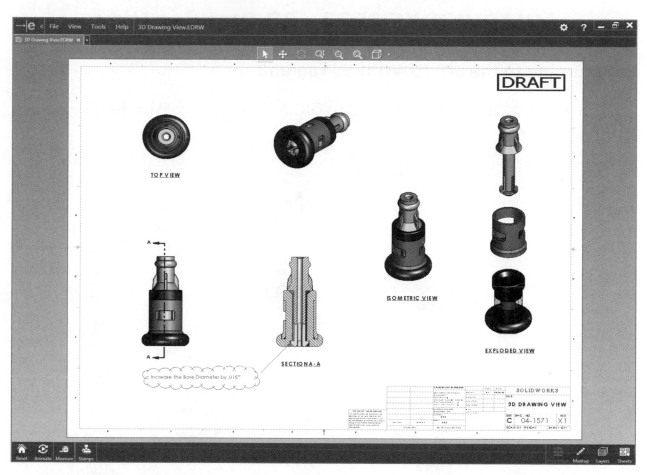 to return to the full drawing mode.

7. Saving as Executable file type:

Click **File / Save As**.

Select **eDrawing Executable Files (*.exe)** from the Save As Type drop-down list.

Click **Save**.

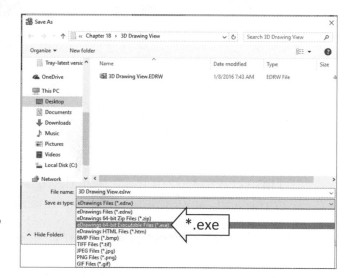

NOTES:
The **.exe** files are much larger in file size than the standard .edrw because the eDrawing-viewer is embedded in the file.

The **.edrw** files are smaller in file size but require eDrawing Viewer or the eDrawing program itself to view.

Continued...

SOLIDWORKS 2023 - 3D Drawing View

8. Returning to the SOLIDWORKS Program:

Switch back to the previous SOLIDWORKS drawing (press **Alt+ Tab**).

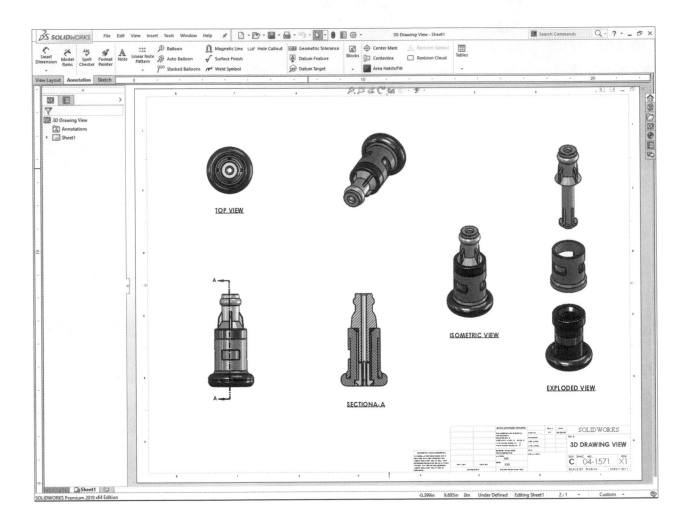

This second half of the lesson discusses the use of the 3D Drawing View command, one of the unique options that allows a flat drawing view to be manipulated and rotated in 3D to make selections of the hidden entities.

3D drawing view mode is particularly helpful when you want to select an obscured edge for the depth of a broken-out section view. Additionally, while in 3D drawing view mode, you can create a new orientation for another model view.

9. Using the 3D Drawing View command:

Create a new Front view and then click the drawing view's border to activate it.

Click **3D Drawing View** icon  from the View toolbar (or under the **View / Modify** menus).

Select the **Rotate** tool  and rotate the Front drawing to a different position approximately as shown.

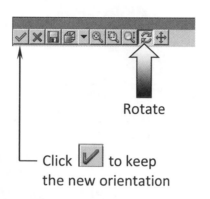

Rotate

Click ✓ to keep
the new orientation

Front Drawing View   **3D Drawing View pop-up toolbar**   *New 3D Drawing View*

10. Saving the New-View orientation:

Click **Save** 🖫 on the 3D Drawing View pop-up toolbar.

Enter **3D Drawing View** in the Named View dialog and click **OK**.

This view orientation will be available in the drawing under the **More Views** section, the next time you insert a model view (arrow).

11. Saving your work:

Click **File / Save As**.

Enter **3D Drawing View** for the name of the file.

Click **Save**.

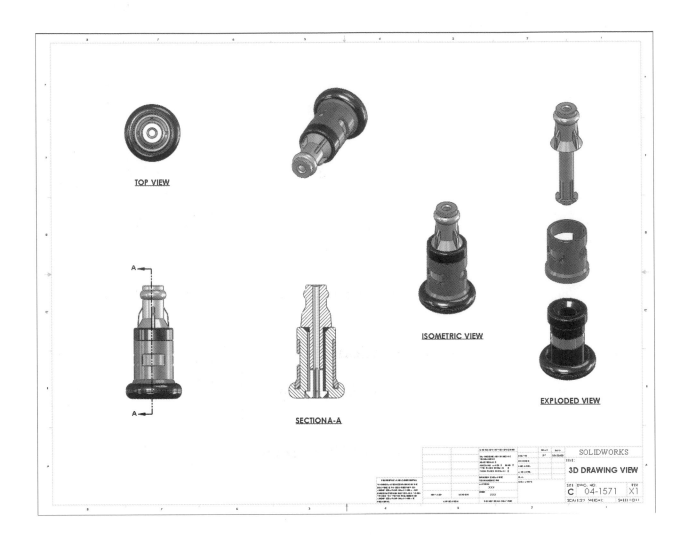

TOP VIEW

ISOMETRIC VIEW

SECTION A-A

EXPLODED VIEW

Reorienting Views

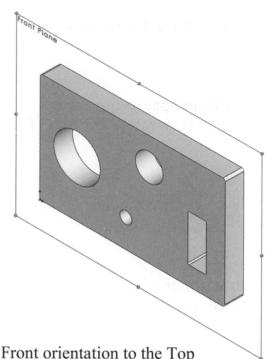

In a part or assembly mode, you can use the Orientation dialog box to select a standard view, activate the View Selector, use Viewports, create Custom Views and save them to SOLIDWORKS. You can also access camera views and snapshots.

To access the Orientation dialog box: select View Orientation on the Standard Views toolbar, press the space bar, or right-click in the graphics area and select View Orientation.

This exercise will show us how to change from the <u>Front</u> orientation to the <u>Top</u> orientation using the options in the View Orientation dialog box.

1. Open the sample model named **View Orientation.sldprt**

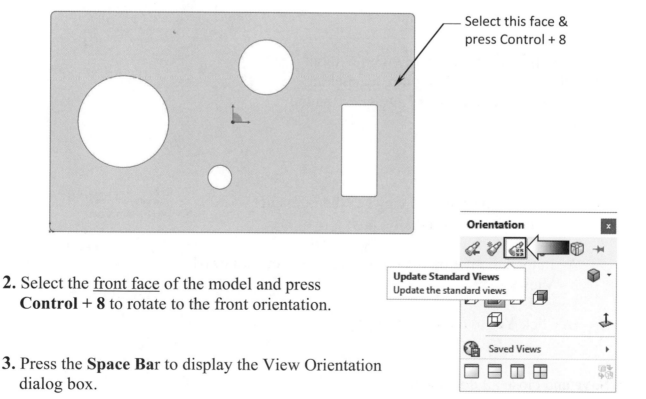

Select this face &
press Control + 8

2. Select the <u>front face</u> of the model and press **Control + 8** to rotate to the front orientation.

Update Standard Views
Update the standard views

3. Press the **Space Bar** to display the View Orientation dialog box.

Click the **Update Standard View** button.

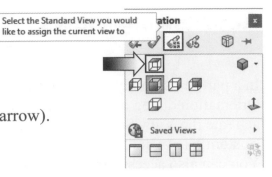

A message appears prompting for a view to assign
the current view to; select the **Top** standard view (arrow).

Click **Yes** to confirm the change.

4. Change to the <u>Top</u> orientation
(Control + 5).

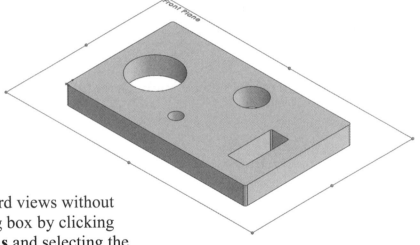

The Front orientation
has been changed to
the Top orientation and
all the standard views
are also updated.

You can also update standard views without
using the Orientation dialog box by clicking
View, **Set Current View As** and selecting the
desired view.

5. To return all standard model views to their default
settings in the Orientation dialog box:

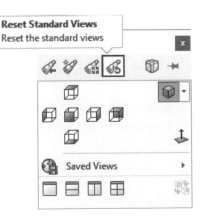

   A. In the Orientation dialog box, click **Reset Standard
      Views**.

   B. Click **Yes** to confirm the update.

6. Save and close all documents.

CHAPTER 19

Configurations

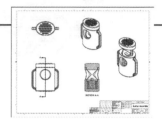

Configurations – Part I
Part, Assembly & Drawing

Configuration is one of the unique options in SOLIDWORKS. It allows you to create multiple variations of a part or an assembly within the same document.

Configurations provide a convenient way to develop and manage families of parts with different dimensions, components, or other parameters.

In a **Part document**, configuration allows you to create families of parts with diffcrent dimensions, features, and custom properties.

In an **Assembly document**, configuration allows you to create:

* Simplified versions of the design by suppressing or hiding the components.

* Families of assemblies with different configurations of the components, parameters, assembly features, dimensions, or configuration-specific custom properties.

In a **Drawing document**, you can display different views of different configurations that you created earlier in the part or assembly documents by accessing the properties of the drawing view.

This chapter will guide you through the basics of creating configurations in the part and the assembly levels. Later on, these configurations will be inserted in a drawing to display the changes that were captured earlier.

Configurations – Part I
Part, Assembly & Drawing

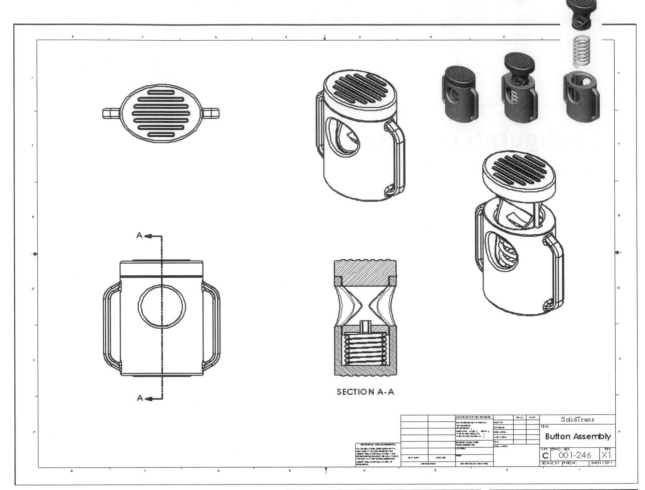

SECTION A-A

SolidTrans

Button Assembly

C | 001-246 | X1

Dimensioning Standards: **ANSI**	Third Angle Projection
Units: **INCHES** – 3 Decimals	

Tools Needed:

 Part
Part document

 Assembly
Assembly document

 Drawing
Drawing document

 Add Configuration...

1. Opening an Assembly Document:

Go to: Training Files folder
      Button Assembly folder
      Open **Button Assembly.sldasm**

2. Using Configurations in the Part mode:

From the FeatureManager tree, right-click on **Button Spring** and select: **Open Part.**

Switch to the **ConfigurationManager** tree.

Right-click the name **Button Spring** and select: **Add Configuration**.

For **Configuration Name**, enter: **Compressed.**

Under **Comment**, enter:

Changed Pitch Dim from .915 to .415.

Click **OK**.

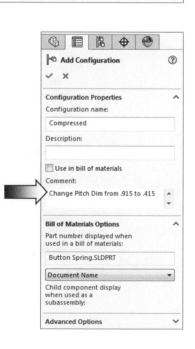

3. Changing the Pitch:

Switch back to the FeatureManager tree (arrow).

Locate the Helix feature on the FeatureManager tree.

Click **Helix/Spiral1** and select **Edit Feature**.

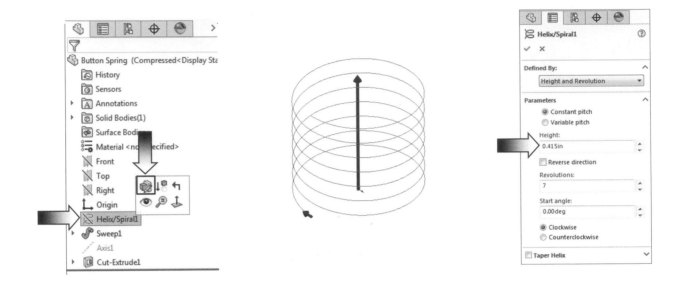

Change the **Height** dimension to **.415** in.

Click **OK**.

Default Configuration
(Pitch = .915)

Compressed Configuration
(Pitch = .415)

4. Creating an Assembly Configuration:

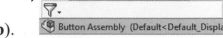

Switch back to the **Assembly** document (**Ctrl+Tab**).

Click the **ConfigurationManager** tab.

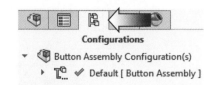

Right-click the name of the assembly and select: **Add Configuration**.

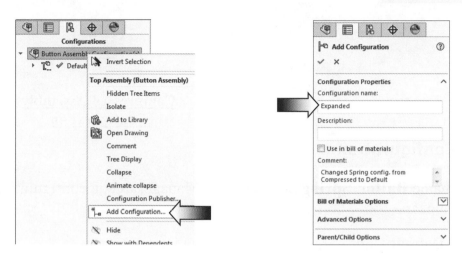

For Configuration Name, enter **Expanded** (also enter a comment for reference).

Click **OK**.

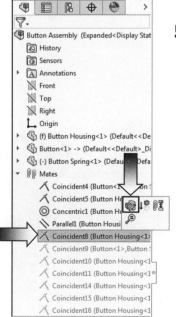

5. Changing the Mate conditions:

Switch back to the **FeatureManager** tree and expand the **Mates** group.

Right-click the mate named **Coincident8** and select: **Suppress**.

By suppressing this mate, the Button (upper part) is no longer locked to the Housing and new Mates can be added to reposition it.

6. Adding new Mates:

Add a **Coincident** mate between the **2 faces** of the two locking features.

Coincident
Mate

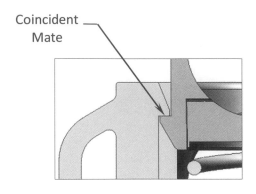

**Compressed Assembly
Configuration**

7. Changing Configuration:

Click the name **Button Spring** on the FeatureManager tree, a configuration selection appears.

Select the **Default** Configuration from the down list (arrow).

Click the green checkmark ☑ to accept the change (arrow).

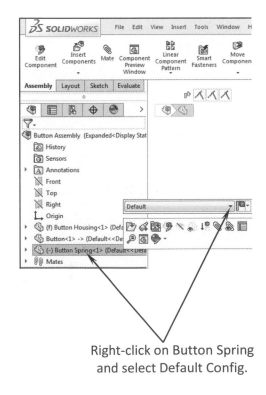

Right-click on Button Spring
and select Default Config.

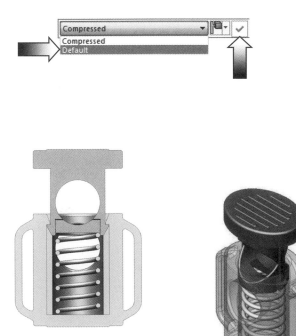

**Default Assembly
Configuration**

8. Using Configurations in a Drawing:

Start a new drawing document; Go to **File / New / Draw** (or Drawing).

Use **C-Landscape** paper size, **Scale: 2:1**.

Using the **View Palette**, drag/drop 4 drawing views shown below (use the **Default** configs).

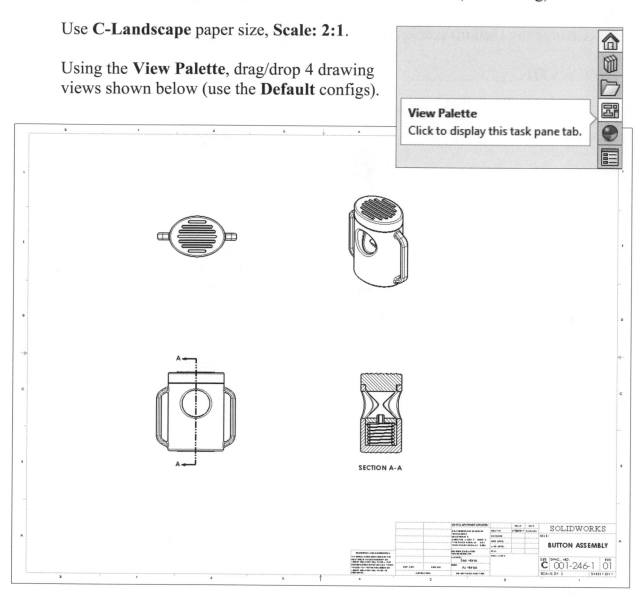

9. Creating a 2nd Isometric view:

Create another Isometric view and place it approximately as shown.

Note: _Copy and Paste also works; first select the Isometric view and press Ctrl+C, then click anywhere in the drawing and press Ctrl+V._

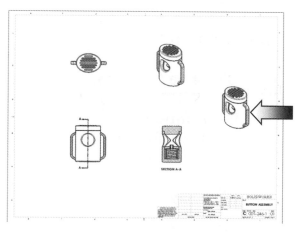

10. Changing the Configuration of a drawing view:

Click the dotted border of the new **Isometric view** to access its configurations.

Change the **Default** configuration to **Expanded** configuration (arrow).

Click **OK**.

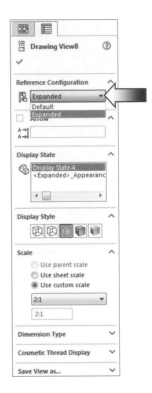

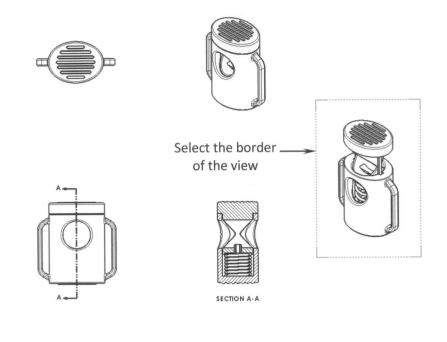

Select the border
of the view

SECTION A-A

The Expanded
configuration is
now the active
configuration for
this specific view.

11. Saving your work:

Click **File / Save As**.

Enter **Button
Assembly** for the
name of the file.

Click **Save**.

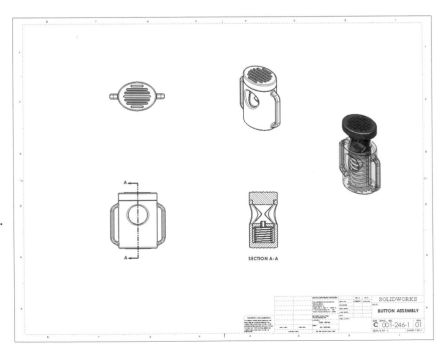

SECTION A-A

SOLIDWORKS

BUTTON ASSEMBLY

Configurations – Part II
Part, Assembly & Drawing

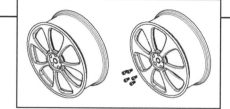

Configurations – Part II
Part, Assembly & Drawing

Configurable Items for Parts:

Part configurations can be used as follows:
* Modify feature dimensions and tolerances.
* Suppress features, equations, and end conditions.
* Assign mass and center of gravity.
* Use different sketch planes, sketch relations, and external sketch relations.
* Set individual face colors.
* Control the configuration of a base part.
* Control the configuration of a split part.
* Control the driving state of sketch dimensions.
* Create derived configurations.
* Define configuration-specific properties.

Configurable Items for Assemblies:

Assembly configurations can be used as follows:
* Change the suppression state (**Suppressed**, **Resolved**) or visibility (**Hide**, **Show**) of components.
* Change the referenced configuration of components.
* Change the dimensions of distance or angle mates, or suppression state of mates.
* Modify the dimensions, tolerances, or other parameters of features that belong to the assembly. This includes assembly feature cuts and holes, component patterns, reference geometry, and sketches that belong to the <u>assembly</u> (not to one of the assembly components).
* Assign mass and center of gravity.
* Suppress features that belong to the assembly.
* Define configuration-specific properties, such as end conditions and sketch relations.
* Create derived configurations.
* Change the suppression state of the Simulation folder in the Feature Manager design tree and its simulation elements (suppressing the folder also suppresses its elements).

Configurations – Part II
Part, Assembly & Drawing

**6 Spokes
Configuration**

**7 Spokes
Configuration**

**7 Spokes with Bolts
Configuration**

Dimensioning Standards: **ANSI**	Third Angle Projection
Units: **INCHES** – 3 Decimals	

Tools Needed:

 Part document
Part

 Assembly document
Assembly

 Drawing document
Drawing

 FeatureManager

ConfigurationManager

 Mate

Part Configurations

This section discusses the use of Configurations at the part level, where the dimensions of the spokes-pattern will be modified to change the number of spokes of the wheel model.

1. Opening a part document:

Go to the Training Files folder, Wheel Assembly folder and open the part document named **Wheel.sldprt**.

Part Configurations:

Switch to the **ConfigurationManager** tree:

A new configuration will be created to capture the change in the number of Spokes.

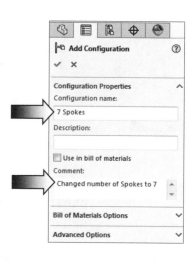

2. Creating a new configuration:

Right-click the name Wheel-Configuration and select **Add Configuration**.

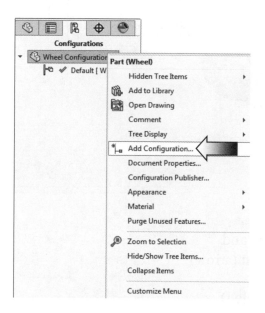

For Configuration Name, enter **7 Spokes** (arrows).

Under Description, enter **Changed number of Spokes to 7** (arrow).

Click **OK**.

NOTE:

> *To display the descriptions next to the file name on the FeatureManager tree, do the following:*
>
> *From the FeatureManager tree, right click on the part name, go to Tree-Display, and select Show Feature Descriptions.*

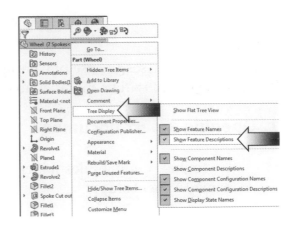

3. Changing the number of the Spokes:

Switch back to the **FeatureManager** tree.

Double-click on the **Spokes Pattern** feature. Locate the number 6 (the number of instances in the pattern) and double-click it.

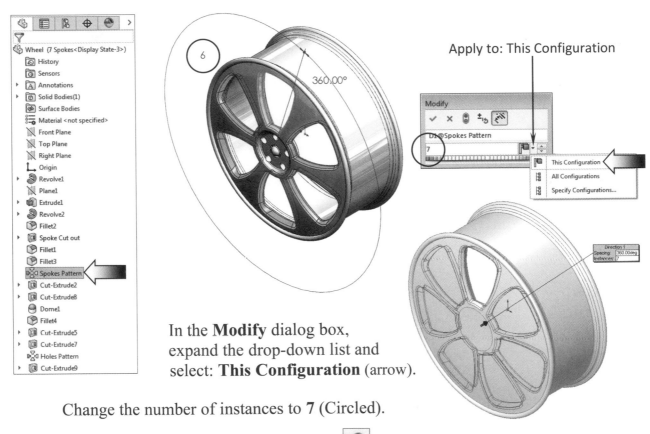

In the **Modify** dialog box, expand the drop-down list and select: **This Configuration** (arrow).

Change the number of instances to **7** (Circled).

Click **OK** and press the **Rebuild** button 🚦 (the stop light) to regenerate the change.

<u>**Important:**</u> *The option **This Configuration** must be selected to prevent other configurations to be affected by this change.*

4. Viewing the configurations:

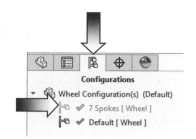

Switch to the **ConfigurationManager** tree.

Double-click on the **Default** configuration to see the 6 Spokes design.

Double-click on the **7 Spokes** configuration to activate it again.

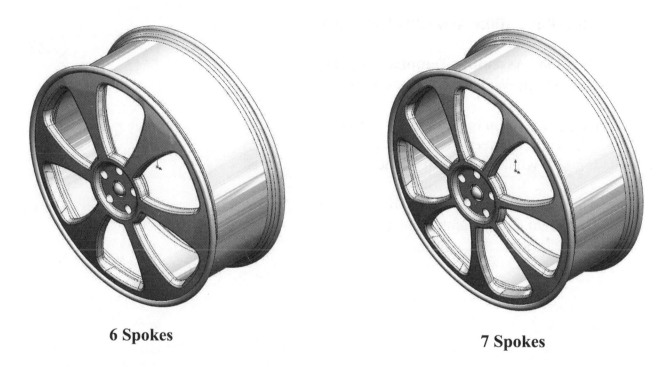

6 Spokes   **7 Spokes**

5. Saving the part:

Save the model as a copy and name it **Wheel.sldprt**.

Keep the part document open. It will be inserted into an assembly document in the next step and used to create new configurations at the assembly level.

Assembly Configurations

This section discusses the use of Configurations at the assembly level, where a Sub-Assembly is inserted and mated to the Wheel as a new configuration. Any configurations created previously at the part level can be toggled and displayed in the top level assembly.

6. Starting a New assembly:

Click **File / Make Assembly from Part**.

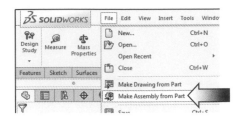

Select the **Assembly Template** from the New SOLIDWORKS Document box.

Click the Origin to place the component. (Clicking the green checkmark on the upper left side will also place it on the Origin.)

The 1st part in the assembly document is the Parent Component, it should be fixed on the origin.

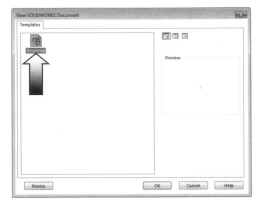

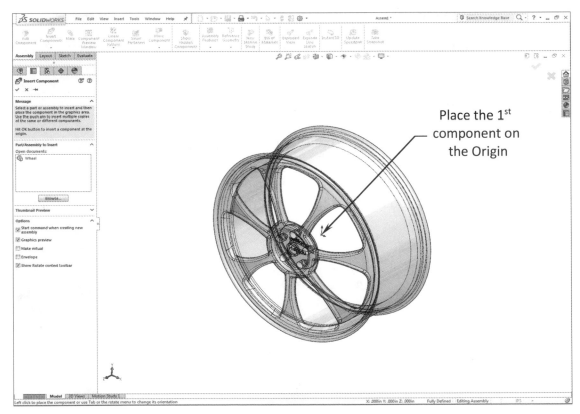

Place the 1st component on the Origin

7. Assembly Configurations:

Switch to the **ConfigurationManager** tree.

Right-click on the Assembly's name and select **Add configuration** (Arrow).

For Configuration Name, enter: **Wheel with Bolts**.

Click **OK**.

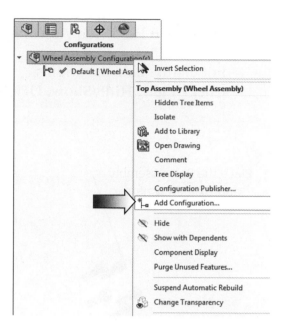

NOTE:

A group of Bolts, which has been saved earlier as an assembly document, is going to be inserted into the Top Level Assembly and becomes a Sub-Assembly.

8. Inserting the Sub-Assembly:

Click **Insert Components.**

Click the **Browse** button.

Browse to the Training Files folder, locate and open the **Bolts Sub-Assembly**.

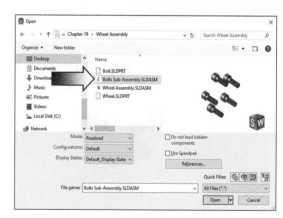

Place the 6 Bolts Sub-Assembly approximately as shown below.

For clarity, hide the origins.
Click: **View, Hide/Show, Origins**.

Place the Bolt-Assembly
approximately here

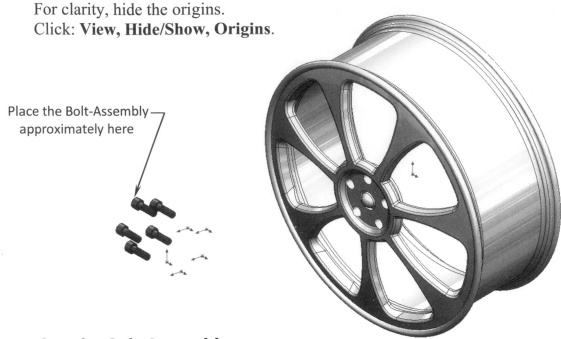

9. Mating the Sub-Assembly:

Enable **Temporary Axis** from the **View / Hide / Show** menus.

Click the **Mate** ⬚ command on the **Assembly** tab or select **Insert / Mate**.

Select the **axis** of one of the <u>Bolts</u> and the **axis** in one of the mating <u>Holes</u> (pictured).

Select 2 Axes

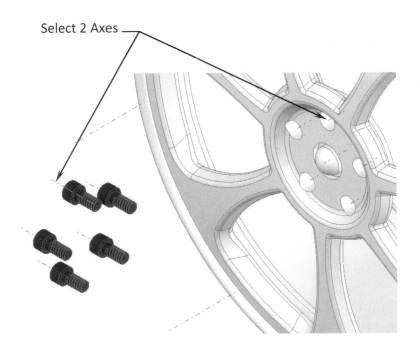

The **Coincident** mate is selected automatically.

Click either **Align** or **Anti-Align** to flip the Bolts to the correct orientation.

Click **OK**.

Add a **Coincident** mate between the <u>bottom face</u> of one of the Bolts and its <u>mating surface</u> (pictured).

Click **OK**.

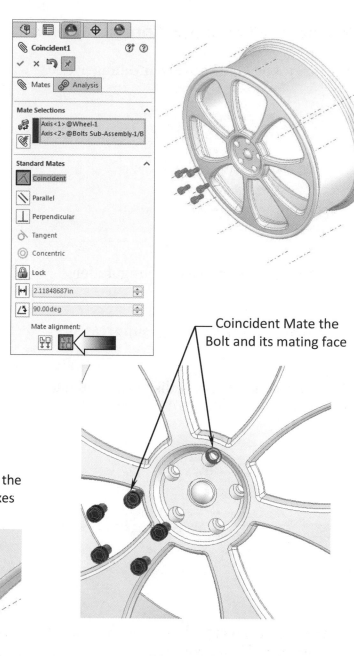

Coincident Mate the
Bolt and its mating face

Coincident Mate the
next 2 center axes

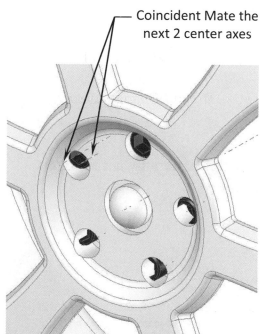

Add another **Coincident** mate to the next **2 axes** to fully centered the 6 bolts.

The Completed Assembly

10. Viewing the Assembly Configurations:

Switch to the **ConfigurationManager** tree.

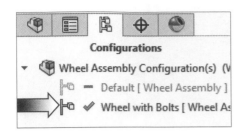

Double-click the **Default** configuration.

Wheel with Bolts Configuration

The Bolts Sub-Assembly is **suppressed**.

The **7 Spokes** pattern is displayed without the bolts.

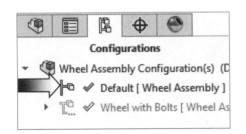

Default Configuration

Make any necessary changes to the Number of Spokes pattern by accessing the Component's Properties and its Configurations (arrow).

An assembly exploded view may need to be created to use in the drawing later.

11. Saving your work:

Save as **Part_Assembly_Configurations** and close all documents.

Drawing Configurations
Changing Configurations in Drawing Views.

This section discusses the use of Configurations at the drawing level, where Configurations created previously in the part and assembly levels can be selected and displayed in different drawing views.

To switch between the configurations in a drawing view: Right-click a drawing view (or hold down **Ctrl** to select multiple drawing views, then right-click) and select **Properties**.

Using the pop-up dialog box, different configurations can now be selected and displayed in the drawing views.

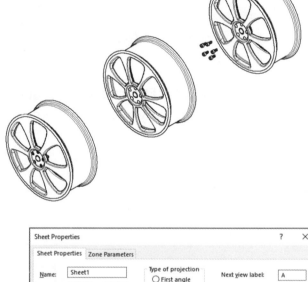

1. Creating an assembly drawing:

Click **File / New / Drawing**.

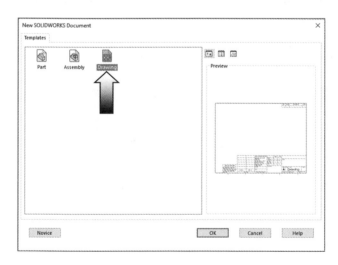

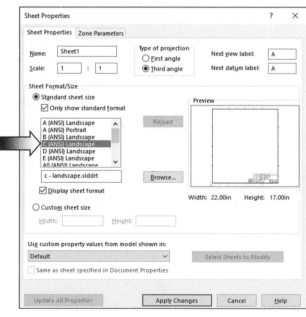

Use **C-Landscape** for paper size.

Set **Scale** to **1 to 1**.

Set the **Projection** to **3rd Angle**.

Click **OK**.

2. Creating the standard drawing views:

Select **Model View** command.

Click **Browse**.

Select the **Wheel Assembly** and click **Open**.

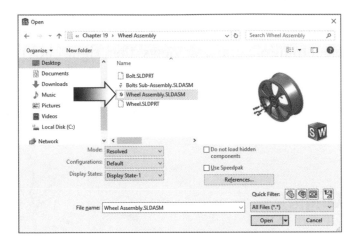

Select the **FRONT** view 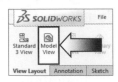 under the Orientation section on the tree and place it approximately as shown below.

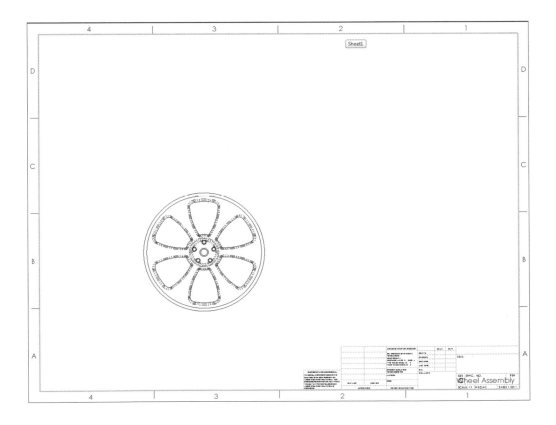

3. Auto-Start the Projected-View:

If the option **Auto Start Projected View** is enabled, SOLIDWORKS will automatically project the next views based on the position of the mouse cursor.

Place the Top view above the Front view.

Click **OK**.

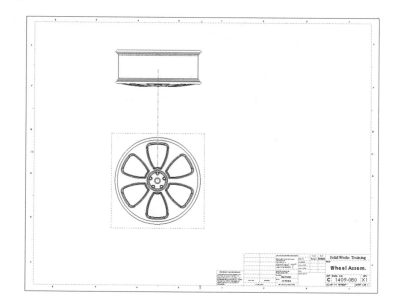

4. Creating the Aligned Section View:

Select the **Front** drawing view dotted border to activate it.

Switch to the **View Layout** tab and click **Section View**.

For Cutting Line, select the **Align** button (arrow) and create the section lines in the order shown below.

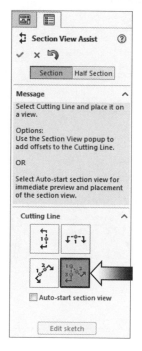

Click **OK** ✔ on the Section Offset toolbar (2nd arrow).

Click **OK** to move to the next step.

<u>**Section Scope**</u>: is an option that allows components to be excluded from the section cut. In other words, when sectioning an assembly drawing view, you will have an option to select which component(s) is going to be affected by this section cut. This option is called Section Scope.

In the Section Scope dialog box, enable **Auto Hatching**, and if needed, click the **Flip Direction** checkbox (arrow).

Place the section view on the right side of the front view as shown below.

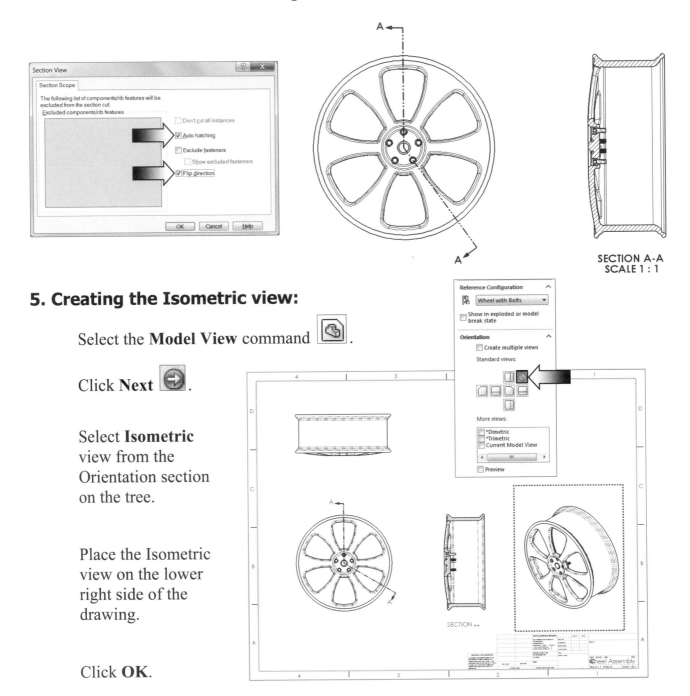

SECTION A-A
SCALE 1 : 1

5. Creating the Isometric view:

Select the **Model View** command .

Click **Next** .

Select **Isometric** view from the Orientation section on the tree.

Place the Isometric view on the lower right side of the drawing.

Click **OK**.

6. Displaying the Exploded View:

An exploded view must be created from the assembly level prior to showing it in the drawing.

Click the **Isometric** drawing view dotted border.

Enable the **Show in Exploded or Model Break State** check box (arrow).

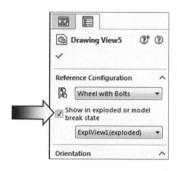

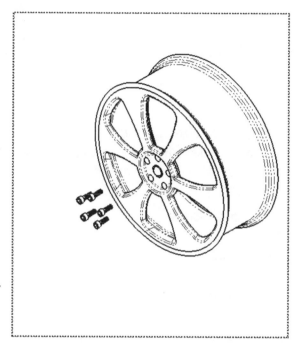

NOTE: The Exploded View option
is also available when
accessing the Properties
of the drawing view.
(Right-click the view's border
and select Properties.)

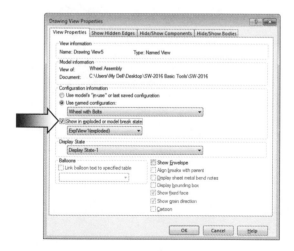

The Assembly Exploded view
is displayed.

7. Changing Configurations:

Use the **Model View** command and create **2 more Isometric views**.

Right-click one of the new Isometric drawing view dotted borders and select: **Properties**.

Select the **Default** configuration.

Set the 2nd Isometric view to the **6-Spokes** configuration (switch to the Assembly document to modify the configurations if needed).

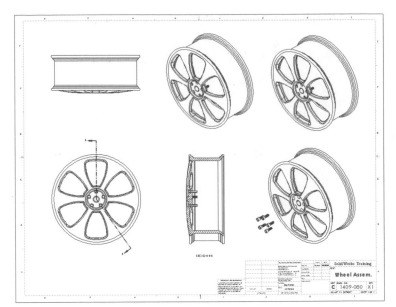

8. Adding Annotations:

Click **Note** 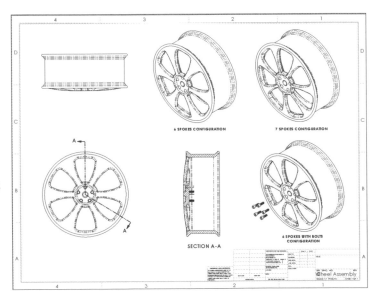 and add the callouts under each drawing view as indicated:

    * 6 SPOKES CONFIGURATION.

    * 7 SPOKES CONFIGURATION.

    * 6 SPOKES WITH BOLTS
      CONFIGURATION.

9. Saving your work:

Click **File / Save As**.

Enter: **Drawing Configurations** for the name of the file.

Click **Save**

Close all documents.

CHAPTER 20

Design Tables

Design Tables in Part, Assembly & Drawing

Design Table Parameters	Description (Legal values)
$PARTNUMBER@_____	For use in a bill of materials
$COMMENT@_____	Any description or text string
$NEVER_EXPAND_IN_BOM@_____	Yes/No to expand in B.O.M.
$STATE@_____	Resolved = R, Suppressed = S
$CONFIGURATION@_____	Configuration name
$SHOW@_____	No longer used
$PRP@_____	Enter any text string
$USER_NOTES@_____	Enter any text string
$COLOR@_____	Specifying 32-bit RGB color
$PARENT@_____	Parent configuration name
$TOLERANCE@_____	Enter tolerance keywords
$SWMASS@_____	Enter any decimal legal value
$SWCOG@_____	Enter any legal X,Y,Z value
$DISPLAYSTATE@_____	Display state name

In a Design Table, you will need to define the names of the configurations, specify the parameters that you want to control, and assign values for each parameter. This chapter will guide you through the use of design tables in both the part and assembly levels.

Design Tables
Part, Assembly & Drawing

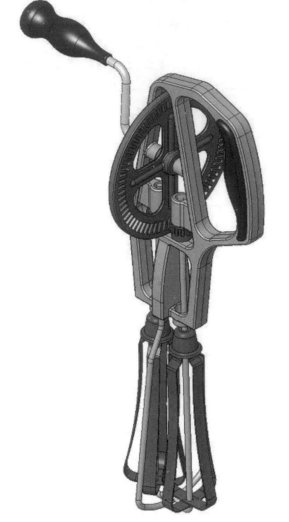

	A	B	C	D	E	F	G	H	I
1	Design Table for: Egg Beater								
2		$state@Egg Beater Handle<1>	$state@Main Gear<1>	$state@Support Rod<1>	$state@Right Spinner<1>	$state@Left Spinner<1>	$state@Right Spinner<2>	$state@Left Spinner<2>	$Configuration@Crank Handle<1>
3	Default	R	R	R	R	R	R	R	Default
4	Config1	R	S	S	S	S	S	S	Oval Handle
5	Config2	R	R	S	S	S	S	S	Default
6	Config3	R	R	R	S	S	S	S	Oval Handle
7	Config4	R	R	R	R	S	S	S	Default
8	Config5	R	R	R	R	R	R	S	Oval Handle
9	Config6	R	R	R	R	R	R	R	Default

Sheet1

Dimensioning Standards: **ANSI**	Third Angle Projection
Units: **INCHES** – 3 Decimals	

Tools Needed:

Design Tables / Microsoft Excel

ConfigurationManager

Part - Design Tables

1. Copying a part document:

<u>Go to:</u> The Training Files
folder, Design Tables folder
Part Design Table.sldprt

Open a copy of the part
document named:
Part Design Table.sldprt

This exercise discusses the
use of **Changing Feature
Dimensions** and **Feature
Suppression-States** in a
Design Table.

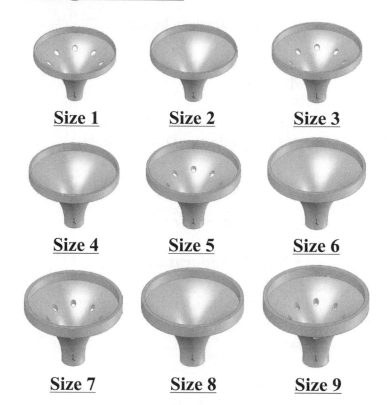

Size 1	Size 2	Size 3
Size 4	Size 5	Size 6
Size 7	Size 8	Size 9

The names of the dimensions will be used as the column's headers in the design
table.

Select **View, Hide/Show** and enable the option **Dimension Names** (arrow).

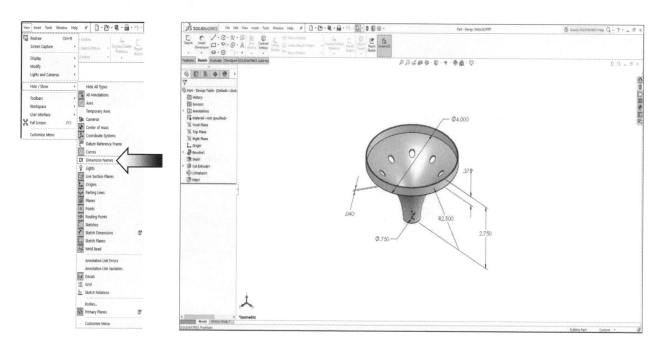

2. Creating a New Design Table:

Click **Insert / Design Table**.

Select the **Blank** option from the Source section.

Enable the option **Allow Model Edits to Update the Design Table**.

Enable the options:

    * **New Parameters**

    * **New Configurations**

    * **Warn When Updating Design Table**

Click **OK**.

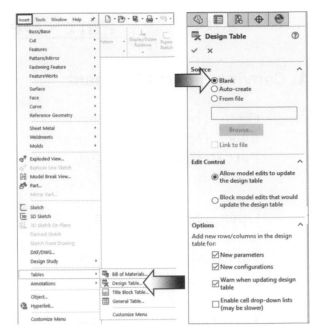

Click **OK** in the Add Rows and Columns Dialog box.

The Microsoft Excel Work Sheet opens.

The name of the part is populated in cell A1 by default.

The cell B2 is selected by default.

The dimensions in the part will be transferred over to the Excel Work Sheet in the next steps.

Notice the names of the dimensions. They should be changed to something more descriptive such as Wall Thickness, Upper Diameter, Height, Radius, Lower Dia., etc.

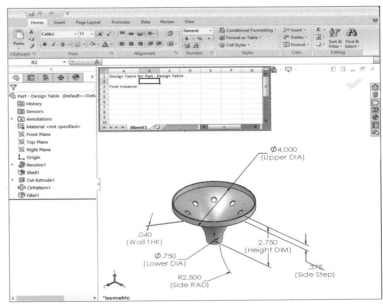

3. Transferring the dimensions to the Design Table:

In Excel, click the cell **B2**.

In SOLIDWORKS, double-click on the Height Dim **2.750**.

The dimension is transferred over to cell **B2**.

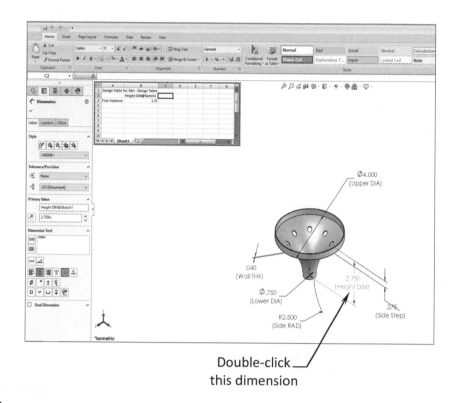

Double-click
this dimension

Repeat the same step and transfer the other dimensions in the order as shown.

	A	B	C	D	E	F	G
1	Design Table for: Part - Design Table						
2		Height DIM@Sketch1	Upper DIA@Sketch1	Lower DIA@Sketch1	Side RAD@Sketch1	Side Step@Sketch1	Wall THK@Shell1
3	Size1	2.75	4	0.75	2.5	0.375	0.04
4	Size2						
5	Size3						
6	Size4						
7	Size5						
8	Size6						
9	Size7						
10	Size8						
11	Size9						
12							

Add configs. names **Size1** thru **Size9**.

4. Using Excel's Addition Formula:

Select the cell **B3**, type the equal sign (=), click the number **2.75** in cell **B2**, and then enter **+.125**.

Copy the formula to the cells below:

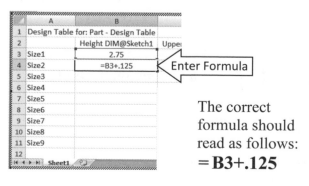

Enter Formula

The correct formula should read as follows:
= B3+.125

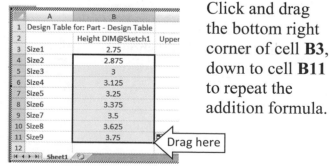

Drag here

Click and drag the bottom right corner of cell **B3**, down to cell **B11** to repeat the addition formula.

	A	B	C	D
1	Design Table for: Part - Design Table			
2		Height DIM@Sketch1	Upper DIA@Sketch1	Lower DIA@
3	Size1	2.75	4	0.75
4	Size2	2.875	=C3+125	
5	Size3	3		
6	Size4	3.125		
7	Size5	3.25		
8	Size6	3.375		
9	Size7	3.5		
10	Size8	3.625		
11	Size9	3.75		
12				

	A	B	C	D
1	Design Table for: Part - Design Table			
2		Height DIM@Sketch1	Upper DIA@Sketch1	Lower DIA@
3	Size1	2.75	4	0.75
4	Size2	2.875	4.125	
5	Size3	3	4.25	
6	Size4	3.125	4.375	
7	Size5	3.25	4.5	
8	Size6	3.375	4.625	
9	Size7	3.5	4.75	
10	Size8	3.625	4.875	
11	Size9	3.75	5	
12				

In cell **C4** type:

$$=C3+.125$$

and copy the formula thru cell **C11**.

	A	B	C	D
1	Design Table for: Part - Design Table			
2		Height DIM@Sketch1	Upper DIA@Sketch1	Lower DIA@Sketch1
3	Size1	2.75	4	0.75
4	Size2	2.875	4.125	=D3+.0625
5	Size3	3	4.25	
6	Size4	3.125	4.375	
7	Size5	3.25	4.5	
8	Size6	3.375	4.625	
9	Size7	3.5	4.75	
10	Size8	3.625	4.875	
11	Size9	3.75	5	
12				

	A	B	C	D
1	Design Table for: Part - Design Table			
2		Height DIM@Sketch1	Upper DIA@Sketch1	Lower DIA@Sketch1
3	Size1	2.75	4	0.75
4	Size2	2.875	4.125	0.8125
5	Size3	3	4.25	0.875
6	Size4	3.125	4.375	0.9375
7	Size5	3.25	4.5	1
8	Size6	3.375	4.625	1.0625
9	Size7	3.5	4.75	1.125
10	Size8	3.625	4.875	1.1875
11	Size9	3.75	5	1.25
12				

Cell **D4** thru cell **C11**, type:

$$=D3+.0625$$

and copy the formula.

	A	B	C	D	E
1	Design Table for: Part - Design Table				
2		Height DIM@Sketch1	Upper DIA@Sketch1	Lower DIA@Sketch1	Side RAD@Sketch1
3	Size1	2.75	4	0.75	2.5
4	Size2	2.875	4.125	0.8125	=E3+.0625
5	Size3	3	4.25	0.875	
6	Size4	3.125	4.375	0.9375	
7	Size5	3.25	4.5	1	
8	Size6	3.375	4.625	1.0625	
9	Size7	3.5	4.75	1.125	
10	Size8	3.625	4.875	1.1875	
11	Size9	3.75	5	1.25	
12					

	A	B	C	D	E
1	Design Table for: Part - Design Table				
2		Height DIM@Sketch1	Upper DIA@Sketch1	Lower DIA@Sketch1	Side RAD@Sketch1
3	Size1	2.75	4	0.75	2.5
4	Size2	2.875	4.125	0.8125	2.5625
5	Size3	3	4.25	0.875	2.625
6	Size4	3.125	4.375	0.9375	2.6875
7	Size5	3.25	4.5	1	2.75
8	Size6	3.375	4.625	1.0625	2.8125
9	Size7	3.5	4.75	1.125	2.875
10	Size8	3.625	4.875	1.1875	2.9375
11	Size9	3.75	5	1.25	3
12					

Cell **E4** thru cell **E11**, type:

$$=E3+.0625$$

and copy the formula.

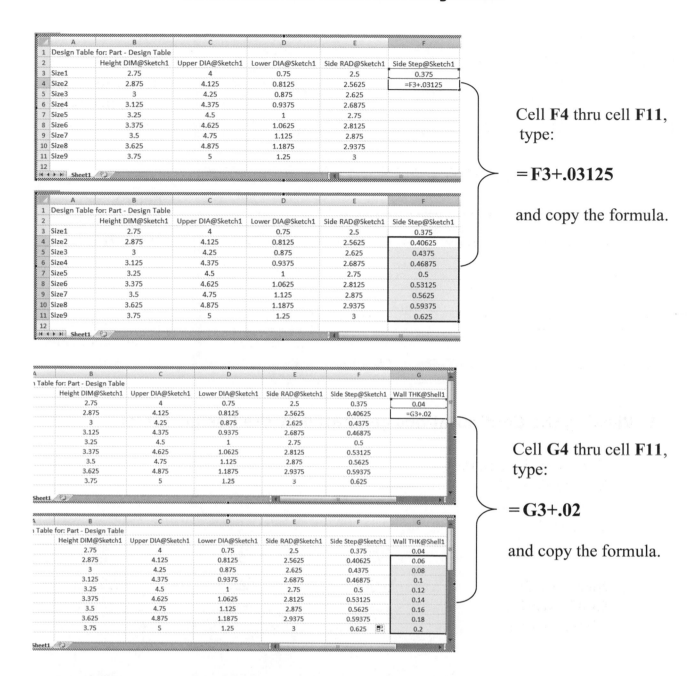

Cell **F4** thru cell **F11**, type:

$$= F3+.03125$$

and copy the formula.

Cell **G4** thru cell **F11**, type:

$$= G3+.02$$

and copy the formula.

5. Controlling the Suppression-States of the holes:

Select the cell **H2** and double-click on the **CutExtrude1** to transfer to cell **H2**.

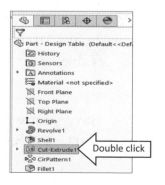

	B	C	D	E	F	G	H
1	for: Part - Design Table						
2	Height DIM@Sketch1	Upper DIA@Sketch1	Lower DIA@Sketch1	Side RAD@Sketch1	Side Step@Sketch1	Wall THK@Shell1	$STATE@Cut-Extrude1
3	2.75	4	0.75	2.5	0.375	0.04	UNSUPPRESSED
4	2.875	4.125	0.8125	2.5625	0.40625	0.06	
5	3	4.25	0.875	2.625	0.4375	0.08	
6	3.125	4.375	0.9375	2.6875	0.46875	0.1	
7	3.25	4.5	1	2.75	0.5	0.12	
8	3.375	4.625	1.0625	2.8125	0.53125	0.14	
9	3.5	4.75	1.125	2.875	0.5625	0.16	
10	3.625	4.875	1.1875	2.9375	0.59375	0.18	
11	3.75	5	1.25	3	0.625	0.2	
12							

Sheet1

In cell **H3**, replace the word **Unsuppressed** with the letter **U**.

	C	D	E	F	G	H	I
1							
2	Upper DIA@Sketch1	Lower DIA@Sketch1	Side RAD@Sketch1	Side Step@Sketch1	Wall THK@Shell1	$STATE@Cut-Extrude1	
3	4	0.75	2.5	0.375	0.04	U	
4	4.125	0.8125	2.5625	0.40625	0.06	S	
5	4.25	0.875	2.625	0.4375	0.08	U	
6	4.375	0.9375	2.6875	0.46875	0.1	S	
7	4.5	1	2.75	0.5	0.12	U	
8	4.625	1.0625	2.8125	0.53125	0.14	S	
9	4.75	1.125	2.875	0.5625	0.16	U	
10	4.875	1.1875	2.9375	0.59375	0.18	S	
11	5	1.25	3	0.625	0.2	U	
12							

Sheet1

Enter **S** for **Suppressed** in cell **H4**.

Enter **S** and **U** for all other cells as shown.

6. Viewing the Configurations generated by the Design Table:

Click anywhere in the SOLIDWORKS graphics area to exit Excel.

Switch to the **Configuration-Manager** tree.

Double-click on **Size2** to see the changes. Verify the other configurations that were created by the Design Table.

Save your work as: **Part - Design Table**.

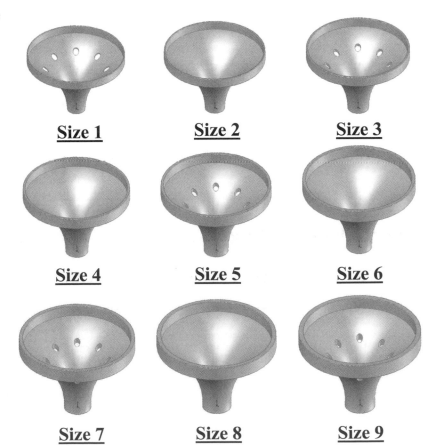

Size 1   Size 2   Size 3

Size 4   Size 5   Size 6

Size 7   Size 8   Size 9

Assembly - Design Tables

1. Copying the Egg Beater Assembly:

<u>Go to:</u> The Training Files folder
Design Tables folder
Egg Beater Assembly Folder.

<u>Copy</u> the **Egg-Beater-Assembly** folder
to your computer desktop.

<u>Open</u> the **Egg Beater Assembly.sldasm**.

This exercise discusses the use of the **STATE**
and **CONFIGURATION** parameters used in a Design Table.

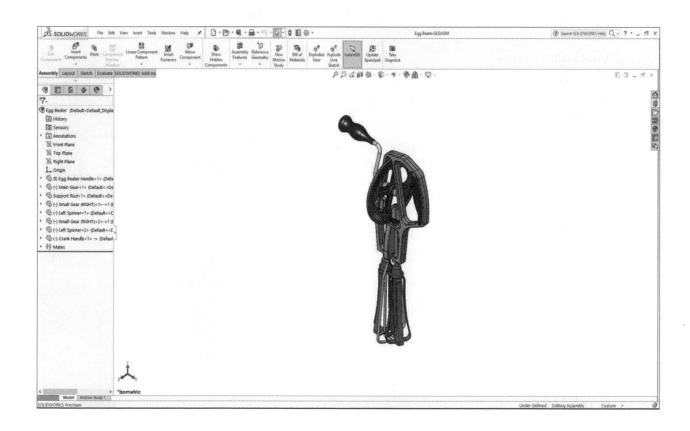

2. Creating a new Assembly Design Table:

Select **Insert / Tables / Design Table** 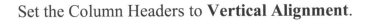.

For Source, select the **Blank** option (arrow).

Under Edit Control, select **Allow Model Edits to Update the Design Table** (arrow).

Under Options, enable **New Parameters**, **New Configs.**, and **Warn When Updating Design Table**.

Click **OK**.

A blank Design Table is loaded on top of SOLIDWORKS screen; the Microsoft Excel application is running side by side with SOLIDWORKS.

Set the Column Headers to **Vertical Alignment**.

Right-click **row #2** (arrow) and select **Format Cells**.

Click the **Alignment** tab and set the Orientation to **90 degrees**.

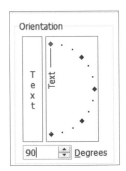

Click **OK**.

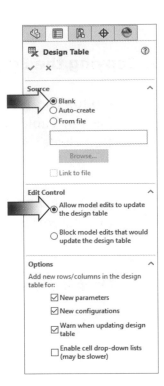

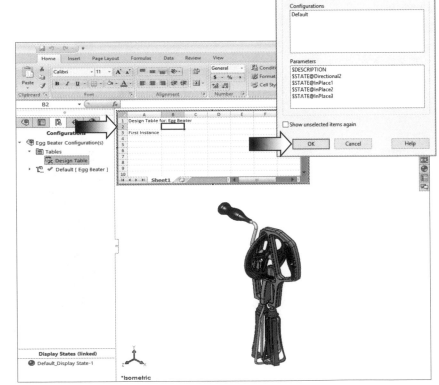

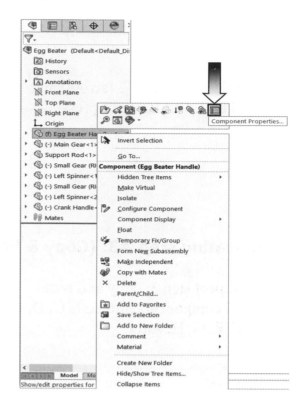

NOTE:
To create a design table, you must define the names of the configurations that you want to create, specify the parameters that you want to control and assign values for each parameter.

3. Defining the column headers:

We are going to copy the name of each component and paste it to the Design Table. They will be used as the Column Headers.

Right-click the part named **Egg Beater Handle** and select **Component Properties** (arrow).

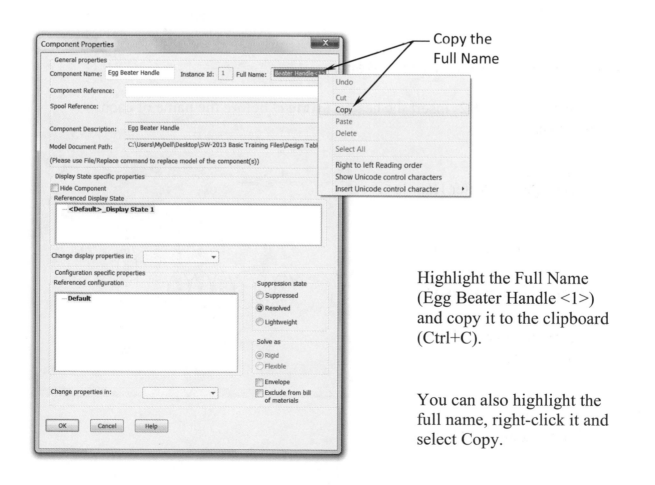

Highlight the Full Name (Egg Beater Handle <1>) and copy it to the clipboard (Ctrl+C).

You can also highlight the full name, right-click it and select Copy.

Select the Cell C2 (arrow) and click **Edit / Paste** or press Ctrl+V.

Click and PASTE here.

4. Repeating step #3: (Copy & Paste all components)

Repeat step 3, copy and paste all components into cells C, D, E, F, G, H, and I.

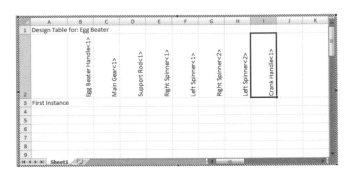

5. Inserting the Control Parameters:

For cells **B2** thru **H2**, insert the header **$state@** before the name of each component.

Example:
$state@Egg Beater Handle<1>

For cell **I2**, insert the header:

$configuration@

before the name of the component.

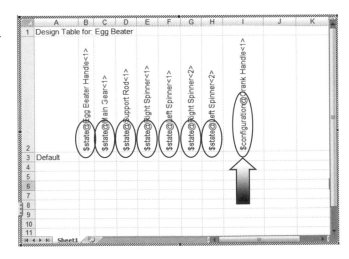

Example:
$configuration@Crank Handle<1>

6. Adding the configuration names:

The actual part numbers can be used in Column A as the name of each configuration.

Starting with **Cell A4**, (below Default), enter **Config1** thru **Config5** as shown.

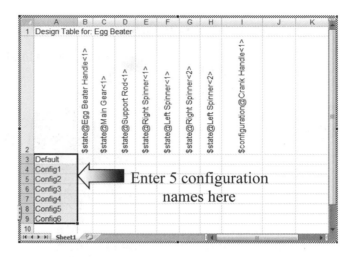

Enter 5 configuration names here

7. Assigning the control values:

To prevent typos, type the letter **R** and letter **S** in each cell.

Enter the values **R** (Resolved) and **S** (Suppressed) into their appropriate cells, from **column B** through **column H**.

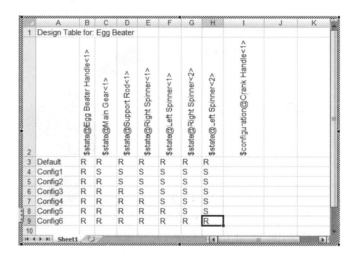

For **column I**, we will enter the names of the configurations instead.

Select the cell **I3** and enter **Default**.

Select the cell **I4** and enter **Oval Handle**.

Repeat the same step for all 6 configurations.

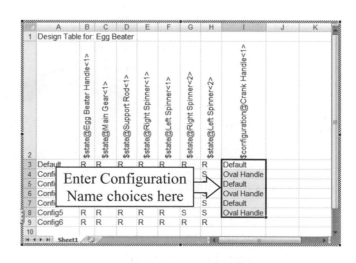

Enter Configuration Name choices here

8. Viewing the new configurations:

Switch to the **ConfigurationManager** Tree.

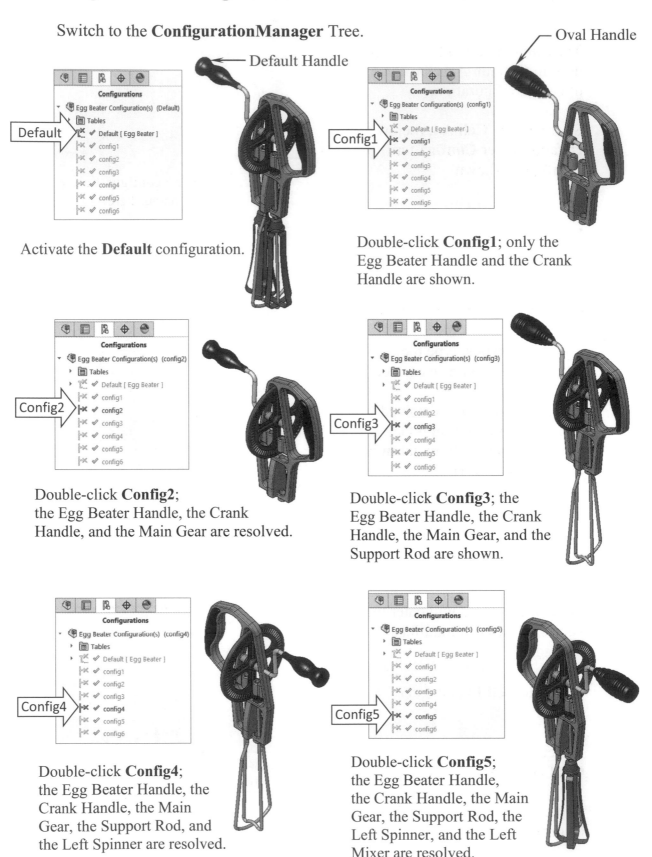

Default Handle

Oval Handle

Activate the **Default** configuration.

Double-click **Config1**; only the Egg Beater Handle and the Crank Handle are shown.

Double-click **Config2**; the Egg Beater Handle, the Crank Handle, and the Main Gear are resolved.

Double-click **Config3**; the Egg Beater Handle, the Crank Handle, the Main Gear, and the Support Rod are shown.

Double-click **Config4**; the Egg Beater Handle, the Crank Handle, the Main Gear, the Support Rod, and the Left Spinner are resolved.

Double-click **Config5**; the Egg Beater Handle, the Crank Handle, the Main Gear, the Support Rod, the Left Spinner, and the Left Mixer are resolved.

<u>Exercise:</u> Part Design Tables

1. Open the existing document named **Part Design Tables_Exe** from the Training folder.
2. Create a design table with 3 different sizes using the dimensions provided in the table.
3. Customize the table by merging the cells, adding colors, and borders.
4. Use the instructions on the following pages, if needed.

Design Table for: Part Design Tables_Exe									
	Lower Boss Thickness@Sketch	Upper Boss Thickness@Sketch	Center Hole@Sketch1	Upper Boss Dia@Sketch1	Lower Boss Dia@Sketch1	D1@Revolve1	Bolt Cicle@Sketch2	Hole on Flange@Sketch2	D3@CirPattern1	D1@CirPattern1
Size 1	0.25	0.25	0.5	1	2	360	1.5	0.25	360	4
Size 2	0.375	0.375	0.625	1.25	2.25	360	1.75	0.275	360	4
Size 3	0.5	0.5	0.75	1.5	2.5	360	2	0.3	360	4

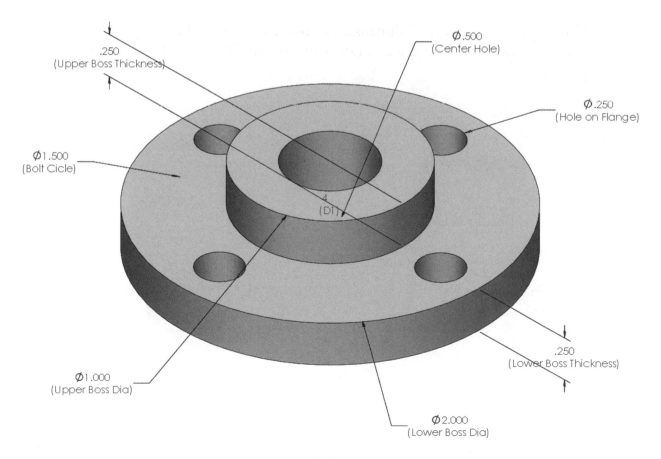

Ø.500
(Center Hole)

.250
(Upper Boss Thickness)

Ø.250
(Hole on Flange)

Ø1.500
(Bolt Cicle)

4
(D1)

.250
(Lower Boss Thickness)

Ø1.000
(Upper Boss Dia)

Ø2.000
(Lower Boss Dia)

1. Opening the main part file:

From the Training Files folder, open the part named **Part Design Tables_Exe**. The dimensions in this part have been renamed for clarity.

2. Inserting a design table:

From the **Insert** menu, click **Tables / Design-Table**.

Click the **Auto Create** button (default).

Leave all other options at their default settings.

Click **OK**.

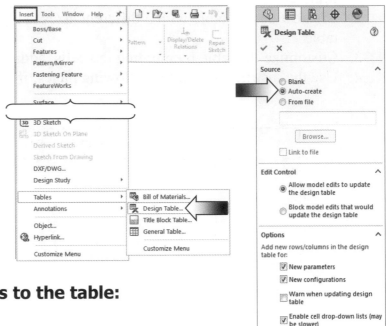

3. Adding model dimensions to the table:

Select **all dimensions** in the Dimensions dialog (arrow).

This option will export all dimensions from the part into the design table. Each dimension will be placed in its own column, in the same order they were created.

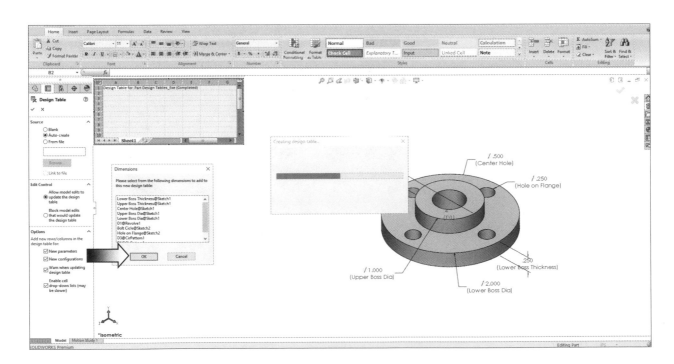

4. Changing the Configuration names:

For clarity, change the name Default to:
Size 1.

Create the next 2 configurations by adding the names **Size 2** and **Size 3**, in cell A4 and A5.

Enter the **new dimensions** for the next 2 sizes.

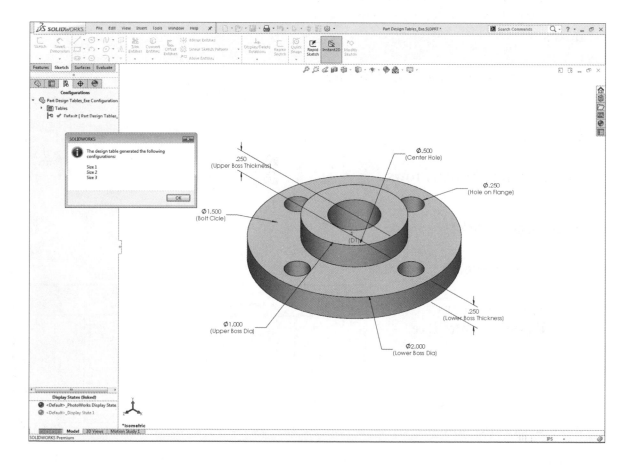

5. Viewing the new configurations:

Click in the SOLIDWORKS background to exit Excel. A dialog box pops up reporting 3 new configurations have been generated by the design table.

Double-click on the names of
the configurations to see the
changes for each size.

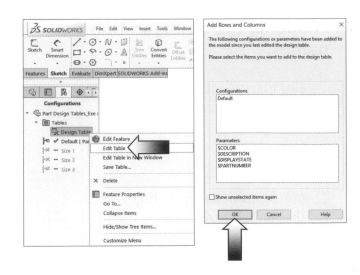

6. Customizing the table:

Expand the **Tables** folder,
right-click on Design Table
and select **Edit Table**.

Click **OK** to close the Rows
and Columns dialog.

Highlight the entire Row1 header, right-
click in the highlighted area and select
Format Cell.

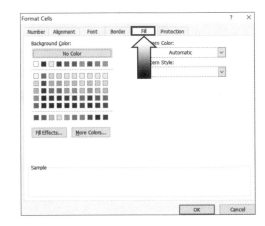

Click the **Fill** tab and select a color for
the Row1.

Change to the **Alignment** tab and enable
the **Wrap Text** and **Merge Cells** checkboxes.

In the **Border** tab, select the **Outline**
button.

Click **OK**.

Repeat step 6 for other rows.

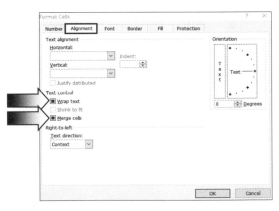

7. Saving your work:

Click **File / Save As**.

For file name, enter
Part Design Table_Exe.

Click **Save**.

Overwrite the original
document if prompted.

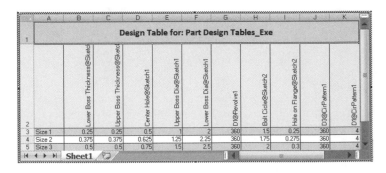

Level 2 Final Exam (1 of 2)

1. Open an assembly document named:
   Assembly Motions - Level 2 Final Exam folder.
2. Create an <u>assembly drawing</u> as shown below.
3. Modify the Bill of Materials to match the information provided in the BOM.

ITEM NO.	PART NUMBER	DESCRIPTION	QTY.
1	1-123-45	Main Housing	1
2	2-123-56	Snap-On Cap	1
3	3-123-67	Main Gear	1
4	4-123-78	Drive Gear	2
5	5-123-89	End Cap	1

SOLIDWORKS

TITLE:

ASSEMBLY MOTIONS

SIZE **B** DWG. NO. 01-12345 REV 01

SCALE: 1:2 WEIGHT: SHEET 1 OF 2

		A	B	C	D
1		ITEM NO.	PART NUMBER	DESCRIPTION	QTY.
2		1	1-123-45	Main Housing	1
3		2	2-123-56	Snap-On Cap	1
4		3	3-123-67	Main Gear	1
5		4	4-123-78	Drive Gear	2
6		5	5-123-89	End Cap	1

4. Change the Balloon Style to Circular Split Line to include the Quantity.
5. Fill out the title block with the information shown in the drawing.
6. Save your drawing as **L2 Final Assembly Drawing**.

Level 2 Final Exam (2 of 2)

1. Open a Part document named:
   Main Housing from the Assembly Motions folder.
2. Create the detailed drawing as shown below.
3. Add the dimensions shown in each drawing view.

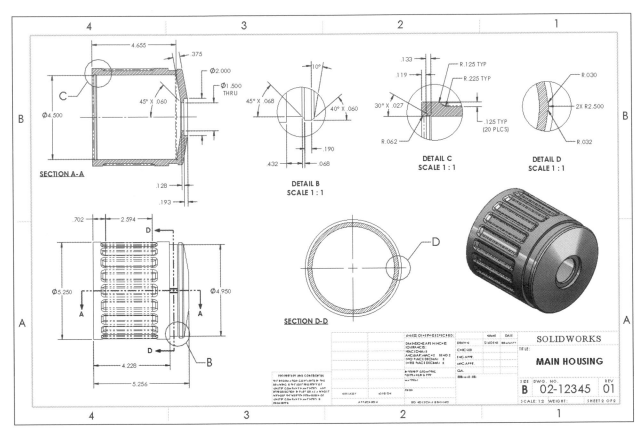

4. Add any missing dimensions as reference dimensions.
5. Fill out the title block with the information shown.
6. Save your drawing as **L2 Final Detailed Drawing**.

TABLE OF U.S. MEASURES

LENGTH

12 inches	=	1 foot
36 inches	=	1 yard (or 3 feet)
5280 feet	=	1 mile (or 1760 yards)

AREA

144 square inches (in)	=	1 square foot (ft)
9 ft	=	1 square yard (yd)
43,560 ft	=	1 acre (A)
640 A	=	1 square mile (mi)

VOLUME

1728 cubic inches (in³)	=	1 cubic foot (ft)
27 ft	=	1 cubic yard (yd)

LIQUID CAPACITY

8 fluid ounces (fl oz)	=	1 cup (c)
2 c	=	1 pint (pt)
2 pt	=	1 quart (qt)
4 pt	=	1 gallon (gal)

WEIGHT

16 ounces (oz)	=	1 pound (lb)
2000 lb	=	1 ton (t)

TEMPERATURE Degrees Fahrenheit (°F)

32° F	=	freezing point of water
98.6° F	=	normal body temperature
212° F	=	boiling point of water

TABLE OF METRIC MEASURES

LENGTH

10 millimeters (mm)	=	1 centimeter (cm)
10 cm	=	1 decimeter (dm)
100 cm	=	1 meter (m)
1000 m	=	1 kilometer (km)

AREA

100 square millimeters (mm)	=	1 square centimeter (cm)
10,000 cm	=	1 square meter (m)
10,000 m	=	1 hectare (ha)
1,000,000 m	=	1 square kilometer (km)

VOLUME

1000 cubic millimeters (ml)	=	1 cubic centimeter (cm)
1 cm	=	1 milliliter (mL)
1,000 m	=	1 Liter (L)
1,000,000 cm	=	1 cubic meter (m)

LIQUID CAPACITY

10 deciliters (dL)	=	1 liter (L) - or 1000 mL
1000 L	=	1 kiloliter (kL)

MASS

1000 milligrams (mg)	=	1 gram (g)
1000 g	=	1 kilogram (kg)
1000 kg	=	1 metric ton (t)

TEMPERATURE Degrees Celsius (°C)

0° C	=	freezing point of water
37° C	=	normal body temperature
100° C	=	boiling point of water

SOLIDWORKS 2023

Certified SOLIDWORKS Associate (CSWA) Mechanical Design

Certification Practice for the Associate Examination

Courtesy of Paul Tran, Sr. Certified SOLIDWORKS Instructor

CSWA – Certified SOLIDWORKS Associate

As a Certified SOLIDWORKS Associate (CSWA), you will stand out from the crowd in today's competitive job market.

The CSWA certification is proof of your SOLIDWORKS® expertise with cutting-edge skills that businesses seek out and reward.

Exam Length: 3 hours - Exam Cost: $99.00

Minimum Passing grade: 70%

Re-test Policy: There is a minimum 14 day waiting period between every attempt of the CSWA exam. Also, a CSWA exam credit must be purchased for each exam attempt.

You must use at least SOLIDWORKS 2011 or newer for this exam. Any use of a previous version will result in the inability to open some of the testing files.

All candidates receive electronic certificates and personal listing on the CSWA directory when they pass *(Courtesy of SOLIDWORKS Corporation)*.

- CSWA Sample Certificate -

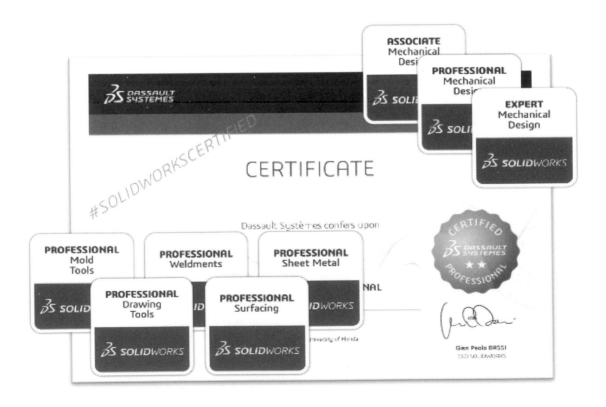

Certified-SOLIDWORKS-Associate program (CSWA)
Certification Practice for the Associate-Examination

Drawings, Parts & Assemblies

Complete all challenges within 180 minutes

(The following examples are intended to assist you in familiarizing yourself with the structures of the exams and the method in which the questions are asked.)

Drafting competencies

Question 1:

Which command was used to create the drawing View B on the lower right? (Circle one)

A. Section View

B. Projected

C. Crop View

D. Detail

Free form shape

View A

View B

(Answers can be found on page 21-5)

Question 2:

Which command was used to create the drawing View B below? (Circle one)

A.   Section View

B.   Crop View

C.   Projected

D.   Detail

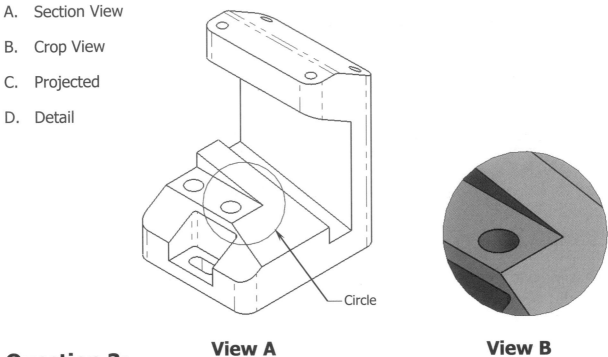

View A   **View B**

Question 3:

Which command was used to create the drawing View B below? (Circle one)

A.   Section View

B.   Aligned Section

C.   Broken-Out Section

D.   Detail

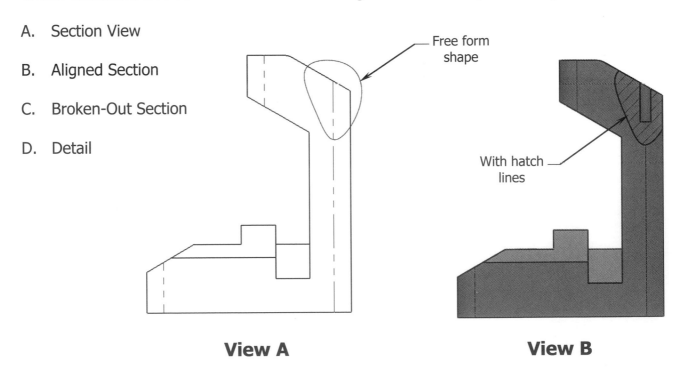

View A   **View B**

Question 4:

Which command was used to create the drawing View B below? (Circle one)

A. Alternate Position View

B. Multiple Positions

C. Exploded View

D. Copy & Paste

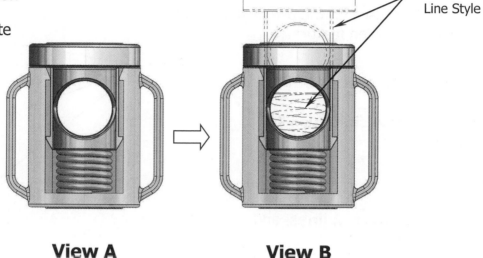

Phantom
Line Style

View A   **View B**

Question 5:

Which command was used to create the drawing View B below? (Circle one)

A. Aligned Section View

B. Horizontal Break

C. Vertical Break

D. Broken Out Section

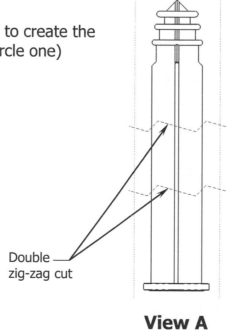

Double
zig-zag cut

View A

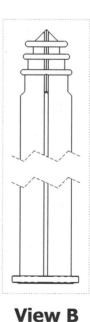

View B

Question 6: Basic Part Modeling (1 of 4)

Create this part in SOLIDWORKS.

Origin: **Arbitrary**

Material: **1060 Alloy Steel**

Unit: **Inches, 3 decimals**

Drafting Standards: **ANSI**

Density: **0.098 lb/in^3**

(This question focuses on the use of sketch tools, relations, and revolved features.)

1. Creating the main body:

Open a **new sketch** on the Front plane.

Sketch a **circle**, **2 lines**, and **2 centerlines**.

Add a **Symmetric** relation between the horizontal centerline and the two lines as noted.

Trim the left portion of the circle and the ends of the two lines as shown.

Add the relations and dimensions shown to fully define the sketch (do not add the reference dimension).

Change the 8.00 diameter dimension to a **R4.00** radius dimension. (Use the options in the Leaders tab, on the Properties tree.)

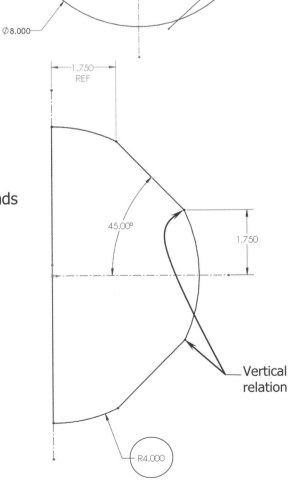

2. Extruding the base feature:

Click **Extruded Boss/Base**.

Use **Mid-Plane** type.

Thickness: **16.00 in**.

Click **OK**.

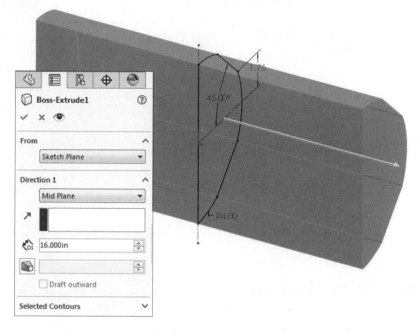

3. Creating the bore hole:

Open a **new sketch** on the <u>front face</u> as noted.

Sketch the profile shown below and add dimensions/relations to fully define the sketch.

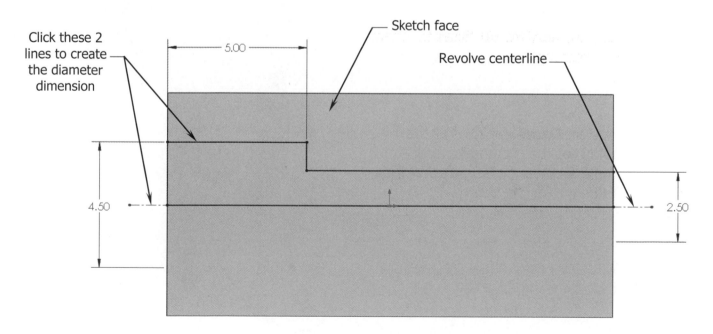

4. Creating a revolved cut:

Click **Revolved-Cut**.

Use the default **360º**.

Click **OK**.

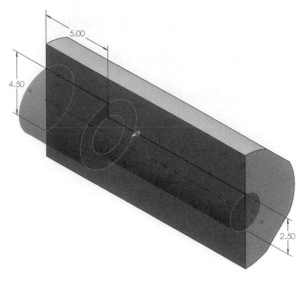

5. Adding the tabs:

Open a **new sketch** on the <u>planar face</u> of the left end.

Select the two angled edges and click **Convert Entities**.

Add the **additional lines** to create two rectangles.

Add an **Equal** relation for the Width of the rectangles, and add the **Parallels** or **Perpendiculars** relations for the others.

Add any other dimensions to fully define the sketch.

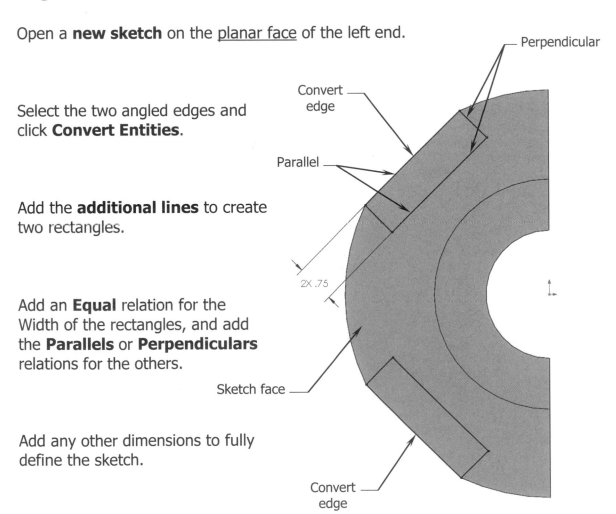

6. Extruding the tabs:

Click **Extruded Boss/Base**.

Use the default **Blind** type.

Enter **1.00 in**. for depth.

Click **OK**.

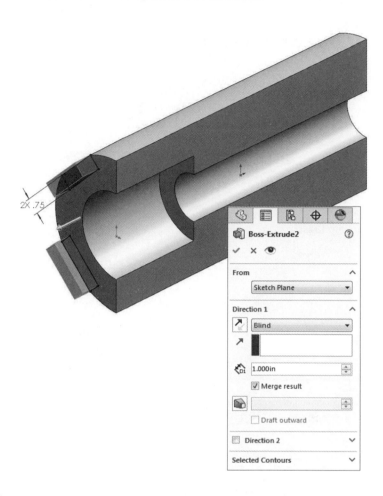

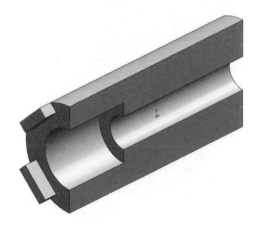

7. Creating the cut features:

Open a **new sketch** on the <u>Top</u> plane.

Sketch **2 rectangles** and make the lines at the bottom **Collinear** with the edge of the model (see next page).

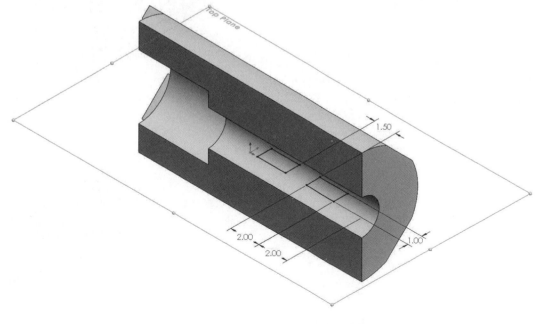

Next⟹

Add the relations/dimensions shown to fully define the sketch.

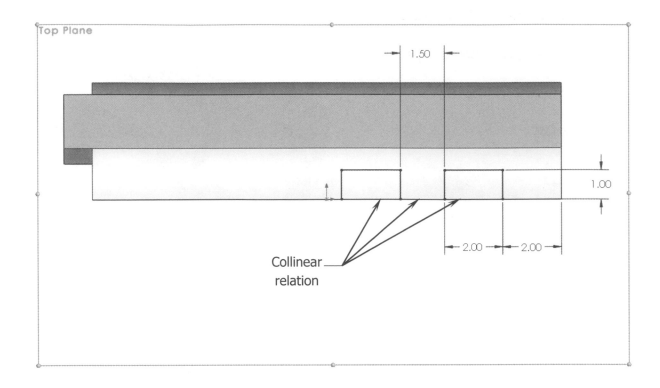

8. Extruding the cuts:

Click **Extruded-Cut**.

Use the **Through All** type.

Click **Reverse** if needed to remove the bottom portions of the half cylinder.

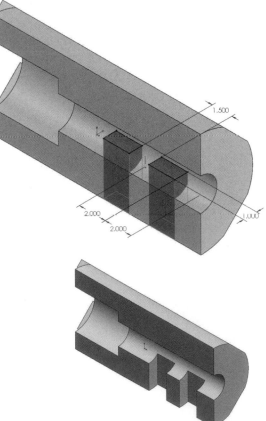

Click **OK**.

9. Adding another cut feature:

Open a **new sketch** on the <u>planar face</u> as noted.

Sketch the profile and add the dimensions below to fully define the sketch.

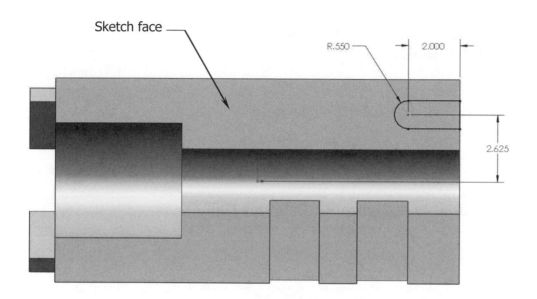

Sketch face

R.550   2.000

2.625

10. Extruding a cut:

Click **Extruded-Cut**.

Use the default **Blind** type.

Enter **.750 in**. for depth.

Click **OK**.

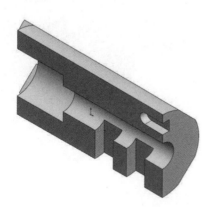

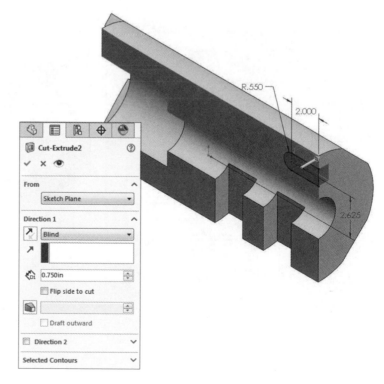

R.550

2.000

2.625

Cut-Extrude2

From

Sketch Plane

Direction 1

Blind

0.750in

Flip side to cut

Draft outward

Direction 2

Selected Contours

11. Calculating the Mass:

Be sure to set the material to **1060 Alloy**.

Switch to the **Evaluate** tab.

Click **Mass Properties**.

If needed, set the Unit of Measure to:
IPS, **3 decimals**.

Enter the final mass of the model here:

_____ lbs.

Save your work as **Hydraulic Cylinder Half.**

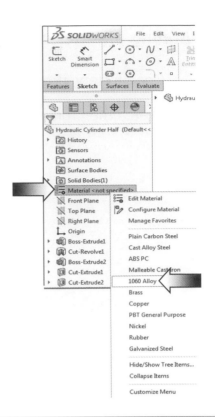

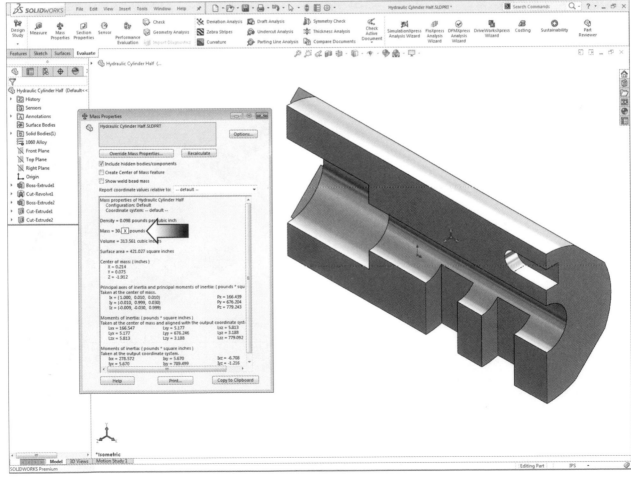

Question 7: Basic Part Modeling (2 of 4)

Create this part in SOLIDWORKS.

Origin: **Arbitrary**

Material: **1060 Alloy**

Unit: **Inches**, **3 decimals**

Drafting Standards: **ANSI**

Density: **0.098 lb/in^3**

(This question focuses on the use of sketch tools, relations, and circular patterns.)

1. Creating the main body:

Open a **new sketch** on the Front plane.

Sketch the profile using the Mirror function to ensure all entities are symmetrical about the vertical centerline.

Add the dimensions and relations needed to fully define the sketch.

*Note: Hold the **Shift** key when adding the 1.00" dimension.

Add a horizontal **centerline** and use it as the revolve line in the next step.

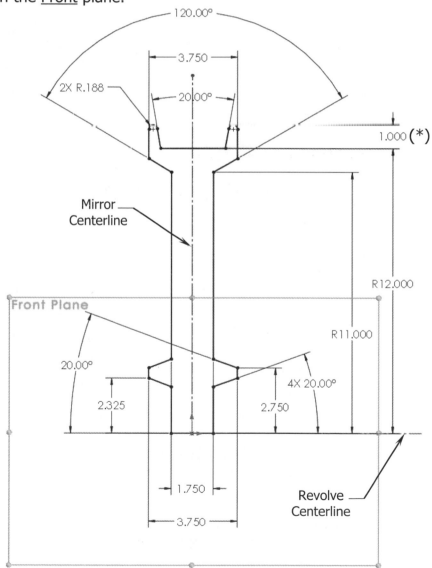

2. Revolving the main body:

Click **Revolved Boss/Base**.

Select the **horizontal centerline** as the Axis of Revolution.

Use the **Blind** type and the default **360°**.

Click **OK**.

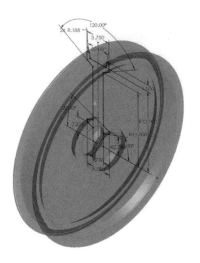

3. Creating the 1st cutout:

Open a **new sketch** on the <u>face</u> as indicated.

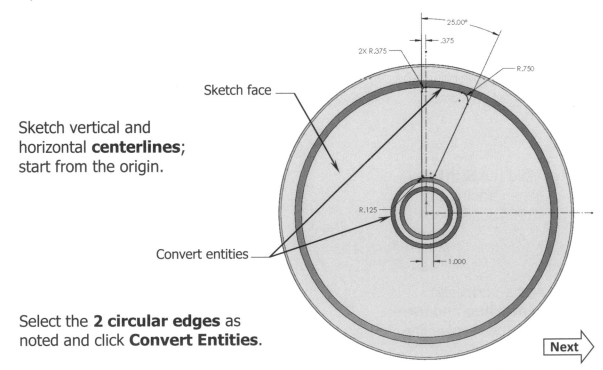

Sketch face

Sketch vertical and horizontal **centerlines**; start from the origin.

Convert entities

Select the **2 circular edges** as noted and click **Convert Entities**.

Next

Trim the sketch entities to create one continuous closed contour.

Add the dimensions shown to fully define the sketch before adding the sketch fillets.

Note:
There are three different fillet sizes in this sketch.

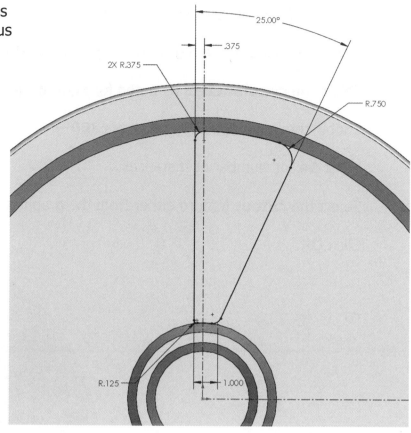

4. Extruding a cut:

Click **Extruded-Cut**.

Use the **Through All** type.

Click **OK**.

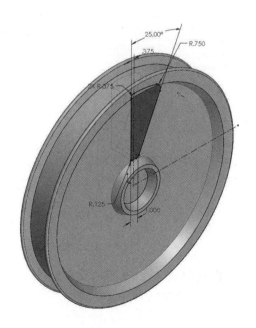

5. Creating a Circular pattern:

Below the Linear Pattern drop-down menu, select **Circular Pattern**.

Select the **circular edge** as noted for Pattern Direction.

Enable the Equal Spacing checkbox **(360°)**.

Enter **14** for Number of Instances.

Select the **cutout** feature either from the graphics area or from the feature tree.

Click **OK**.

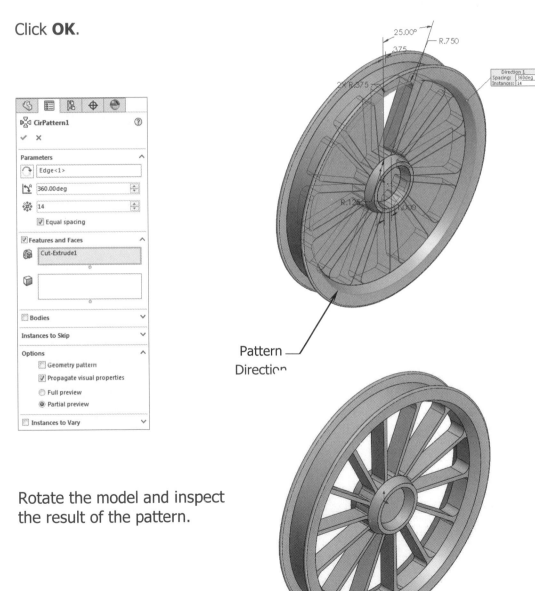

Pattern
Direction

Rotate the model and inspect
the result of the pattern.

Press **Control+1** to change to
the **Front** view orientation.

6. Calculating the Mass:

Be sure to set the material to **1060 Alloy**.

Switch to the **Evaluate** tab.

Click **Mass Properties**.

If needed, set the Unit of Measure to:
IPS, **3 decimals**.

Enter the final mass of the model here:

_____ lbs.

Save your work as **CSWA_Wheel**.

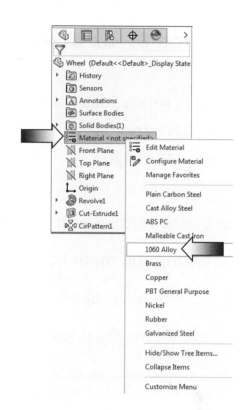

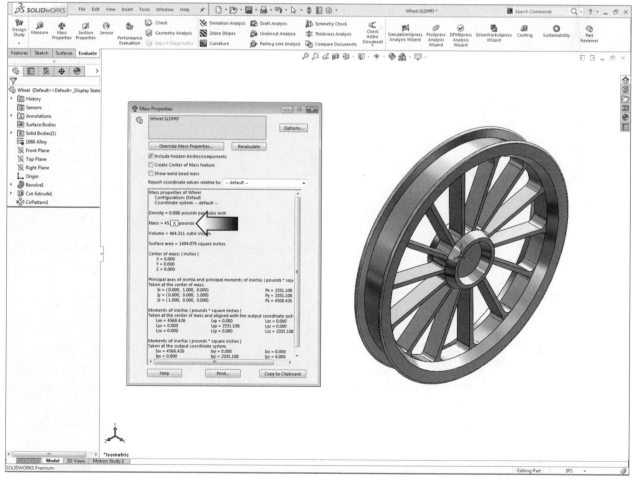

Question 8: Basic Part Modeling (3 of 4)

Create this part in SOLIDWORKS.

Origin: **Arbitrary**

Material: **AISI 1020**

Unit: **Inches, 3 decimals**

Drafting Standards: **ANSI**

Density: **0.285 lb/in^3**

(This question focuses on the use of sketch tools, relations, and extrude features.)

1. Creating the main body:

Open a **new sketch** on the <u>Front</u> plane.

Sketch the profile below and only add the corner fillets <u>after</u> the sketch is fully defined.

Add the dimensions and relations as indicated. Do not add the 2 reference dimensions (36 and 45.43); use them to check or measure the geometry only.

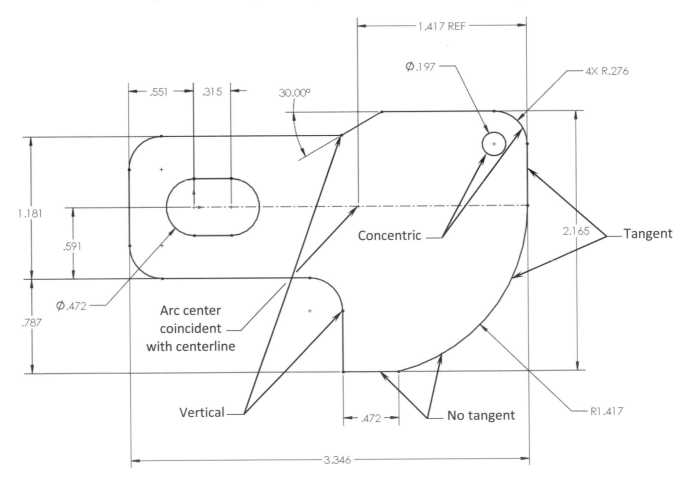

2. Extruding the base:

Click **Extruded Boss/Base**.

Select the **Mid Plane** type.

Enter **1.00 in**. for depth.

Click **OK**.

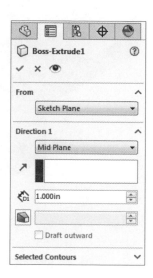

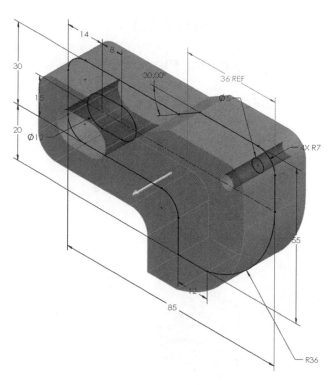

3. Creating the center cutout:

Open a **new sketch** on the Front plane.

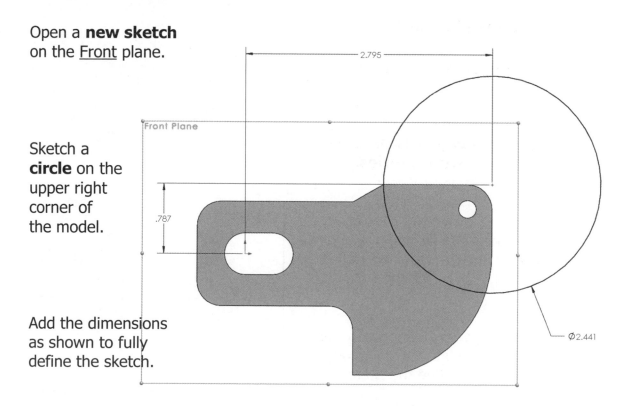

Sketch a **circle** on the upper right corner of the model.

Add the dimensions as shown to fully define the sketch.

4. Extruding a cut:

Click **Extruded Cut**.

Use the **Mid-Plane** type.

Enter **.475 in**. for depth.

Click **OK**.

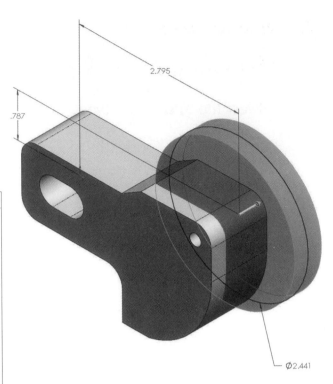

5. Creating the side cut:

Open a **new sketch** on the planar face as noted.

Select the **3 edges** as indicated and click **Convert Entities**.

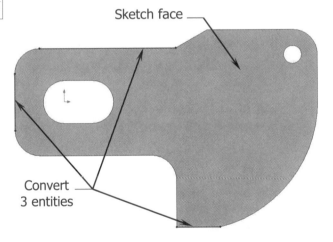

Extend the converted lines to merge their end points, and add **3 other lines** to close-off the sketch.

Add a **Collinear** relation between the horizontal line and the model edge to fully define the sketch.

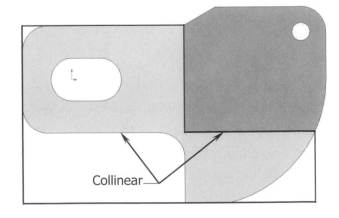

6. Extruding a cut:

Click **Extruded Cut**.

Use the default **Blind** type.

Enter **.492 in**. for depth.

Click **OK**.

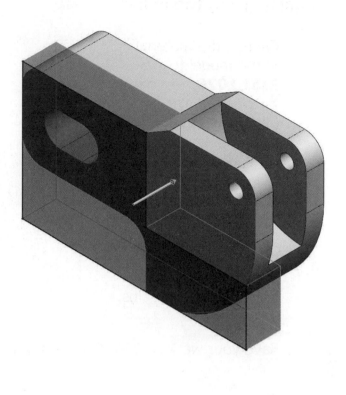

Rotate the model to verify the result of the cut.

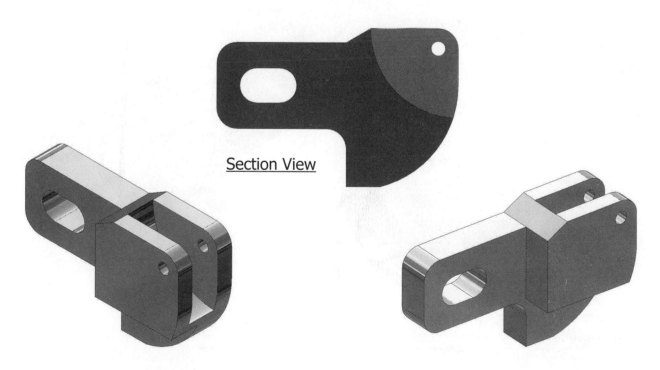

Section View

7. Calculating the Mass:

Change the material
of the model to
AISI 1020.

Click **Mass Properties**.

If needed, set the Unit
of Measure to:
IPS, 3 decimals.

Enter the final mass of
the model here:

_____ lbs.

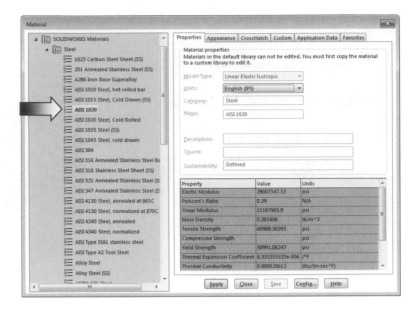

Save your work as **CSWA Tool Block Lever.**

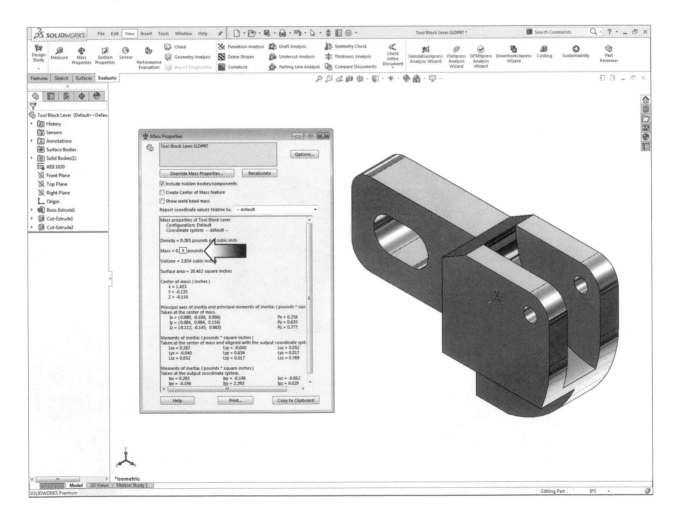

Question 9: Basic Part Modeling (4 of 4)

Create this part in SOLIDWORKS.

Origin: **Arbitrary**

Material: **AISI 1020**

Unit: **Inches**, 3 decimals

Drafting Standards: **ANSI**

Density: **0.285 lb/in^3**

(This question focuses on the use of sketch tools, relations, and extrude features.)

1. Creating the main body:

Open a **new sketch** on the Front plane.

Sketch the profile below and keep the origin in the center of the large hole.

Add the dimensions/relations as shown below to fully define the sketch.

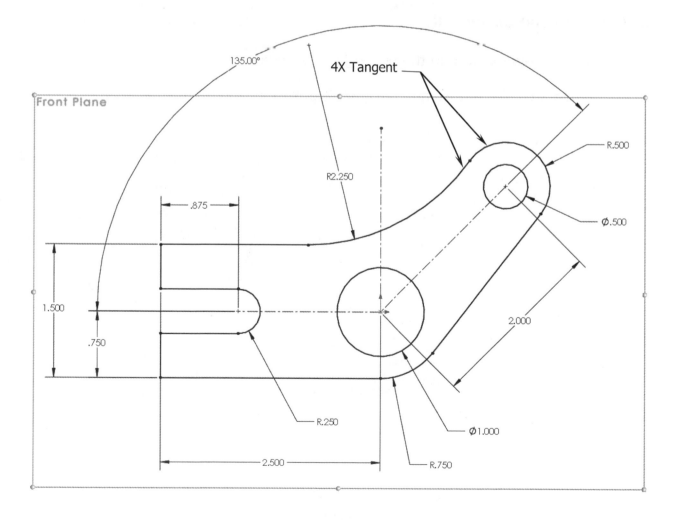

2. Extruding the base:

Click **Extruded Boss/Base**.

Use the **Mid-Plane** type.

Enter **1.00 in**. for depth.

Click **OK**.

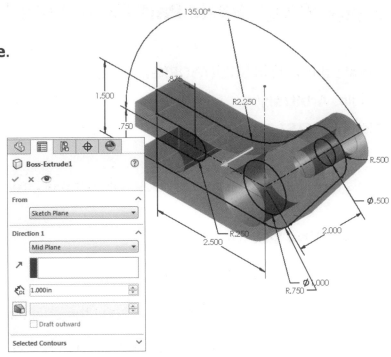

3. Creating the upper cut:

Open a **new sketch** on the <u>planar face</u> as noted.

Sketch a **circle**, a **line**, and **convert 2 entities** as indicated.

Sketch a circle
and a line

Convert entity

Sketch face

The center of the circle is coincident with the center of the existing hole.

Trim the entities and add the dimensions shown to fully define the sketch.

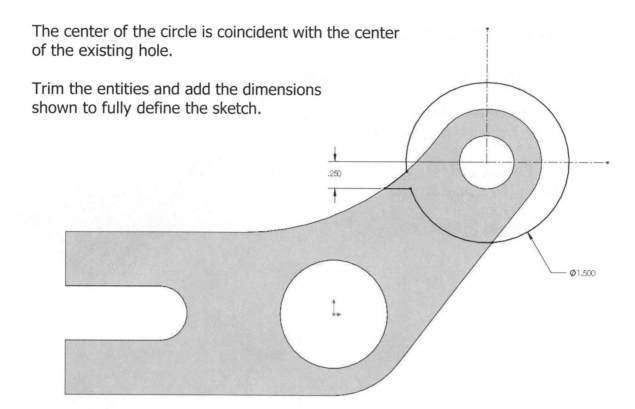

4. Extruding a blind cut:

Click **Extruded Cut**.

Use the default **Blind** type.

Enter **.250 in**. for depth.

Click **OK**.

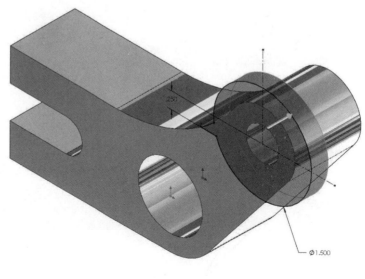

5. Mirroring the cut feature:

Click **Mirror** from the feature toolbar.

For Mirror Face/Plane, select the **Front** plane from the feature tree.

For Features to Mirror, select the **Cut-Extrude** either from the tree or from the graphics area.

Click **OK**.

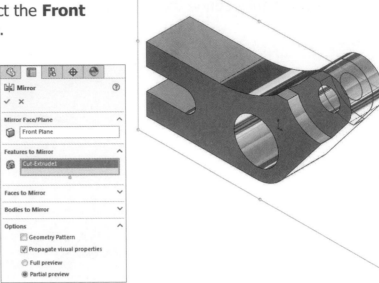

6. Creating the center cut:

Open a **new sketch** on the <u>Top</u> plane.

Sketch a **rectangle** approximately as shown.

Add the dimensions/relations needed to fully define the sketch.

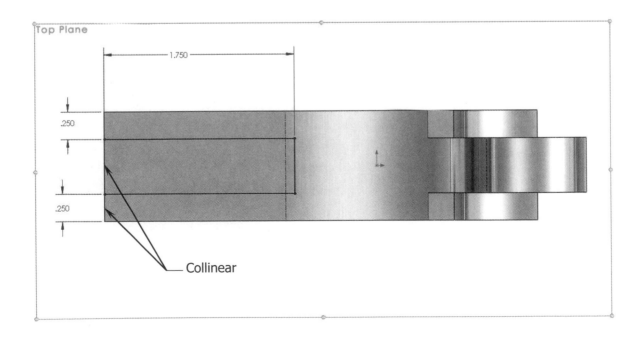

7. Extruding a through cut:

Click **Extruded Cut**.

Use the **Through All-Both** for Direction 1.

The 2nd direction is selected automatically.

Click **OK**.

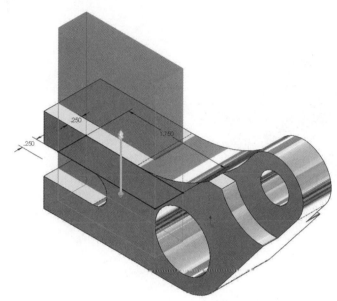

8. Creating a recess feature:

Open a **new sketch** on the <u>planar face</u> as noted.

While the selected face is still highlighted, click **Offset Entities**.

Enter **.080 in**. for offset distance and click the **Reverse** checkbox if needed to place the offset entities on the **inside**.

Close the Offset Entities command.

Sketch face

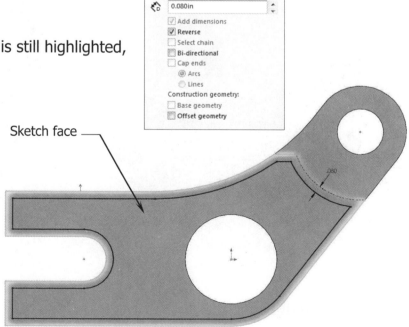

9. Extruding a blind cut:

Click **Extruded Cut**.

Use the default **Blind** type.

Enter **.125 in**. for depth.

Click **OK**.

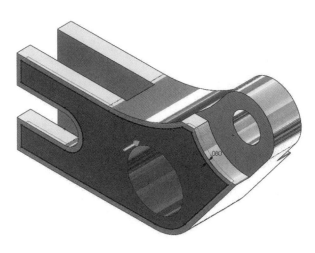

Rotate the model to verify the result of the cut feature.

Note: The recess feature is only added to one side. Do not mirror it.

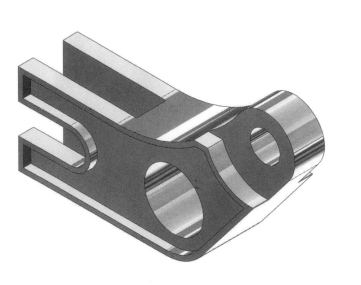

Front Isometric

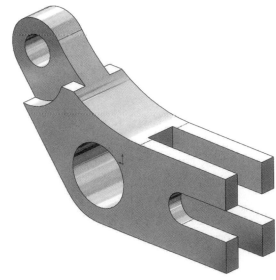

Rear Isometric

10. Calculating the Mass:

Change the material
of the model to
AISI 1020.

Click **Mass Properties**.

If needed, set the Unit
of Measure to:
IPS, **3 decimals**.

Enter the final mass of
the model here:

_____ lbs.

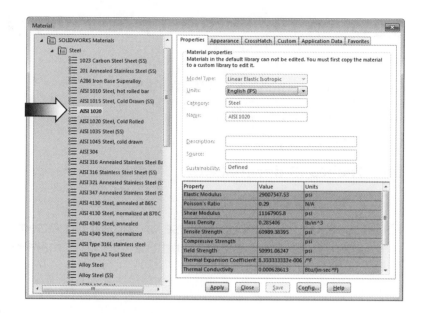

Save your work as **CSWA Bracket**.

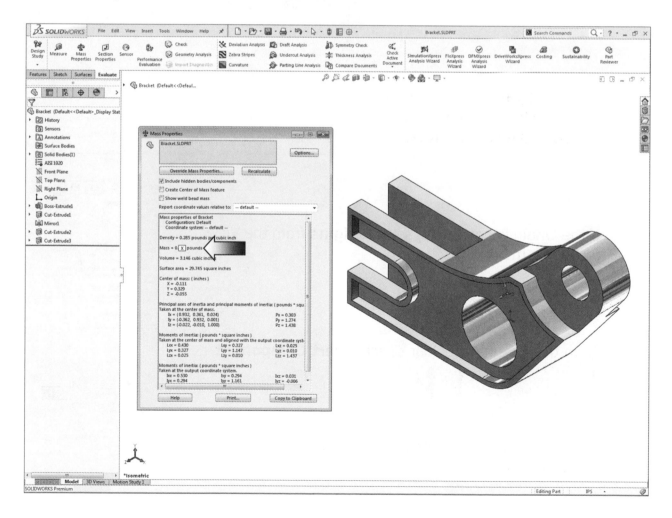

Question 10: Bottom Up Assembly (1 of 2)

Create this assembly in SOLIDWORKS.

Drafting Standards: **ANSI**

Origin: **Arbitrary**

Unit: **Inches**, **3 decimals**

(This question focuses on the use of Bottom Up Assembly method, mating components, and changing the mate conditions.)

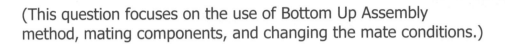

1. Creating a new assembly:

Select: **File, New, Assembly**.

Click the **Cancel** button and set the Units to **IPS** and the Drafting Standard to **ANSI**.

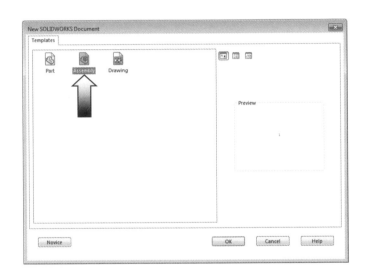

If the Origin is not visible, select **Origins** from the View pull-down menu.

From the Assembly toolbar select **Insert Component**.

Locate the part named **Base** from the CSWA Training Folder and open it.

Place the 1st component on the assembly's origin.

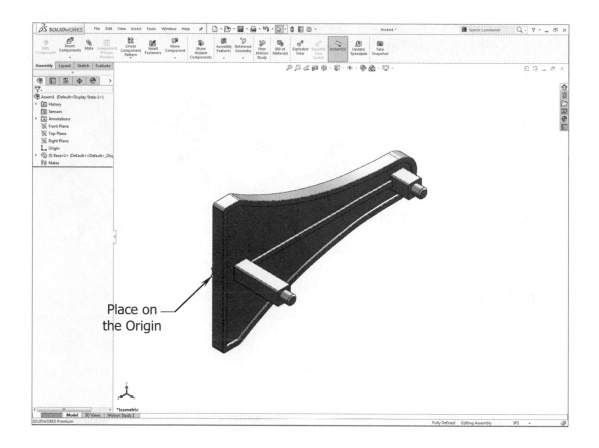

Place on the Origin

2. Inserting other components:

Insert the rest of the components into the assembly as labeled.

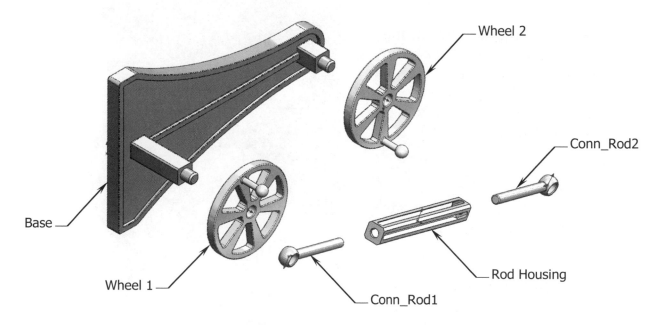

Wheel 2

Conn_Rod2

Base

Wheel 1

Conn_Rod1

Rod Housing

3. Mating the components:

The components that need to be moved or rotated will get only 2 mates assigned to them, but the ones that are fixed will get 3 mates instead.

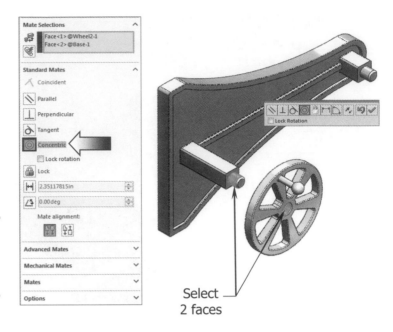

For clarity, hide all components except for the **Base** and the **Wheel1**.

Click the **Mate** command from the Assembly toolbar.

Select —
2 faces

Select the **cylindrical face** of the Pin on the left side of the Base and the **hole** in the middle of the Wheel1.

A **Concentric** mate is selected automatically.

Click **OK** to accept the mate.

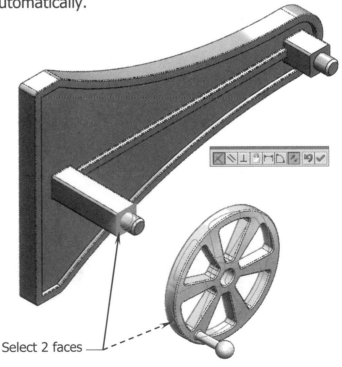

Select the **planar face** at the end of the Pin and the **planar face** on the far side of the Wheel1. (The dotted line indicates the surface in the back of the component.)

When 2 planar faces are selected, a **Coincident** mate is added automatically.

Select 2 faces —

Click **OK** to accept the mate.

4. Mating the Wheel2 to the Base:

Show the component **Wheel2**.

Click the **Mate** command if it is no longer active.

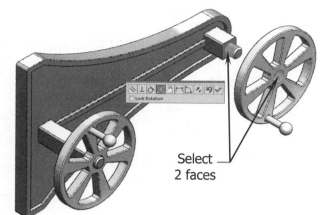

Select 2 faces

Add a **Concentric** mate between the **cylindrical face** of the 2nd Pin and the **hole** in the middle of the Wheel2.

Click **OK**.

Next, select the **planar face** at the end of the Pin and the **planar face** on the far side of the Wheel1.

Select 2 faces

A **Coincident** mate is added automatically.

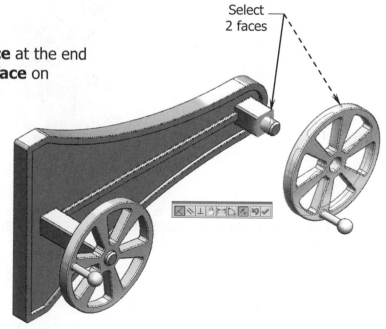

Click **OK**.

5. Assembling the Conn_Rod1 to the Wheel1:

Show the component **Conn-Rod1**.

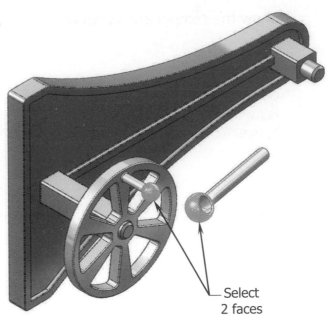

Select the **cylindrical faces** of both components, the **Wheel1** and the **Conn-Rod1**.

A **Concentric** mate is added to the 2 selected faces.

Click **OK**.

Select 2 faces

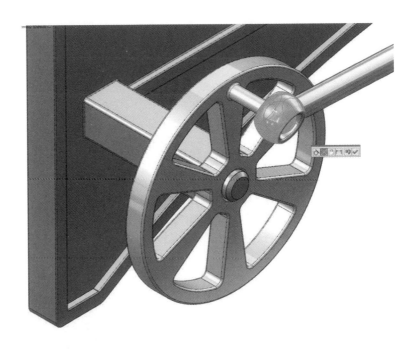

Test the degrees of freedom of each component by dragging them back and forth.

Move the Conn-Rod1 to a position where it does not interfere with the knob in the Wheel1 (as shown in the image above).

6. Mating the Conn_Rod2 to the Wheel2:

Show the component **Conn_Rod2**.

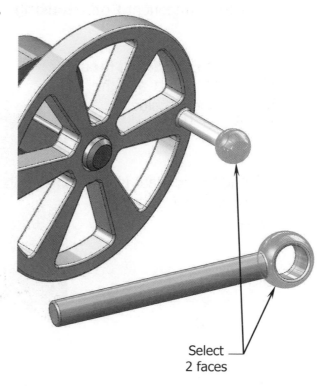

Select the **cylindrical faces** of both components, the **Conn-Rod2** and the **Wheel2** as pictured.

Another **Concentric** mate is added to the 2 selected faces.

Click **OK**.

Select 2 faces

Drag the Conn_Rod2 back and forth to test it.

Move the Conn_Rod2 to the position similar to the one pictured above.

7. Mating the Rod_Housing to the Conn_Rods:

Show the component **Rod_Housing**.

Select the **cylindrical face** of the Conn-Rod1 and the **hole** on the left side of the Rod_Housing.

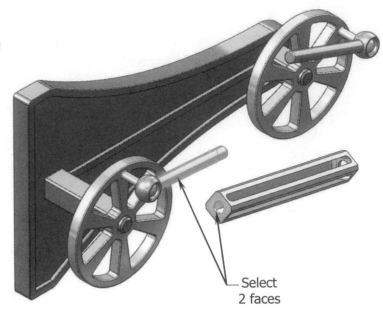

A **Concentric** mate is added to the 2 selected faces.

Click **OK**.

Select 2 faces

Test the Rod_Housing by dragging it back and forth. It should be constrained to move only along the longitudinal axis of the Conn-Rod.

Move the Rod_Housing to a position similar to the one pictured above.

Zoom in to the right side of the assembly. We will assemble the Conn_Rod2 to its housing.

Click the **Mate** command if it is no longer selected.

Select the **cylindrical face** of the Conn-Rod1 and the **hole** on the right side of the Rod_Housing.

Another **Concentric** mate is added to the 2 selected faces.

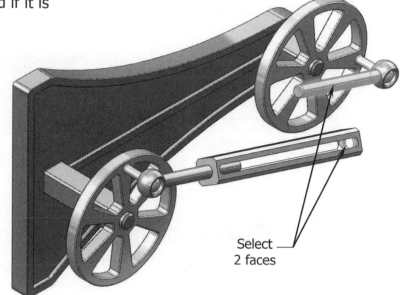

Select
2 faces

Click **OK**.

8. Adding a Symmetric mate:

Using the Feature tree, expand the Rod-Housing and select its **Front** Plane (A).

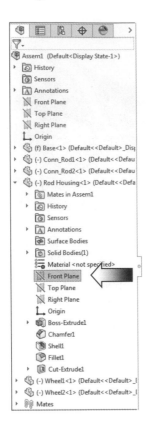

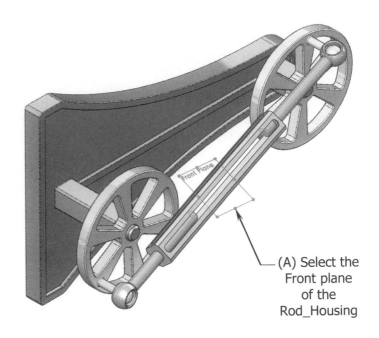

(A) Select the Front plane of the Rod_Housing

Change to the **Advanced Mates** section and select the **Symmetric** button (B).

The 2 ends are spaced evenly. Click **OK**.

(B) Select the 2 end faces of the 2 Conn_Rods

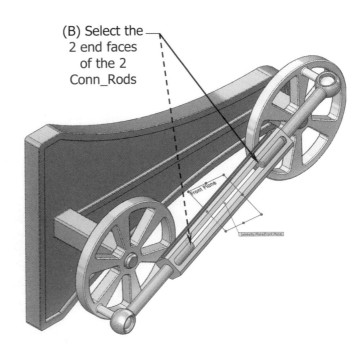

9. Testing the assembly motions:

Exit out of the Mate mode.

Drag the handle on the Wheel1 as indicated.

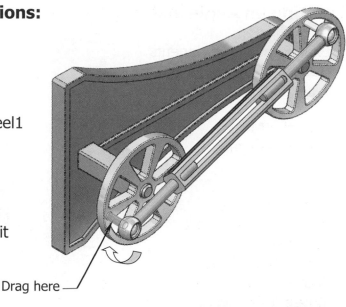

When the Wheel1 is turned it moves the Conn-Rod1 with it, and at the same time the Rod_Housing and the Conn_Rod2 is also moved along.

Drag here

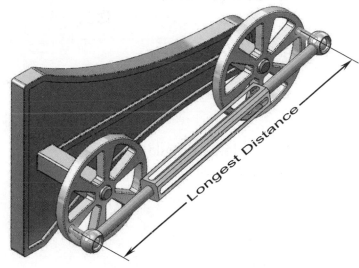

Longest Distance

Without a **Limit** mate the 2 Conn-Rods will collide with each other, but since Limit mate is not part of the question, we are going to use an alternate method to test the motion of this assembly.

Drag the handles on both wheels to fully extend them to their longest distance (approximately as shown).

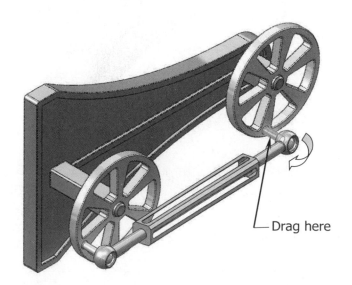

Drag here

Drag the handle on the Wheel2 in either direction. The Wheel2 should move both Con_Rods and the Rod_Housing with it but without any collisions.

10. Creating an Angle mate:

Expand both components **Base** and **Conn_Rod1**.

Click the **Mate** command again.

Using the Feature Manager tree, select the **Top** plane of the **Base** and the **Front** plane of the **Wheel1** (arrows).

Select the **Angle** button and enter **90°**. Click **OK**.

The 2 planes should be perpendicular to each other.

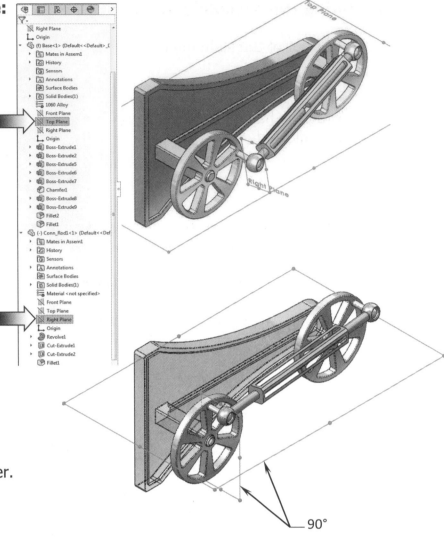

Also create a **45° Angle** mate between the **Front** plane of the **Base** and the **Front** plane of the **Wheel2**.

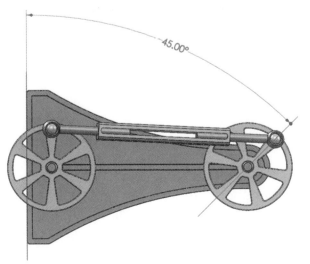

11. Measuring the distance:

Switch to the **Evaluate** tab.

Select the **Measure** command and measure the distance between the **left end** of the Rod_Housing and the **end face** of the Conn_Rod1.

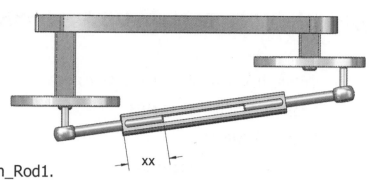

Enter the distance (in inches) here: _____

12. Changing the mate Angle:

Expand the **Mates Group** at the bottom of the Feature tree (arrow).

<u>Edit</u> the **Angle mate** and change the angle to **180º**. Click **OK**.

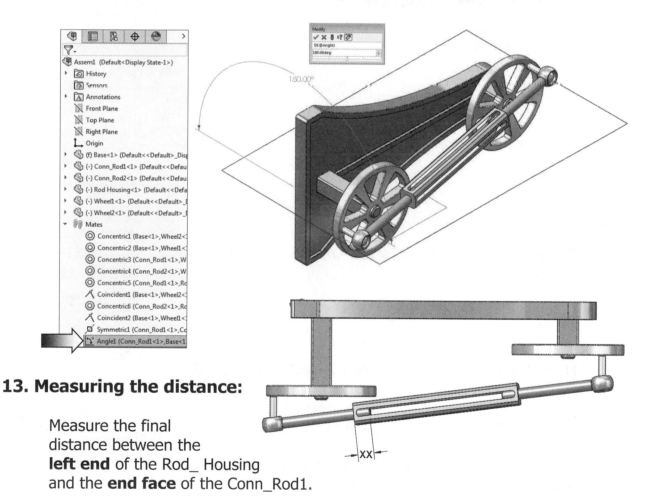

13. Measuring the distance:

Measure the final distance between the **left end** of the Rod_ Housing and the **end face** of the Conn_Rod1.

Enter the distance (in inches) here: _____

Question 11: Bottom Up Assembly (2 of 2)

Create this assembly in SOLIDWORKS.

Drafting Standards: **ANSI**

Origin: **Arbitrary**

Unit: **Inches, 3 decimals**

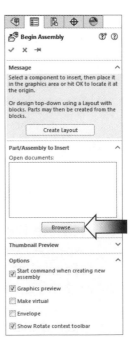

(This question focuses on the use of Bottom Up Assembly method, mating components, and changing the mate conditions.)

1. Creating a new assembly:

Select: **File, New, Assembly**.

Click the **Cancel** button and set the Units to **IPS** and the Drafting Standard to **ANSI**.

If the Origin is not visible, select **Origins** from the View pull-down menu.

From the Assembly toolbar select **Insert Component**.

Locate the part named **Base_Exe** from the **CSWA Assembly Exercise** folder and open it.

Place the 1st component on the assembly's origin.

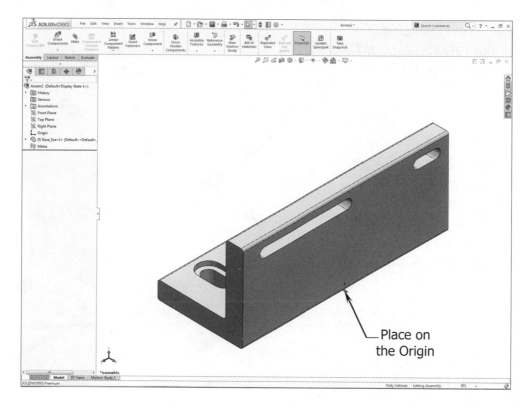

Place on
the Origin

2. Inserting other components:

Insert the rest of the components into the assembly as labeled.

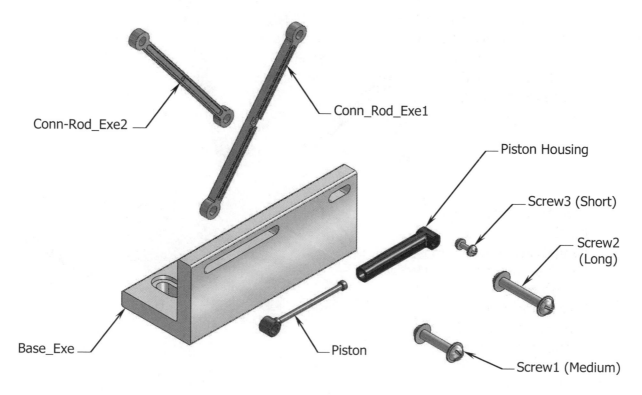

Conn-Rod_Exe2

Conn_Rod_Exe1

Piston Housing

Screw3 (Short)

Screw2 (Long)

Base_Exe

Piston

Screw1 (Medium)

3. Changing configurations:

The Screw contains 3 different configurations, a Long, a Medium, and a Short.

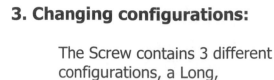

To change configuration simply click the Screw, in the pop-up menu select the configuration from the list and click the check mark (arrow).

To quickly make a copy of a component, simply hold the Control key and drag it aside.

Create a total of 3 instances of the Screw and change their configurations as labeled. Click the **green check** after each change.

Use the left mouse button to move a component and then use the right button to rotate. Rearrange the components similar to the image shown.

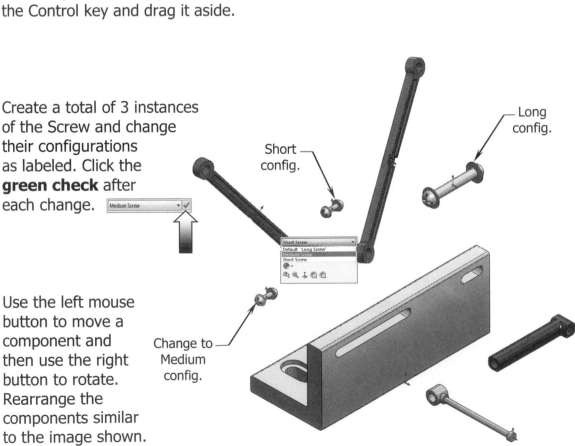

4. Mating the Piston Housing:

Click the **Mate** command again, to reactivate it.

Select the **face of the slot** of the Base_Exe and the **hole** in the Piston Housing.

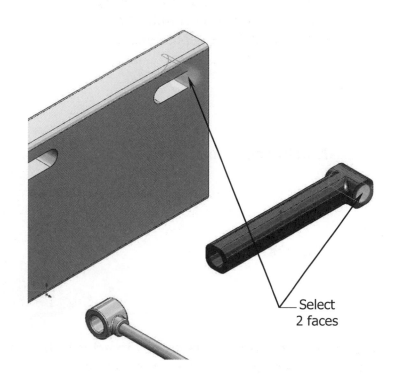

A **Concentric** mate is selected automatically for the selection.

Select 2 faces

Click **OK** to accept the mate.

Concentric

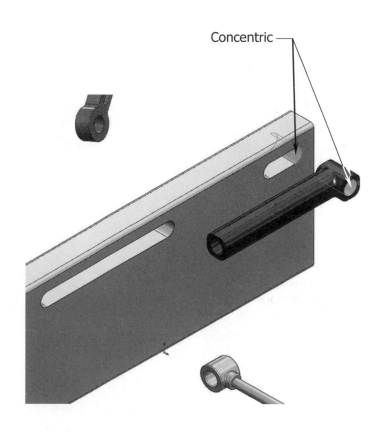

Select the **planar face** in the back of the
Base_Exe and the **planar face** on the
far side of the Piston Housing.
(The dashed line represents a
hidden face.)

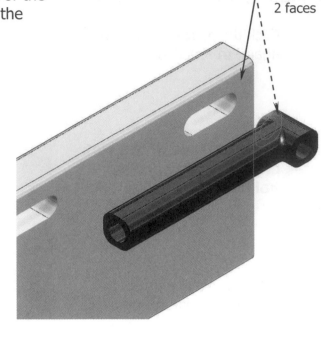

Select
2 faces

A **Coincident** mate is selected.

Click **OK**.

NOTE: Press the F5 function key to activate the Selection-Filters and use the **Filter-Faces** to assist you with selecting the faces more precisely and easily.

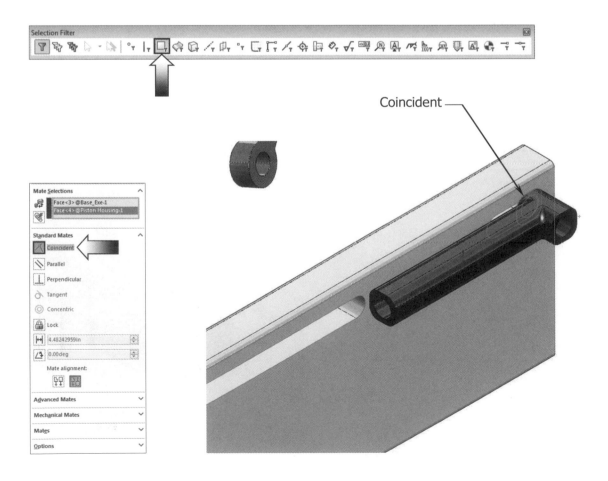

Coincident

5. Mating the Conn-Rod_Exe1:

Move the Conn-Rod_Exe1 closer to the Base_Exe as pictured; it will be mated to the Piston Housing.

Click the **Mate** command if it is no longer active.

Select the **hole** in the Conn-Rod_Exe1 and the **hole** in the Piston Housing.

A **Concentric** mate is selected.

Click **OK**.

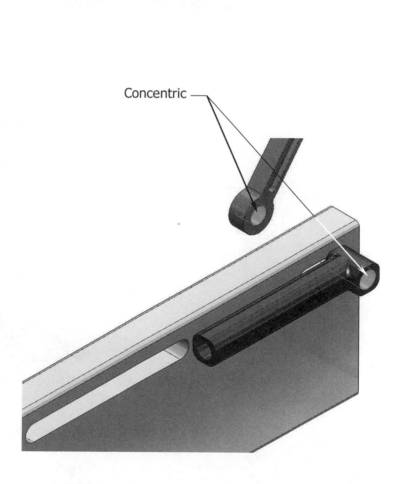

Select 2 faces

Concentric

Select the **planar face** in the front of the Conn-Rod_exe1 and the **planar face** on the far side of the Piston Housing.

Select 2 faces

A **Coincident** mate is selected.

Click **OK**.

Usually there will be a Washer placed between the 2 components, but we are going to skip it because this exercise does not need one.

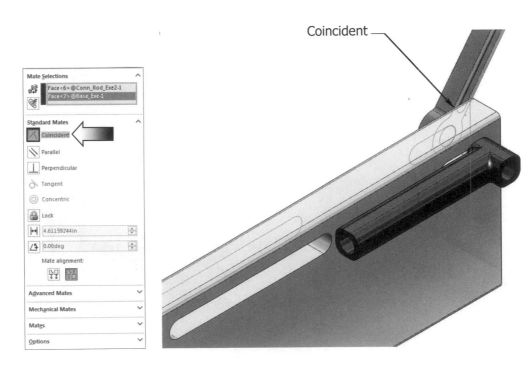

Coincident

Mate Selections	
Face<6> @Conn_Rod_Exe2-1	
Face<7> @Base_Exe-1	

Standard Mates	
Coincident	←
Parallel	
Perpendicular	
Tangent	
Concentric	
Lock	
4.61159244in	
0.00deg	

Mate alignment:

Advanced Mates	∨
Mechanical Mates	∨
Mates	∨
Options	∨

6. Mating the Long Screw:

Move the **Long screw** closer as pictured. It will be mated to the Conn-Rod_Exe1.

Ensure that the mate command is still active.

Select the **cylindrical face** of the Long Screw and the **hole** in the Piston Housing.

A **Concentric** mate is selected automatically.

Click **OK**.

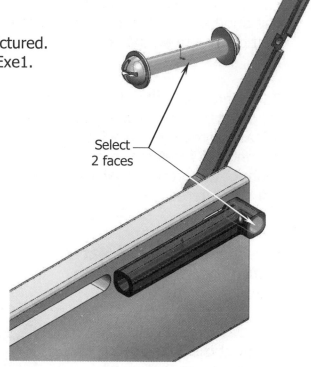

Select 2 faces

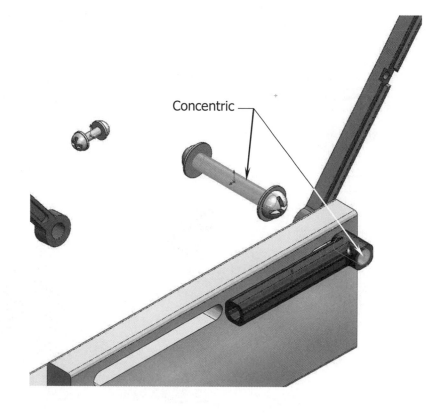

Concentric

When 2 components need to be centered with one another, Width mate is one of the best options to do that with.

A. Select 2 opposing faces for Width

To specify the width of each part, you will need to select two opposing faces of each one such as the faces of the tab and the faces that represent the width of the groove in which the tab will be centered.

Expand the **Advanced Mates** section and select the **Width** option (arrow).

B. Select 2 opposing faces for Tab

A. Select the **2 opposing faces** of the Long Screw.

B. Select the **2 opposing faces** of the Piston Housing and the Conn-Rod_exe1.

The 2 selected components are centered automatically.

Click **OK**.

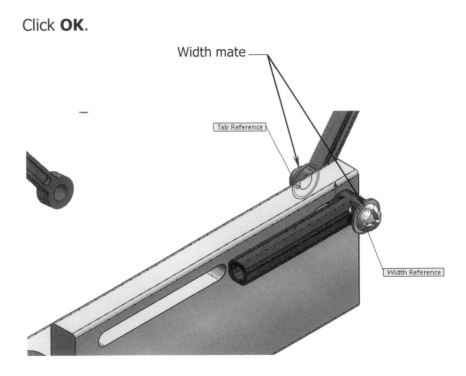

Width mate

Tab Reference

Width Reference

7. Mating the Piston to its Housing:

Move the Piston closer to the Piston Housing as pictured.

Select the **cylindrical face** of the Piston and the **hole** in the Piston Housing as indicated.

A **Concentric** mate is selected but the alignment is incorrect.

Locate the **Align/Anti Align** buttons (arrow) at the bottom of the Standard Mates section.

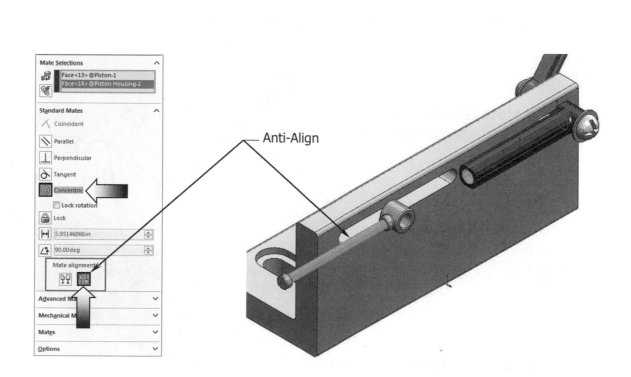

Toggle the **Align/Anti Align** buttons (arrow) to flip the Piston to the correct orientation.

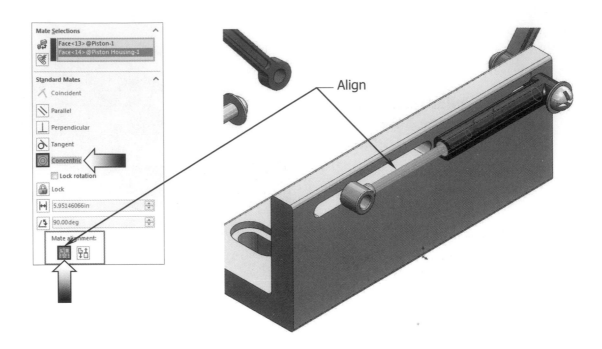

Align

8. Mating the Conn_Rod_Exe2 to the Piston:

Move the Conn-Rod_Exe2 closer to the Piston as shown.

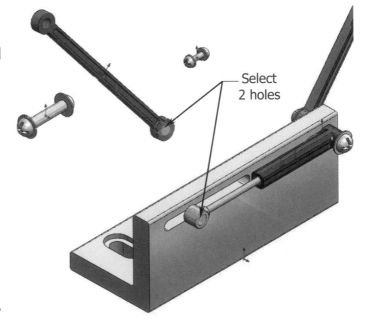

The **Mate** command should still be active; select it otherwise.

Select 2 holes

Select the **hole** on the bottom of the Conn-Rod_exe2 and the **hole** in the Piston.

A **Concentric** mate is selected.

OK to accept the concentric mate.

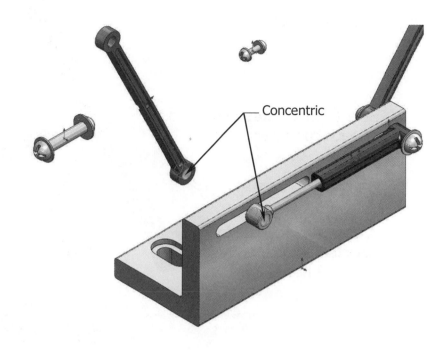

Concentric

9. Rearranging the components:

Drag the top of the Conn_Rod_Exe2 to the right, and drag the top of the Conn_Rod_Exe1 to the left.

From the right view orientation the 2 Conn-Rods should look similar to the image shown below.

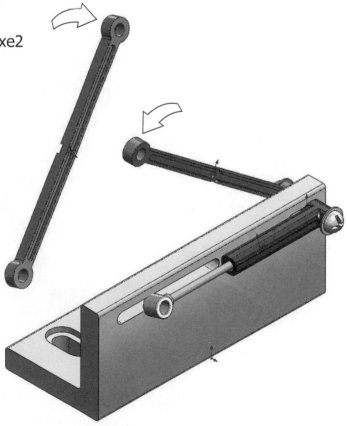

10. Mating the two Connecting Rods:

Click the **Mate** command.

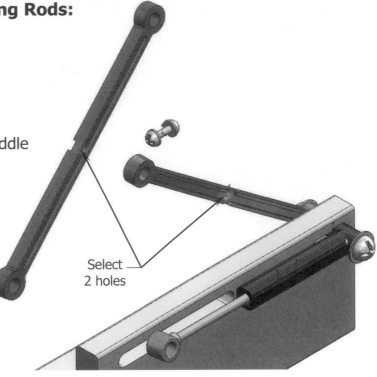

Select the **2 holes** in the middle of each connecting rod.

A **Concentric** mate is selected automatically for the 2 holes.

Select
2 holes

Click **OK**.

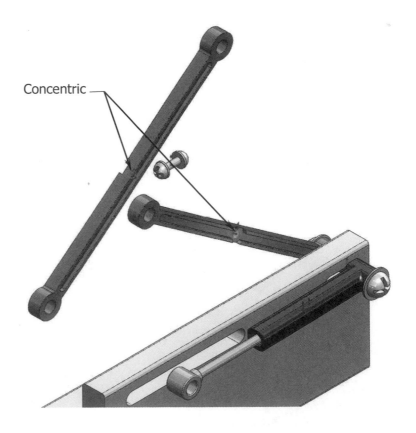

Concentric

For the coincident mate, select **the face in the front** of the Conn_Rod_Exe2 and **the face on the far side** of the Conn_Rod_Exe1.

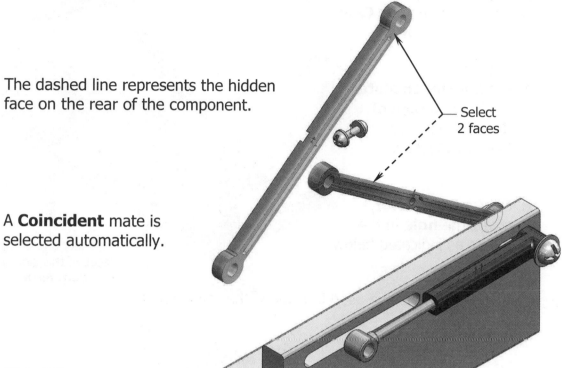

The dashed line represents the hidden face on the rear of the component.

Select 2 faces

A **Coincident** mate is selected automatically.

Click **OK**.

The front face of the Conn_Rod2 moves forward and touches the back face of the Conn-Rod1.

Coincident

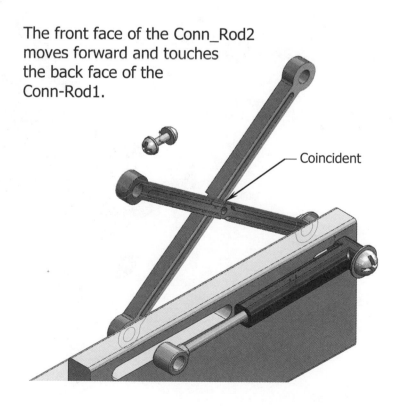

11. Adding a Cam mate: (Note: The Slot mate option is only available in SW-2014 or newer.)

Expand the **Mechanical Mates** section and select the **Cam** mate option (arrow).

A. For **Entities to Mate**, right click **a face of the slot** and pick Select Tangency.

B. For **Cam Follower**, select the **hole** in the Piston as indicated below.

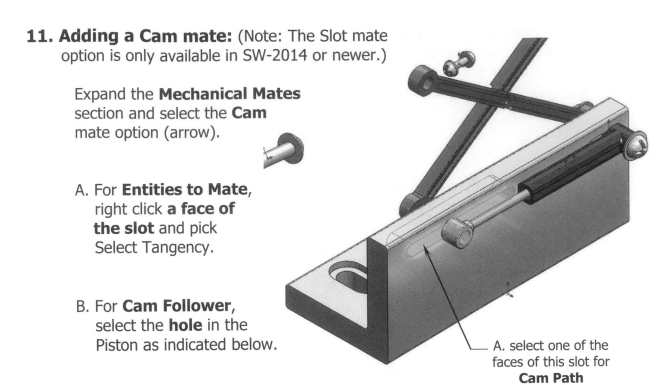

A. select one of the faces of this slot for **Cam Path**

Note: A Slot Mate can also be used to achieve the same result.

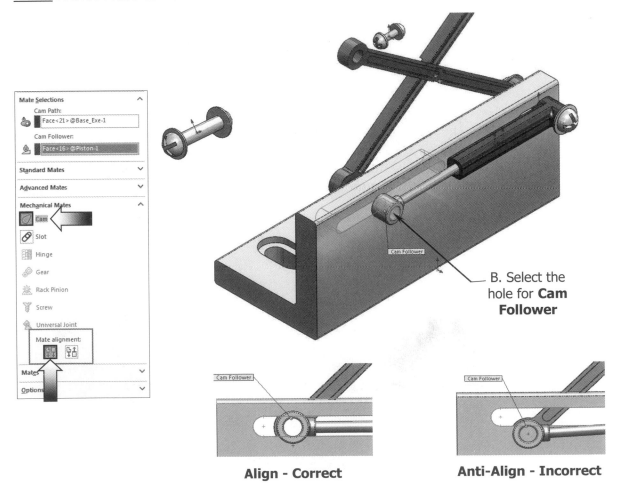

B. Select the hole for **Cam Follower**

Align - Correct

Anti-Align - Incorrect

12. Constraining the Medium Screw:

The **Mate** command should still be active.

Select the **body** of the Medium Screw and the **hole** in the Piston as indicated.

A **Concentric** mate is selected.

Click **OK**.

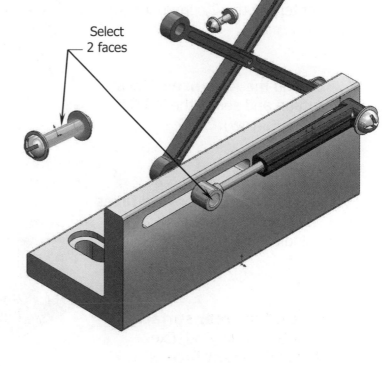

Select 2 faces

The cylindrical body of the Medium Screw is rotated and constrained to the center axis of the hole.

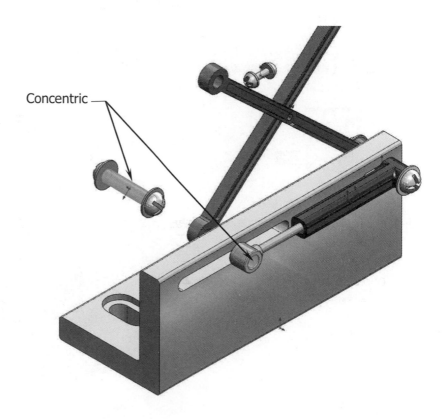

Concentric

The **Width** mate is used to align the centers of the 2 components.

Expand the **Advanced Mates** section and select the **Width** option (arrow).

A. Select the **2 opposing faces** of the Medium Screw as indicated.

B. Select the **rear surface** of the Conn_Rod_Exe2 and the **front face** of the Piston for Tab.

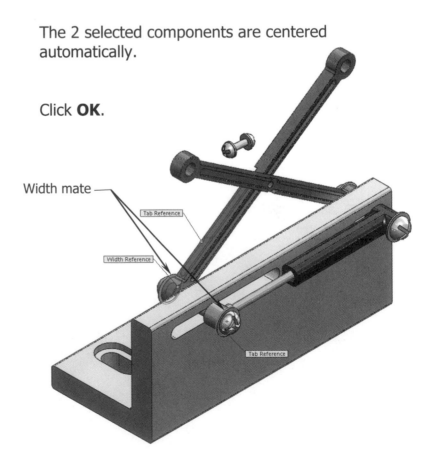

A. Select 2 opposing faces for Width

B. Select 2 opposing faces for Tab

The 2 selected components are centered automatically.

Click **OK**.

Width mate

Tab Reference

Width Reference

Tab Reference

13. Mating the Small Screw:

Click the **Mate** command if it is no longer active.

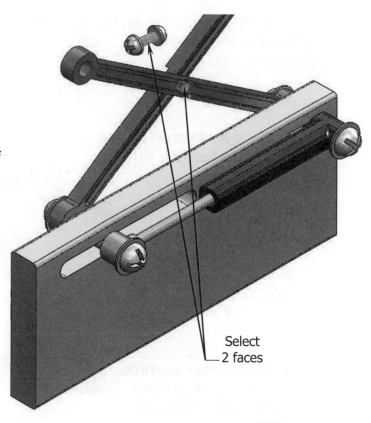

Select the **cylindrical face** of the Small Screw and the **hole** in the Conn-Rod_Exe1.

A **Concentric** mate is selected.

Click **OK** to accept the mate.

Select 2 faces

Concentric

Use the **Width** mate again to align the centers of the Small Screw and the 2 Conn-Rods.

Expand the **Advanced Mates** section and select the **Width** option (arrow).

A. Select the **2 opposing faces** of the Small Screw as indicated.

B. Select the **rear surface** of the Conn_Rod_Exe2 and the **front face** of the Conn-Rod_Exe1 for Tab.

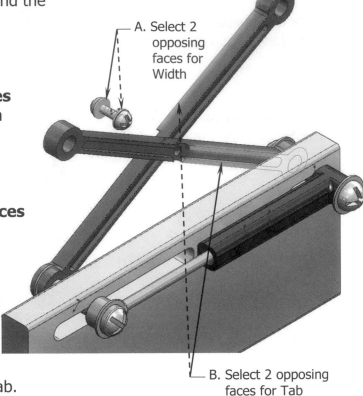

A. Select 2 opposing faces for Width

B. Select 2 opposing faces for Tab

Click **OK**.

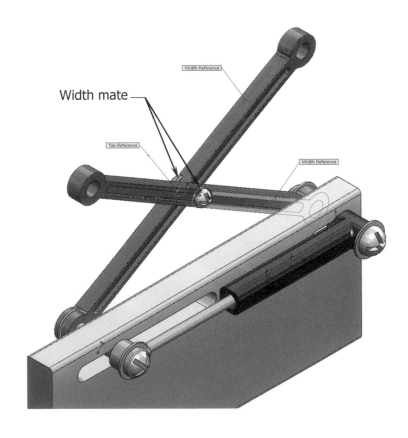

Width mate

14. Adding an Angle Mate:

After all components have been assembled, a reference location needs to be established so that the center of mass can be measured from it.

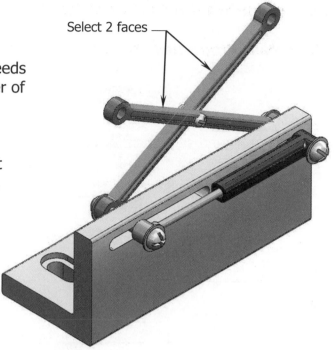

Select 2 faces

An Angle mate is used at this point to establish the reference location.

Click the **Mate** command if it is not active.

Select the **2 faces** of the 2 arms as indicated.

Click the **Angle** button (arrow) and enter **60°**.

The 2 arms move to the position similar to the image below. Click **OK**.

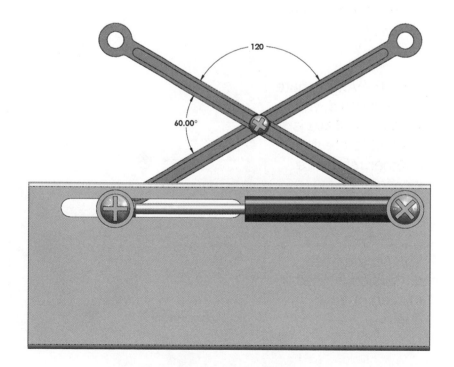

15. Measuring the Center of Mass of the assembly:

Click **Mass Properties** from the Evaluate tool tab.

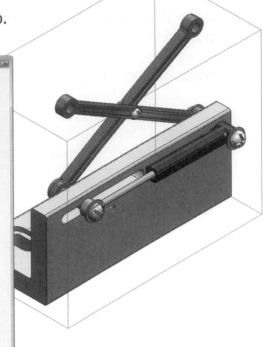

Enter the **Center of Mass** below (in Inches)

X = _____

Y = _____

Z = _____

16. Suppressing a mate:

Expand the **Mates Group** and suppress the Angle mate that was done in step number 14.

After the Angle mate is suppressed, a **Distance Mate** is needed to create another reference location.

This time a linear dimension is used instead of an angle.

17. Creating a Section View:

Click the **Section View** command from the View Heads Up tool bar.

Section View
Displays a cutaway of a part or assembly using one or more cross section planes.

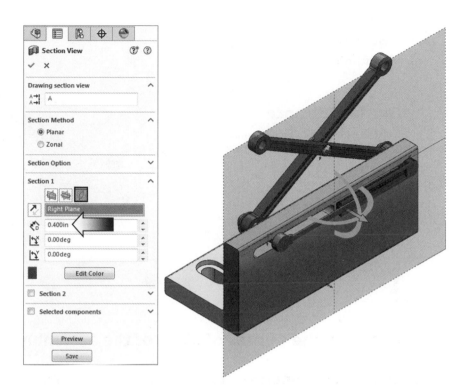

Select the **Right Plane** as cutting Plane and enter **.400"** for **Offset Distance**.

Click the Reverse button if needed to remove the front portion of the assembly.

Click **OK** to close out of the section command.

18. Measuring the distance:

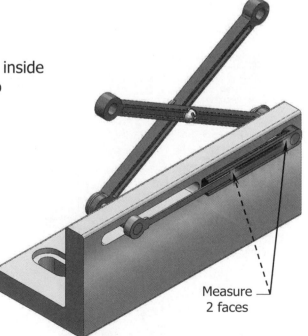

The Section View allows us to see the inside of the Piston Housing. We will need to create a distance Mate between the end of the Piston and the inside face of the Piston Housing.

Switch back to the Evaluate tab and click the **Measure** command.

Measure the <u>distance</u> between the **2 faces** as indicated.

Push **Esc** to exit the measure tool.

Measure 2 faces

19. Creating a Distance Mate:

Click the **Mate** command.

Select the **2 faces** that were used in the last step.

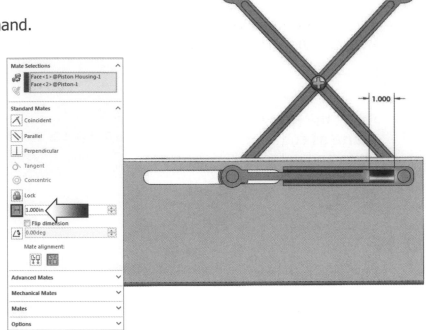

Click the **Distance** button (arrow) and enter **1.00"**.

The 2 Arms move to a new position (which is equivalent to about 95.34deg).

20. Measuring the Center of Mass of the assembly:

Click **Mass Properties** from the Evaluate tool tab.

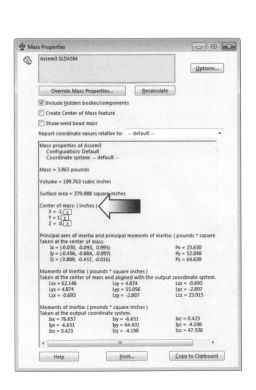

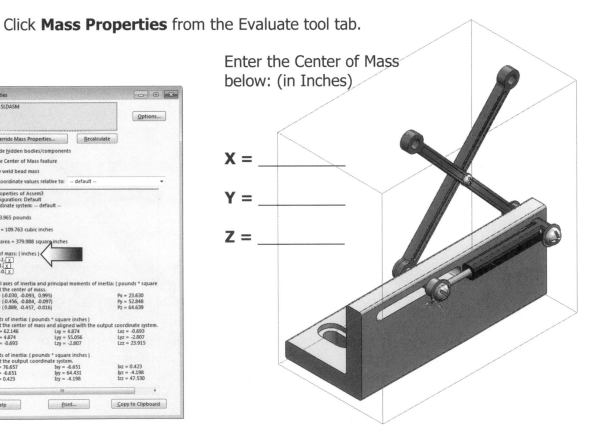

Enter the Center of Mass below: (in Inches)

X = _____

Y = _____

Z = _____

Practice this material two or three times and time yourself to see if you could complete all challenges within a three-hour time frame.

When you are ready to take the CSWA exam, go to the web link below and purchase the exam using a credit card:
http://www.solidworks.com/sw/support/796_ENU_HTML.htm

The test costs $99 for a student or customer without a maintenance subscription, but the exam is free of charge for customers who have purchased the maintenance subscription. For more information, go to the quick link within the same webpage above and select *Certifications offers for subscription service customers*.

As of the writing of this text, there are a total of **428,825 Certified SOLIDWORKS Associate (CSWA)** world-wide.
Go to the link below and enter your state to find out how many CSWA are in your state:
https://solidworks.virtualtester.com/#userdir_button

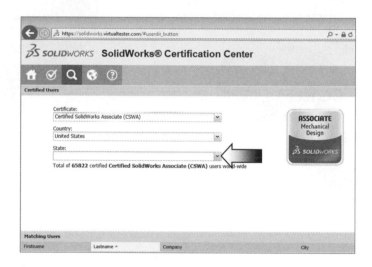

Exam Length: 3 hours – Exam cost: $99.00

Minimum passing grade: 70%

Re-test Policy: There is a minimum 14 day waiting period between every attempt of the CSWA exam. Also, a CSWA exam credit must be purchased for each exam attempt.

All candidates receive electronic certificates and personal listing on the CSWA directory when they pass. You can also update or change your log in information afterward.

Dual monitors are recommended but not required. You could save up to 15 minutes for <u>not</u> having to switch back and forth between the exam and the SOLIDWORKS application.

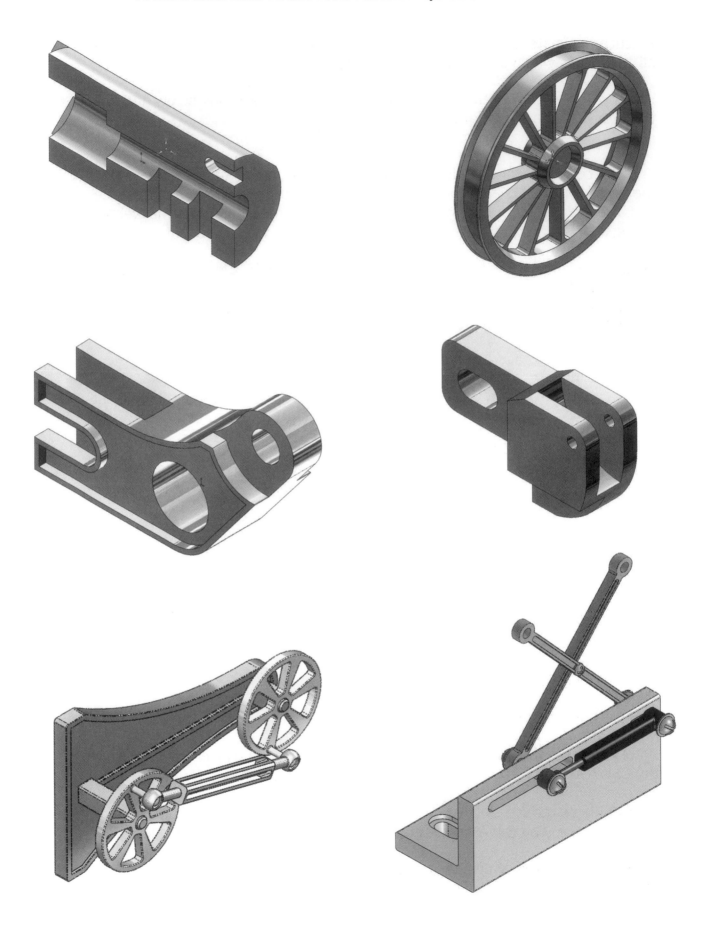

Glossary

Alloys:

An Alloy is a mixture of two or more metals (and sometimes a non-metal). The mixture is made by heating and melting the substances together.

Example of alloys are Bronze (Copper and Tin), Brass (Copper and Zinc), and Steel (Iron and Carbon).

Gravity and Mass:

Gravity is the force that pulls everything on earth toward the ground and makes things feel heavy. Gravity makes all falling bodies accelerate at a constant 32ft. per second (9.8 m/s). In the earth's atmosphere, air resistance slows acceleration. Only on the airless Moon would a feather and a metal block fall to the ground together.

The mass of an object is the amount of material it contains.

A body with greater mass has more inertia; it needs a greater force to accelerate. Weight depends on the force of gravity, but mass does not.

When an object spins around another (for example: a satellite orbiting the earth) it is pushed outward. Two forces are at work here: Centrifugal (pushing outward) and Centripetal (pulling inward). If you whirl a ball around you on a string, you pull it inward (Centripetal force). The ball seems to pull outward (Centrifugal force) and if released will fly off in a straight line.

Heat:

Heat is a form of energy and can move from one substance to another in one of three ways: by Convection, by Radiation, and by Conduction.

Convection takes place only in liquids like water (for example: water in a kettle) and gases (for example: air warmed by a heat source such as a fire or radiator). When liquid or gas is heated, it expands and becomes less dense. Warm air above the radiator rises and cool air moves in to take its place, creating a convection current.

Radiation is movement of heat through the air. Heat forms match set molecules of air moving, and rays of heat spread out around the heat source.

Conduction occurs in solids such as metals. The handle of a metal spoon left in boiling liquid warms up as molecules at the heated end move faster and collide with their neighbors, setting them moving. The heat travels through the metal, which is a good conductor of heat.

Inertia:

A body with a large mass is harder to start and also to stop. A heavy truck traveling at 50mph needs more power breaks to stop its motion than a smaller car traveling at the same speed.

Inertia is the tendency of an object either to stay still or to move steadily in a straight line, unless another force (such as a brick wall stopping the vehicle) makes it behave differently.

Joules:

The Joule is the SI unit of work or energy.
One Joule of work is done when a force of one Newton moves through a distance of one meter. The Joule is named after the English scientist James Joule (1818-1889).

Materials:

Stainless steel is an alloy of steel with chromium or nickel.

Steel is made by the basic oxygen process. The raw material is about three parts melted iron and one-part scrap steel. Blowing oxygen into the melted iron raises the temperature and gets rid of impurities.

All plastics are chemical compounds called polymers. Glass is made by mixing and heating sand, limestone, and soda ash. When these ingredients melt they turn into glass, which is hardened when it cools. Glass is in fact not a solid but a "supercooled" liquid, it can be shaped by blowing, pressing, drawing, casting into molds, rolling, and floating across molten tin, to make large sheets.

Ceramic objects, such as pottery and porcelain, electrical insulators, bricks, and roof tiles are all made from clay. The clay is shaped or molded when wet and soft, and heated in a kiln until it hardens.

Machine Tools:

Machine tools are powered tools used for shaping metal or other materials by drilling holes, chiseling, grinding, pressing, or cutting. Often the material (the work piece) is moved while the tool stays still (lathe), or vice versa, the work piece stays while the tool moves (mill). Most common machine tools are Mill, Lathe, Saw, Broach, Punch press, Grind, Bore and Stamp break.

CNC

Computer Numerical Control is the automation of machine tools that are operated by precisely programmed commands encoded on a storage medium, as opposed to controlled manually via hand wheels or levers, or mechanically automated via cams alone. Most CNC today is computer numerical control in which computers play an integral part of the control.

3D Printing

All methods work by working in layers, adding material, etc. different to other techniques, which are subtractive. Support is needed because almost all methods could support multi material printing, but it is currently only available in certain top tier machines.

A method of turning digital shapes into physical objects. Due to its nature, it allows us to accurately control the shape of the product. The drawback is size restraints and materials are often not durable.

While FDM does not seem like the best method for instrument manufacturing, it is one of the cheapest and most universally available methods.

EDM
Electric Discharge Machining.

FDM
Fused Deposition Modeling.

SLA
Stereo Lithography.

SLS
Selective Laser Sintering.

SLM
Selective Laser Melting.

J-P
Jetted Photopolymer (or Polyjet).

Newton's Law:
1. Every object remains stopped or goes on moving at a steady rate in a straight line unless acted upon by another force. This is the inertia principle.
2. The amount of force needed to make an object change its speed depends on the mass of the object and the amount of the acceleration or deceleration required.
3. To every action there is an equal and opposite reaction. When a body is pushed by a force, another force pushes back with equal strength.

Polymers:
A polymer is made of one or more large molecules formed from thousands of smaller molecules. Rubber and Wood are natural polymers. Plastics are synthetic (artificially made) polymers.

Speed and Velocity:
Speed is the rate at which a moving object changes position (how far it moves in a fixed time).
Velocity is speed in a particular direction.
If either speed or direction is changed, velocity also changes.

Absorbed
A feature, sketch, or annotation that is contained in another item (usually a feature) in the FeatureManager design tree. Examples are the profile sketch and profile path in a base-sweep, or a cosmetic thread annotation in a hole.

Align

Tools that assist in lining up annotations and dimensions (left, right, top, bottom, and so on). For aligning parts in an assembly.

Alternate position view

A drawing view in which one or more views are superimposed in phantom lines on the original view. Alternate position views are often used to show range of motion of an assembly.

Anchor point

The end of a leader that attaches to the note, block, or other annotation. Sheet formats contain anchor points for a bill of materials, a hole table, a revision table, and a weldment cut list.

Annotation

A text note or a symbol that adds specific design intent to a part, assembly, or drawing. Specific types of annotations include note, hole callout, surface finish symbol, datum feature symbol, datum target, geometric tolerance symbol, weld symbol, balloon, and stacked balloon. Annotations that apply only to drawings include center mark, annotation centerline, area hatch, and block.

Appearance callouts

Callouts that display the colors and textures of the face, feature, body, and part under the entity selected and are a shortcut to editing colors and textures.

Area hatch

A crosshatch pattern or fill applied to a selected face or to a closed sketch in a drawing.

Assembly

A document in which parts, features, and other assemblies (sub-assemblies) are mated together. The parts and sub-assemblies exist in documents separate from the assembly. For example, in an assembly, a piston can be mated to other parts, such as a connecting rod or cylinder. This new assembly can then be used as a sub-assembly in an assembly of an engine. The extension for a SOLIDWORKS assembly file name is .SLDASM.

Attachment point

The end of a leader that attaches to the model (to an edge, vertex, or face, for example) or to a drawing sheet.

Axis

A straight line that can be used to create model geometry, features, or patterns. An axis can be made in a number of different ways, including using the intersection of two planes.

Balloon

Labels parts in an assembly, typically including item numbers and quantity. In drawings, the item numbers are related to rows in a bill of materials.

Base

The first solid feature of a part.

Baseline dimensions

Sets of dimensions measured from the same edge or vertex in a drawing.

Bend

A feature in a sheet metal part. A bend generated from a filleted corner, cylindrical face, or conical face is a round bend; a bend generated from sketched straight lines is a sharp bend.

Bill of materials

A table inserted into a drawing to keep a record of the parts used in an assembly.

Block

A user-defined annotation that you can use in parts, assemblies, and drawings. A block can contain text, sketch entities (except points), and area hatch, and it can be saved in a file for later use as, for example, a custom callout or a company logo.

Bottom-up assembly

An assembly modeling technique where you create parts and then insert them into an assembly.

Broken-out section

A drawing view that exposes inner details of a drawing view by removing material from a closed profile, usually a spline.

Cavity

The mold half that holds the cavity feature of the design part.

Center mark

A cross that marks the center of a circle or arc.

Centerline

A centerline marks, in phantom font, an axis of symmetry in a sketch or drawing.

Chamfer

Bevels a selected edge or vertex. You can apply chamfers to both sketches and features.

Child
A dependent feature related to a previously built feature. For example, a chamfer on the edge of a hole is a child of the parent hole.

Click-release
As you sketch, if you click and then release the pointer, you are in click-release mode. Move the pointer and click again to define the next point in the sketch sequence.

Click-drag
As you sketch, if you click and drag the pointer, you are in click-drag mode. When you release the pointer, the sketch entity is complete.

Closed profile
Also called a closed contour, it is a sketch or sketch entity with no exposed endpoints; for example, a circle or polygon.

Collapse
The opposite of explode. The collapse action returns an exploded assembly's parts to their normal positions.

Collision Detection
An assembly function that detects collisions between components when components move or rotate. A collision occurs when an entity on one component coincides with any entity on another component.

Component
Any part or sub-assembly within an assembly.

Configuration
A variation of a part or assembly within a single document. Variations can include different dimensions, features, and properties. For example, a single part such as a bolt can contain different configurations that vary the diameter and length.

ConfigurationManager
Located on the left side of the SOLIDWORKS window, it is a means to create, select, and view the configurations of parts and assemblies.

Constraint
The relations between sketch entities, or between sketch entities and planes, axes, edges, or vertices.

Construction geometry
The characteristic of a sketch entity that the entity is used in creating other geometry but is not itself used in creating features.

Coordinate system

A system of planes used to assign Cartesian coordinates to features, parts, and assemblies. Part and assembly documents contain default coordinate systems; other coordinate systems can be defined with reference geometry. Coordinate systems can be used with measurement tools and for exporting documents to other file formats.

Cosmetic thread

An annotation that represents threads.

Crosshatch

A pattern (or fill) applied to drawing views such as section views and broken-out sections.

Curvature

Curvature is equal to the inverse of the radius of the curve. The curvature can be displayed in different colors according to the local radius (usually of a surface).

Cut

A feature that removes material from a part by such actions as extrude, revolve, loft, sweep, thicken, cavity, and so on.

Dangling

A dimension, relation, or drawing section view that is unresolved. For example, if a piece of geometry is dimensioned, and that geometry is later deleted, the dimension becomes dangling.

Degrees of freedom

Geometry that is not defined by dimensions or relations is free to move. In 2D sketches, there arc three degrees of freedom: movement along the X and Y axes, and rotation about the Z axis (the axis normal to the sketch plane). In 3D sketches and in assemblies, there are six degrees of freedom: movement along the X, Y, and Z axes, and rotation about the X, Y, and Z axes.

Derived part

A derived part is a new base, mirror, or component part created directly from an existing part and linked to the original part such that changes to the original part are reflected in the derived part.

Derived sketch

A copy of a sketch, in either the same part or the same assembly that is connected to the original sketch. Changes in the original sketch are reflected in the derived sketch.

Design Library
Located in the Task Pane, the Design Library provides a central location for reusable elements such as parts, assemblies, and so on.

Design table
An Excel spreadsheet that is used to create multiple configurations in a part or assembly document.

Detached drawing
A drawing format that allows opening and working in a drawing without loading the corresponding models into memory. The models are loaded on an as-needed basis.

Detail view
A portion of a larger view, usually at a larger scale than the original view.

Dimension line
A linear dimension line references the dimension text to extension lines indicating the entity being measured. An angular dimension line references the dimension text directly to the measured object.

DimXpertManager
Located on the left side of the SOLIDWORKS window, it is a means to manage dimensions and tolerances created using DimXpert for parts according to the requirements of the ASME Y.14.41-2003 standard.

DisplayManager
The DisplayManager lists the appearances, decals, lights, scene, and cameras applied to the current model. From the DisplayManager, you can view applied content, and add, edit, or delete items. When PhotoView 360 is added in, the DisplayManager also provides access to PhotoView options.

Document
A file containing a part, assembly, or drawing.

Draft
The degree of taper or angle of a face usually applied to molds or castings.

Drawing
A 2D representation of a 3D part or assembly. The extension for a SOLIDWORKS drawing file name is .SLDDRW.

Drawing sheet
A page in a drawing document.

Driven dimension
Measurements of the model, but they do not drive the model and their values cannot be changed.

Driving dimension
Also referred to as a model dimension, it sets the value for a sketch entity. It can also control distance, thickness, and feature parameters.

Edge
A single outside boundary of a feature.

Edge flange
A sheet metal feature that combines a bend and a tab in a single operation.

Equation
Creates a mathematical relation between sketch dimensions, using dimension names as variables, or between feature parameters, such as the depth of an extruded feature or the instance count in a pattern.

Exploded view
Shows an assembly with its components separated from one another, usually to show how to assemble the mechanism.

Export
Save a SOLIDWORKS document in another format for use in other CAD/CAM, rapid prototyping, web, or graphics software applications.

Extension line
The line extending from the model indicating the point from which a dimension is measured.

Extrude
A feature that linearly projects a sketch to either add material to a part (in a base or boss) or remove material from a part (in a cut or hole).

Face
A selectable area (planar or otherwise) of a model or surface with boundaries that help define the shape of the model or surface. For example, a rectangular solid has six faces.

Fasteners
A SOLIDWORKS Toolbox library that adds fasteners automatically to holes in an assembly.

Feature

An individual shape that, combined with other features, makes up a part or assembly. Some features, such as bosses and cuts, originate as sketches. Other features, such as shells and fillets, modify a feature's geometry. However, not all features have associated geometry. Features are always listed in the FeatureManager design tree.

FeatureManager design tree

Located on the left side of the SOLIDWORKS window, it provides an outline view of the active part, assembly, or drawing.

Fill

A solid area hatch or crosshatch. Fill also applies to patches on surfaces.

Fillet

An internal rounding of a corner or edge in a sketch, or an edge on a surface or solid.

Forming tool

Dies that bend, stretch, or otherwise form sheet metal to create such form features as louvers, lances, flanges, and ribs.

Fully defined

A sketch where all lines and curves in the sketch, and their positions, are described by dimensions or relations, or both, and cannot be moved. Fully defined sketch entities are shown in black.

Geometric tolerance

A set of standard symbols that specify the geometric characteristics and dimensional requirements of a feature.

Graphics area

The area in the SOLIDWORKS window where the part, assembly, or drawing appears.

Guide curve

A 2D or 3D curve used to guide a sweep or loft.

Handle

An arrow, square, or circle that you can drag to adjust the size or position of an entity (a feature, dimension, or sketch entity, for example).

Helix

A curve defined by pitch, revolutions, and height. A helix can be used, for example, as a path for a swept feature cutting threads in a bolt.

Hem

A sheet metal feature that folds back at the edge of a part. A hem can be open, closed, double, or teardrop.

HLR

(Hidden lines removed) a view mode in which all edges of the model that are not visible from the current view angle are removed from the display.

HLV

(Hidden lines visible) A view mode in which all edges of the model that are not visible from the current view angle are shown gray or dashed.

Import

Open files from other CAD software applications into a SOLIDWORKS document.

In-context feature

A feature with an external reference to the geometry of another component; the in-context feature changes automatically if the geometry of the referenced model or feature changes.

Inference

The system automatically creates (infers) relations between dragged entities (sketched entities, annotations, and components) and other entities and geometry. This is useful when positioning entities relative to one another.

Instance

An item in a pattern or a component in an assembly that occurs more than once. Blocks are inserted into drawings as instances of block definitions.

Interference detection

A tool that displays any interference between selected components in an assembly.

Jog

A sheet metal feature that adds material to a part by creating two bends from a sketched line.

Knit

A tool that combines two or more faces or surfaces into one. The edges of the surfaces must be adjacent and not overlapping, but they cannot ever be planar. There is no difference in the appearance of the face or the surface after knitting.

Layout sketch

A sketch that contains important sketch entities, dimensions, and relations. You reference the entities in the layout sketch when creating new sketches, building new geometry, or positioning components in an assembly. This allows for easier updating of your model because changes you make to the layout sketch propagate to the entire model.

Leader
A solid line from an annotation (note, dimension, and so on) to the referenced feature.

Library feature
A frequently used feature, or combination of features, that is created once and then saved for future use.

Lightweight
A part in an assembly or a drawing has only a subset of its model data loaded into memory. The remaining model data is loaded on an as-needed basis. This improves performance of large and complex assemblies.

Line
A straight sketch entity with two endpoints. A line can be created by projecting an external entity such as an edge, plane, axis, or sketch curve into the sketch.

Loft
A base, boss, cut, or surface feature created by transitions between profiles.

Lofted bend
A sheet metal feature that produces a roll form or a transitional shape from two open profile sketches. Lofted bends often create funnels and chutes.

Mass properties
A tool that evaluates the characteristics of a part or an assembly such as volume, surface area, centroid, and so on.

Mate
A geometric relationship, such as coincident, perpendicular, tangent, and so on, between parts in an assembly.

Mate reference
Specifies one or more entities of a component to use for automatic mating. When you drag a component with a mate reference into an assembly, the software tries to find other combinations of the same mate reference name and mate type.

Mates folder
A collection of mates that are solved together. The order in which the mates appear within the Mates folder does not matter.

Mirror
a) A mirror feature is a copy of a selected feature, mirrored about a plane or planar face.
b) A mirror sketch entity is a copy of a selected sketch entity that is mirrored about a centerline.

Miter flange
A sheet metal feature that joins multiple edge flanges together and miters the corner.

Model
3D solid geometry in a part or assembly document. If a part or assembly document contains multiple configurations, each configuration is a separate model.

Model dimension
A dimension specified in a sketch or a feature in a part or assembly document that defines some entity in a 3D model.

Model item
A characteristic or dimension of feature geometry that can be used in detailing drawings.

Model view
A drawing view of a part or assembly.

Mold
A set of manufacturing tooling used to shape molten plastic or other material into a designed part. You design the mold using a sequence of integrated tools that result in cavity and core blocks that are derived parts of the part to be molded.

Motion Study
Motion Studies are graphical simulations of motion and visual properties with assembly models. Analogous to a configuration, they do not actually change the original assembly model or its properties. They display the model as it changes based on simulation elements you add.

Multibody part
A part with separate solid bodies within the same part document. Unlike the components in an assembly, multibody parts are not dynamic.

Native format
DXF and DWG files remain in their original format (are not converted into SOLIDWORKS format) when viewed in SOLIDWORKS drawing sheets (view only).

Open profile

Also called an open contour, it is a sketch or sketch entity with endpoints exposed. For example, a U-shaped profile is open.

Ordinate dimensions

A chain of dimensions measured from a zero ordinate in a drawing or sketch.

Origin

The model origin appears as three gray arrows and represents the (0,0,0) coordinate of the model. When a sketch is active, a sketch origin appears in red and represents the (0,0,0) coordinate of the sketch. Dimensions and relations can be added to the model origin, but not to a sketch origin.

Out-of-context feature

A feature with an external reference to the geometry of another component that is not open.

Over defined

A sketch is over defined when dimensions or relations are either in conflict or redundant.

Parameter

A value used to define a sketch or feature (often a dimension).

Parent

An existing feature upon which other features depend. For example, in a block with a hole, the block is the parent to the child hole feature.

Part

A single 3D object made up of features. A part can become a component in an assembly, and it can be represented in 2D in a drawing. Examples of parts are bolt, pin, plate, and so on. The extension for a SOLIDWORKS part file name is .SLDPRT.

Path

A sketch, edge, or curve used in creating a sweep or loft.

Pattern

A pattern repeats selected sketch entities, features, or components in an array, which can be linear, circular, or sketch driven. If the seed entity is changed, the other instances in the pattern update.

Physical Dynamics

An assembly tool that displays the motion of assembly components in a realistic way. When you drag a component, the component applies a force to other components it touches. Components move only within their degrees of freedom.

Pierce relation
Makes a sketch point coincident to the location at which an axis, edge, line, or spline pierces the sketch plane.

Planar
Entities that can lie on one plane. For example, a circle is planar, but a helix is not.

Plane
Flat construction geometry. Planes can be used for a 2D sketch, section view of a model, a neutral plane in a draft feature, and others.

Point
A singular location in a sketch, or a projection into a sketch at a single location of an external entity (origin, vertex, axis, or point in an external sketch).

Predefined view
A drawing view in which the view position, orientation, and so on can be specified before a model is inserted. You can save drawing documents with predefined views as templates.

Profile
A sketch entity used to create a feature (such as a loft) or a drawing view (such as a detail view). A profile can be open (such as a U shape or open spline) or closed (such as a circle or closed spline).

Projected dimension
If you dimension entities in an isometric view, projected dimensions are the flat dimensions in 2D.

Projected view
A drawing view projected orthogonally from an existing view.

PropertyManager
Located on the left side of the SOLIDWORKS window, it is used for dynamic editing of sketch entities and most features.

RealView graphics
A hardware (graphics card) support of advanced shading in real time; the rendering applies to the model and is retained as you move or rotate a part.

Rebuild
Tool that updates (or regenerates) the document with any changes made since the last time the model was rebuilt. Rebuild is typically used after changing a model dimension.

Reference dimension

A dimension in a drawing that shows the measurement of an item but cannot drive the model and its value cannot be modified. When model dimensions change, reference dimensions update.

Reference geometry

Includes planes, axes, coordinate systems, and 3D curves. Reference geometry is used to assist in creating features such as lofts, sweeps, drafts, chamfers, and patterns.

Relation

A geometric constraint between sketch entities or between a sketch entity and a plane, axis, edge, or vertex. Relations can be added automatically or manually.

Relative view

A relative (or relative to model) drawing view is created relative to planar surfaces in a part or assembly.

Reload

Refreshes shared documents. For example, if you open a part file for read-only access while another user makes changes to the same part, you can reload the new version, including the changes.

Reorder

Reordering (changing the order of) items is possible in the FeatureManager design tree. In parts, you can change the order in which features are solved. In assemblies, you can control the order in which components appear in a bill of materials.

Replace

Substitutes one or more open instances of a component in an assembly with a different component.

Resolved

A state of an assembly component (in an assembly or drawing document) in which it is fully loaded in memory. All the component's model data is available, so its entities can be selected, referenced, edited, and used in mates, and so on.

Revolve

A feature that creates a base or boss, a revolved cut, or revolved surface by revolving one or more sketched profiles around a centerline.

Rip

A sheet metal feature that removes material at an edge to allow a bend.

Rollback

Suppresses all items below the rollback bar.

Section

Another term for profile in sweeps.

Section line

A line or centerline sketched in a drawing view to create a section view.

Section scope

Specifies the components to be left uncut when you create an assembly drawing section view.

Section view

A section view (or section cut) is (1) a part or assembly view cut by a plane, or (2) a drawing view created by cutting another drawing view with a section line.

Seed

A sketch or an entity (a feature, face, or body) that is the basis for a pattern. If you edit the seed, the other entities in the pattern are updated.

Shaded

Displays a model as a colored solid.

Shared values

Also called linked values, these are named variables that you assign to set the value of two or more dimensions to be equal.

Sheet format

Includes page size and orientation, standard text, borders, title blocks, and so on. Sheet formats can be customized and saved for future use. Each sheet of a drawing document can have a different format.

Shell

A feature that hollows out a part, leaving open the selected faces and thin walls on the remaining faces. A hollow part is created when no faces are selected to be open.

Sketch

A collection of lines and other 2D objects on a plane or face that forms the basis for a feature such as a base or a boss. A 3D sketch is non-planar and can be used to guide a sweep or loft, for example.

Smart Fasteners
Automatically adds fasteners (bolts and screws) to an assembly using the SOLIDWORKS Toolbox library of fasteners.

SmartMates
An assembly mating relation that is created automatically.

Solid sweep
A cut sweep created by moving a tool body along a path to cut out 3D material from a model.

Spiral
A flat or 2D helix, defined by a circle, pitch, and number of revolutions.

Spline
A sketched 2D or 3D curve defined by a set of control points.

Split line
Projects a sketched curve onto a selected model face, dividing the face into multiple faces so that each can be selected individually. A split line can be used to create draft features, to create face blend fillets, and to radiate surfaces to cut molds.

Stacked balloon
A set of balloons with only one leader. The balloons can be stacked vertically (up or down) or horizontally (left or right).

Standard 3 views
The three orthographic views (front, right, and top) that are often the basis of a drawing.

StereoLithography
The process of creating rapid prototype parts using a faceted mesh representation in STL files.

Sub-assembly
An assembly document that is part of a larger assembly. For example, the steering mechanism of a car is a sub-assembly of the car.

Suppress
Removes an entity from the display and from any calculations in which it is involved. You can suppress features, assembly components, and so on. Suppressing an entity does not delete the entity; you can unsuppress the entity to restore it.

Surface

A zero-thickness planar or 3D entity with edge boundaries. Surfaces are often used to create solid features. Reference surfaces can be used to modify solid features.

Sweep

Creates a base, boss, cut, or surface feature by moving a profile (section) along a path. For cut sweeps, you can create solid sweeps by moving a tool body along a path.

Tangent arc

An arc that is tangent to another entity, such as a line.

Tangent edge

The transition edge between rounded or filleted faces in hidden lines visible or hidden lines removed modes in drawings.

Task Pane

Located on the right-side of the SOLIDWORKS window, the Task Pane contains SOLIDWORKS Resources, the Design Library, and the File Explorer.

Template

A document (part, assembly, or drawing) that forms the basis of a new document. It can include user-defined parameters, annotations, predefined views, geometry, and so on.

Temporary axis

An axis created implicitly for every conical or cylindrical face in a model.

Thin feature

An extruded or revolved feature with constant wall thickness. Sheet metal parts are typically created from thin features.

TolAnalyst

A tolerance analysis application that determines the effects that dimensions and tolerances have on parts and assemblies.

Top-down design

An assembly modeling technique where you create parts in the context of an assembly by referencing the geometry of other components. Changes to the referenced components propagate to the parts that you create in context.

Triad

Three axes with arrows defining the X, Y, and Z directions. A reference triad appears in part and assembly documents to assist in orienting the viewing of models. Triads also assist when moving or rotating components in assemblies.

Under defined
A sketch is under defined when there are not enough dimensions and relations to prevent entities from moving or changing size.

Vertex
A point at which two or more lines or edges intersect. Vertices can be selected for sketching, dimensioning, and many other operations.

Viewports
Windows that display views of models. You can specify one, two, or four viewports. Viewports with orthogonal views can be linked, which links orientation and rotation.

Virtual sharp
A sketch point at the intersection of two entities after the intersection itself has been removed by a feature such as a fillet or chamfer. Dimensions and relations to the virtual sharp are retained even though the actual intersection no longer exists.

Weldment
A multibody part with structural members.

Weldment cut list
A table that tabulates the bodies in a weldment along with descriptions and lengths.

Wireframe
A view mode in which all edges of the part or assembly are displayed.

Zebra stripes
Simulate the reflection of long strips of light on a very shiny surface. They allow you to see small changes in a surface that may be hard to see with a standard display.

Zoom
To simulate movement toward or away from a part or an assembly.

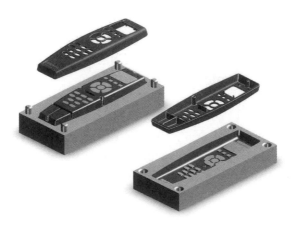

Index

SOLIDWORKS Quick Guide
Command Icons & Toolbars

 Creates a new document.

 Opens an existing document.

 Saves an active document.

 Make Drawing from Part/Assembly.

 Make Assembly from Part/Assembly.

 Prints the active document.

 Print preview.

 Cuts the selection & puts it on the clipboard.

 Copies the selection & puts it on the clipboard.

 Inserts the clipboard contents.

 Deletes the selection.

 Reverses the last action.

 Rebuilds the part / assembly / drawing.

 Redo the last action that was undone.

 Saves all documents.

 Edits material.

 Closes an existing document.

 Shows or hides the Selection Filter toolbar.

 Shows or hides the Web toolbar.

 Properties.

 File properties.

 Loads or unloads the 3D instant website add-in.

 Select tool.

 Select the entire document.

 Checks read-only files.

 Options.

 Help.

 Full screen view.

 OK.

 Cancel.

 Magnified selection.

 Select.

 Sketch.

 3D Sketch.

 Sketches a rectangle from the center.

 Sketches a CenterPoint arc slot.

 Sketches a 3-point arc slot.

 Sketches a straight slot.

 Sketches a CenterPoint straight slot.

 Sketches a 3-point arc.

 Creates sketched ellipses.

Quick Reference Guide to SOLIDWORKS Command Icons & Toolbars

SKETCH TOOLS Toolbar

 3D sketch on plane.

 Sets up Grid parameters.

 Creates a sketch on a selected plane or face.

 Equation driven curve.

 Modifies a sketch.

 Copies sketch entities.

 Scales sketch entities.

 Rotates sketch entities.

 Sketches 3-point rectangle from the center.

 Sketches 3-point corner rectangle.

 Sketches a line.

 Creates a center point arc: center, start, end.

 Creates an arc tangent to a line.

 Sketches splines on a surface or face.

 Sketches a circle.

 Sketches a circle by its perimeter.

 Makes a path of sketch entities.

Mirrors entities dynamically about a centerline.

Insert a plane into the 3D sketch.

Instant 2D.

Sketch numeric input.

Detaches segment on drag.

Sketch picture.

 Partial ellipses.

 Adds a Parabola.

 Adds a spline.

 Sketches a polygon.

 Sketches a corner rectangle.

 Sketches a parallelogram.

 Creates points.

 Creates sketched centerlines.

 Adds text to sketch.

 Converts selected model edges or sketch entities to sketch segments.

 Creates a sketch along the intersection of multiple bodies.

 Converts face curves on the selected face into 3D sketch entities.

 Mirrors selected segments about a centerline.

 Fillets the corner of two lines.

 Creates a chamfer between two sketch entities.

 Creates a sketch curve by offsetting model edges or sketch entities at a specified distance.

 Trims a sketch segment.

 Extends a sketch segment.

 Splits a sketch segment.

 Construction Geometry.

 Creates linear steps and repeat of sketch entities.

 Creates circular steps and repeat of sketch entities.

SHEET METAL Toolbar

 Add a bend from a selected sketch in a Sheet Metal part.

 Shows flat pattern for this sheet metal part.

 Shows part without inserting any bends.

 Inserts a rip feature to a sheet metal part.

 Create a Sheet Metal part or add material to existing Sheet Metal part.

 Inserts a Sheet Metal Miter Flange feature.

 Folds selected bends.

 Unfolds selected bends.

 Inserts bends using a sketch line.

 Inserts a flange by pulling an edge.

 Inserts a sheet metal corner feature.

 Inserts a Hem feature by selecting edges.

 Breaks a corner by filleting/chamfering it.

 Inserts a Jog feature using a sketch line.

 Inserts a lofted bend feature using 2 sketches.

 Creates inverse dent on a sheet metal part.

 Trims out material from a corner, in a sheet metal part.

 Inserts a fillet weld bead.

 Converts a solid/surface into a sheet metal part.

 Adds a Cross Break feature into a selected face.

 Sweeps an open profile along an open/closed path.

 Adds a gusset/rib across a bend.

 Corner relief.

 Welds the selected corner.

SURFACES Toolbar

 Creates mid surfaces between offset face pairs.

 Patches surface holes and external edges.

 Creates an extruded surface.

 Creates a revolved surface.

 Creates a swept surface.

 Creates a lofted surface.

 Creates an offset surface.

 Radiates a surface originating from a curve, parallel to a plane.

 Knits surfaces together.

 Creates a planar surface from a sketch or a set of edges.

 Creates a surface by importing data from a file.

 Extends a surface.

 Trims a surface.

 Surface flattens.

 Deletes Face(s).

 Replaces Face with Surface.

 Patches surface holes and external edges by extending the surfaces.

 Creates parting surfaces between core & cavity surfaces.

 Inserts ruled surfaces from edges.

WELDMENTS Toolbar

 Creates a weldment feature.

 Creates a structure member feature.

 Adds a gusset feature between 2 planar adjoining faces.

 Creates an end cap feature.

 Adds a fillet weld bead feature.

 Trims or extends structure members.

Weld bead.

DIMENSIONS/RELATIONS Toolbar

 Inserts dimension between two lines.

 Creates a horizontal dimension between selected entities.

 Creates a vertical dimension between selected entities.

 Creates a reference dimension between selected entities.

 Creates a set of ordinate dimensions.

 Creates a set of Horizontal ordinate

 Creates a set of Vertical ordinate dimensions.

 Creates a chamfer dimension.

 Adds a geometric relation.

 Automatically Adds Dimensions to the current sketch.

 Displays and deletes geometric relations.

 Fully defines a sketch.

 Scans a sketch for elements of equal length or radius.

 Angular Running dimension.

 Display / Delete dimension.

 Isolate changed dimension.

 Path length dimension.

BLOCK Toolbar

 Makes a new block.

 Edits the selected block.

 Inserts a new block to a sketch or drawing.

 Adds/Removes sketch entities to/from blocks.

 Updates parent sketches affected by this block.

 Saves the block to a file.

 Explodes the selected block.

 Inserts a belt.

STANDARD VIEWS Toolbar

 Front view.

 Back view.

 Left view.

 Right view.

 Top view.

 Bottom view.

 Isometric view.

 Trimetric view.

 Dimetric view.

 Normal to view.

 Links all views in the viewport together.

 Displays viewport with front & right

 Displays a 4-view viewport with 1st or 3rd angle of projection.

 Displays viewport with front & top views.

 Displays viewport with a single view.

 View selector.

 New view.

FEATURES Toolbar

 Creates a boss feature by extruding a sketched profile.

 Creates a revolved feature based on profile and angle parameter.

 Creates a cut feature by extruding a sketched profile.

 Creates a cut feature by revolving a sketched profile.

 Thread.

 Creates a cut by sweeping a closed profile along an open or closed path.

 Loft cut.

 Creates a cut by thickening one or more adjacent surfaces.

 Adds a deformed surface by push or pull on points.

 Creates a lofted feature between two or more profiles.

 Creates a solid feature by thickening one or more adjacent surfaces.

 Creates a filled feature.

 Chamfers an edge or a chain of tangent edges.

 Inserts a rib feature.

 Combine.

 Creates a shell feature.

 Applies draft to a selected surface.

 Creates a cylindrical hole.

 Inserts a hole with a pre-defined cross section.

 Puts a dome surface on a face.

 Model break view.

 Applies global deformation to solid or surface bodies.

 Wraps closed sketch contour(s) onto a face.

 Curve Driven pattern.

 Suppresses the selected feature or component.

 Un-suppresses the selected feature or component.

 Flexes solid and surface bodies.

 Intersect.

 Variable Patterns.

 Live Section Plane.

 Mirrors.

 Scale.

 Creates a Sketch Driven pattern.

 Creates a Table-Driven Pattern.

 Inserts a split Feature.

 Hole series.

 Joins bodies from one or more parts into a single part in the context of an assembly.

 Deletes a solid or a surface.

 Instant 3D.

 Inserts apart from file into the active part document.

 Moves/Copies solid and surface bodies or moves graphics bodies.

 Merges short edges on faces.

 Pushes solid / surface model by another solid / surface model.

 Moves face(s) of a solid.

 FeatureWorks Options.

 Linear Pattern.

 Fill Pattern.

 Cuts a solid model with a

 Boundary Boss/Base.

 Boundary Cut.

 Circular Pattern.

 Recognize Features.

 Grid System.

MOLD TOOLS Toolbar

 Extracts core(s) from existing tooling split.

 Constructs a surface patch.

 Moves face(s) of a solid.

 Creates offset surfaces.

 Inserts cavity into a base part.

 Scales a model by a specified factor.

 Applies draft to a selected surface.

 Inserts a split line feature.

 Creates parting lines to separate core & cavity surfaces.

 Finds & creates mold shut-off surfaces.

 Creates a planar surface from a sketch or a set of edges.

 Knits surfaces together.

 Inserts ruled surfaces from edges.

 Creates parting surfaces between core & cavity surfaces.

 Creates multiple bodies from a single body.

 Inserts a tooling split feature.

 Creates parting surfaces between the core & cavity.

 Inserts surface body folders for mold operation.

SELECTION FILTERS Toolbar

 Turns selection filters on and off.

 Clears all filters.

 Selects all filters.

 Inverts current selection.

 Allows selection of edges only.

 Allows selection filter for vertices only.

 Allows selection of faces only.

 Adds filter for Surface Bodies.

 Adds filter for Solid Bodies.

 Adds filter for Axes.

 Adds filter for Planes.

 Adds filter for Sketch Points.

 Allows selection for sketch only.

 Adds filter for Sketch Segments.

 Adds filter for Midpoints.

 Adds filter for Center Marks.

 Adds filter for Centerline.

 Adds filter for Dimensions and Hole Callouts.

 Adds filter for Surface Finish Symbols.

 Adds filter for Geometric Tolerances.

 Adds filter for Notes / Balloons.

 Adds filter for Weld Symbols.

 Adds filter for Weld beads.

 Adds filter for Datum Targets.

 Adds filter for Datum feature only.

 Adds filter for blocks.

 Adds filter for Cosmetic Threads.

Adds filter for Dowel pin symbols.

 Adds filter for connection points.

 Adds filter for routing points.

SOLIDWORKS Add-Ins Toolbar

 Loads/unloads CircuitWorks add-in.

 Loads/unloads the Design Checker add-in.

 Loads/unloads the PhotoView 360 add-in.

 Loads/unloads the Scan-to-3D add-in.

 Loads/unloads the SOLIDWORKS Motions add-in.

 Loads/unloads the SOLIDWORKS Routing add-in.

 Loads/unloads the SOLIDWORKS Simulation add-in.

 Loads/unloads the SOLIDWORKS Toolbox add-in.

 Loads/unloads the SOLIDWORKS TolAnalysis add-in.

 Loads/unloads the SOLIDWORKS Flow Simulation add-in.

 Loads/unloads the SOLIDWORKS Plastics add-in.

 Loads/unloads the SOLIDWORKS MBD SNL license.

FASTENING FEATURES Toolbar

 Creates a parameterized mounting boss.

 Creates a parameterized snap hook.

 Creates a groove to mate with a hook feature.

 Uses sketch elements to create a vent for air flow.

 Creates a lip/groove feature.

SCREEN CAPTURE Toolbar

 Copies the current graphics window to the clipboard.

 Records the current graphics window to an AVI file.

 Stops recording the current graphics window to an AVI file.

EXPLODE LINE SKETCH Toolbar

 Adds a route line that connects entities.

 Adds a jog to the route lines.

LINE FORMAT Toolbar

 Changes layer properties.

 Changes the current document layer.

 Changes line color.

 Changes line thickness.

 Changes line style.

 Hides / Shows a hidden edge.

 Changes line display mode.

Did you know??

* Ctrl+Q will force a rebuild on all features of a part.

* Ctrl+B will rebuild the feature being worked on and its dependents.

2D-To-3D Toolbar

 Makes a Front sketch from the selected entities.

 Makes a Top sketch from the selected entities.

 Makes a Right sketch from the selected entities.

Makes a Left sketch from the selected entities.

Makes a Bottom sketch from the selected entities.

Makes a Back sketch from the selected entities.

Makes an Auxiliary sketch from the selected entities.

Creates a new sketch from the selected entities.

Repairs the selected sketch.

Aligns a sketch to the selected point.

Creates an extrusion from the selected sketch segments, starting at the selected sketch point.

Creates a cut from the selected sketch segments, optionally starting at the selected sketch point.

ALIGN Toolbar

Aligns the left side of the selected annotations with the leftmost annotation.

Aligns the right side of the selected annotations with the rightmost annotation.

Aligns the top side of the selected annotations with the topmost annotation.

Aligns the bottom side of the selected annotations with the lowermost annotation.

Evenly spaces the selected annotations horizontally.

Evenly spaces the selected annotations vertically.

Centrally aligns the selected annotations horizontally.

Centrally aligns the selected annotations vertically.

Compacts the selected annotations horizontally.

Compacts the selected annotations vertically.

Creates a group from the selected items.

Deletes the grouping between these items.

Aligns & groups selected dimensions along a line or an arc.

Aligns & groups dimensions at uniform distances.

Evenly spaces selected dimensions.

Aligns collinear selected dimensions.

Aligns stagger selected dimensions.

SOLIDWORKS MBD Toolbar

Captures 3D view.

Manages 3D PDF templates.

Creates shareable 3D PDF presentations.

Toggles dynamic annotation views.

MACRO Toolbar

Runs a Macro.

Stops Macro recorder.

Records (or pauses recording of) actions to create a Macro.

Launches the Macro Editor and begins editing a new macro.

Opens a Macro file for editing.

Creates a custom macro.

SMARTMATES icons

Concentric & Coincident 2 circular edges.

Concentric 2 cylindrical faces.

Coincident 2 linear edges.

Coincident 2 planar faces.

Coincident 2 vertices.

Coincident 2 origins or coordinate systems.

TABLE Toolbar

 Adds a hole table of selected holes from a specified origin datum.

 Adds a Bill of Materials.

 Adds a revision table.

 Displays a Design table in a drawing.

 Adds a weldments cuts list table.

 Adds an Excel based Bill of Materials.

 Adds a weldment cut list table.

REFERENCE GEOMETRY Toolbar

 Adds a reference plane.

 Creates an axis.

 Creates a coordinate system.

 Adds the center of mass.

 Specifies entities to use as references using SmartMates.

SPLINE TOOLS Toolbar

 Inserts a point to a spline.

 Displays all points where the concavity of selected spline changes.

 Displays minimum radius of selected spline.

 Displays curvature combs of selected spline.

 Reduces numbers of points in a selected spline.

 Adds a tangency control.

 Adds a curvature control.

 Adds a spline based on selected sketch entities & edges.

 Displays the spline control polygon.

ANNOTATIONS Toolbar

 Inserts a note.

 Inserts a surface finish symbol.

 Inserts a new geometric tolerancing symbol.

 Attaches a balloon to the selected edge or face.

 Adds balloons for all components in selected view.

 Inserts a stacked balloon.

 Attaches a datum feature symbol to a selected edge / detail.

 Inserts a weld symbol on the selected edge / face / vertex.

 Inserts a datum target symbol and / or point attached to a selected edge / line.

 Selects and inserts block.

 Inserts annotations & reference geometry from the part / assembly into the selected.

 Adds center marks to circles on model.

 Inserts a Centerline.

 Inserts a hole callout.

 Adds a cosmetic thread to the selected cylindrical feature.

 Inserts a Multi-Jog leader.

 Selects a circular edge or an arc for Dowel pin symbol insertion.

 Adds a view location symbol.

 Inserts latest version symbol.

 Adds cross hatch patterns or solid fill.

 Adds a weld bead caterpillar on an edge.

 Adds a weld symbol on a selected entity.

 Inserts a revision cloud.

 Inserts a magnetic line.

 Hides/shows annotation.

DRAWINGS Toolbar

 Updates the selected view to the model's current stage.

 Creates a detail view.

 Creates a section view.

 Inserts an Alternate Position view.

 Unfolds a new view from an existing view.

 Generates a standard 3-view drawing (1st or 3rd angle).

 Inserts an auxiliary view of an inclined surface.

 Adds an Orthogonal or Named view based on an existing part or assembly.

 Adds a Relative view by two orthogonal faces or planes.

 Adds a Predefined orthogonal projected or Named view with a model.

 Adds an empty view.

 Adds vertical break lines to selected view.

 Crops a view.

 Creates a Broken-out section.

QUICK SNAP Toolbar

 Snap to points.

 Snap to center points.

 Snap to midpoints.

 Snap to quadrant points.

 Snap to intersection of 2 curves.

 Snap to nearest curve.

 Snap tangent to curve.

 Snap perpendicular to curve.

 Snap parallel to line.

 Snap horizontally / vertically to points.

 Snap horizontally / vertically.

 Snap to discrete line lengths.

 Snap to angle.

LAYOUT Toolbar

 Creates the assembly layout sketch.

 Sketches a line.

 Sketches a corner rectangle.

 Sketches a circle.

 Sketches a 3-point arc.

 Rounds a corner.

 Trims or extends a sketch.

 Adds sketch entities by offsetting faces, edges and curves.

 Mirrors selected entities about a centerline.

 Adds a relation.

 Creates a dimension.

 Displays / Deletes geometric relations.

 Makes a new block.

 Edits the selected block.

 Inserts a new block to the sketch or drawing.

 Adds / Removes sketch entities to / from a block.

 Saves the block to a file.

 Explodes the selected block.

 Creates a new part from a layout sketch block.

 Positions 2 components relative to one another.

CURVES Toolbar

 Projects sketch onto selected surface.

 Inserts a split line feature.

 Creates a composite curve from selected edges, curves and sketches.

 Creates a curve through free points.

 Creates a 3D curve through reference points.

 Helical curve defined by a base sketch and shape parameters.

VIEW Toolbar

 Displays a view in the selected orientation.

 Reverts to previous view.

 Redraws the current window.

 Zooms out to see entire model.

 Zooms in by dragging a bounding box.

 Zooms in or out by dragging up or down.

 Zooms to fit all selected entities.

 Dynamic view rotation.

 Scrolls view by dragging.

 Displays image in wireframe mode.

 Displays hidden edges in gray.

 Displays image with hidden lines removed.

 Controls the visibility of planes.

 Controls the visibility of axis.

 Controls the visibility of parting lines.

 Controls the visibility of temporary axis.

 Controls the visibility of origins.

 Controls the visibility of coordinate systems.

 Controls the visibility of reference curves.

 Controls the visibility of sketches.

 Controls the visibility of 3D sketch planes.

 Controls the visibility of 3D sketch.

 Controls the visibility of all annotations.

 Controls the visibility of reference points.

 Controls the visibility of routing points.

 Controls the visibility of lights.

 Controls the visibility of cameras.

 Controls the visibility of sketch relations.

 Changes the display state for the current configuration.

 Rolls the model view.

 Turns the orientation of the model view.

 Dynamically manipulate the model view in 3D to make selection.

 Changes the display style for the active view.

 Displays a shade view of the model with its edges.

 Displays a shade view of the model.

 Toggles between draft quality & high quality HLV.

 Cycles through or applies a specific scene.

 Views the models through one of the model's cameras.

 Displays a part or assembly w/different colors according to the local radius of curvature.

 Displays zebra stripes.

 Displays a model with hardware accelerated shades.

 Applies a cartoon effect to model edges & faces.

 Views simulations symbols.

TOOLS Toolbar

 Calculates the distance between selected items.

 Adds or edits equation.

 Calculates the mass properties of the model.

 Checks the model for geometry errors.

 Inserts or edits a Design Table.

 Evaluates section properties for faces and sketches that lie in parallel planes.

 Reports Statistics for this Part/Assembly.

 Deviation Analysis.

 Runs the SimulationXpress analysis wizard Powered by SOLIDWORKS Simulation.

 Checks the spelling.

 Import diagnostics.

 Runs the DFMXpress analysis wizard.

 Runs the SOLIDWORKS FloXpress analysis wizard.

ASSEMBLY Toolbar

 Creates a new part & inserts it into the assembly.

 Adds an existing part or sub-assembly to the assembly.

 Creates a new assembly & inserts it into the assembly.

 Turns on/off large assembly mode for this document.

 Hides / shows model(s) associated with the selected model(s).

 Toggles the transparency of components.

 Changes the selected components to suppressed or resolved.

 Inserts a belt.

 Toggles between editing part and assembly.

 Smart Fasteners.

 Positions two components relative to one another.

 External references will not be created.

 Moves a component.

 Rotates an un-mated component around its center point.

 Replaces selected components.

 Replaces mate entities of mates of the selected components on the selected Mates group.

 Creates a New Exploded view.

 Creates or edits explode line sketch.

 Interference detection.

 Shows or Hides the Simulation toolbar.

 Patterns components in one or two linear directions.

 Patterns components around an axis.

 Sets the transparency of the components other than the one being edited.

 Sketch driven component pattern.

 Pattern driven component pattern.

 Curve driven component pattern.

 Chain driven component pattern.

 SmartMates by dragging & dropping components.

 Checks assembly hole alignments.

 Mirrors subassemblies and parts.

To add or remove an icon
to or from the toolbar, first select:

Tools/Customize/Commands

Next, select a **Category**, click a button to see its description and then drag/drop the command icon into any toolbar.

SOLIDWORKS Quick-Guide©
STANDARD Keyboard Shortcuts

Rotate the model

* Horizontally or Vertically:_____ Arrow keys

* Horizontally or Vertically 90°:_____ Shift + Arrow keys

* Clockwise or Counterclockwise:_____ Alt + left or right Arrow

* Pan the model: _____ Ctrl + Arrow keys

* Zoom in:_____ Z (shift + Z or capital Z)

* Zoom out: _____ z (lower case z)

* Zoom to fit: _____ F

* Previous view: _____ Ctrl+Shift+Z

View Orientation

* View Orientation Menu: _____ Space bar

* Front:_____ Ctrl+1

* Back:_____ Ctrl+2

* Left: _____ Ctrl+3

* Right: _____ Ctrl+4

* Top: _____ Ctrl+5

* Bottom:_____ Ctrl+6

* Isometric:_____ Ctrl+7

Selection Filter & Misc.

* Filter Edges:_____ e

* Filter Vertices: _____ v

* Filter Faces: _____ x

* Toggle Selection filter toolbar:_____ F5

* Toggle Selection Filter toolbar (on/off): _____ F6

* New SOLIDWORKS document:_____ F1

* Open Document: _____ Ctrl+O

* Open from Web folder:_____ Ctrl+W

* Save: _____ Ctrl+S

* Print:_____ Ctrl+P

* Magnifying Glass Zoom _____ g

* Switch between the SOLIDWORKS documents _____ Ctrl + Tab

SOLIDWORKS Sample Customized Hot Keys

Function Keys

F1	SW-Help
F2	2D Sketch
F3	3D Sketch
F4	Modify
F5	Selection Filters
F6	Move (2D Sketch)
F7	Rotate (2D Sketch)
F8	Measure
F9	Extrude
F10	Revolve
F11	Sweep
F12	Loft

Sketch

C	Circle
P	Polygon
E	Ellipse
O	Offset Entities
Alt + C	Convert Entities
M	Mirror
Alt + M	Dynamic Mirror
Alt + F	Sketch Fillet
T	Trim
Alt + X	Extend
D	Smart Dimension
Alt + R	Add Relation
Alt + P	Plane
Control + F	Fully Define Sketch
Control + Q	Exit Sketch